Visualization and Optimization

OPERATIONS RESEARCH/COMPUTER SCIENCE INTERFACES SERIES

Ramesh Sharda, Series Editor
Conoco/DuPont Chair of Management of Technology
Oklahoma State University
Stillwater, Oklahoma U.S.A.

Other published titles in the series:

Brown, Donald/Scherer, William T.
University of Virginia
Intelligent Scheduling Systems

Nash, Stephen G./Sofer, Ariela
George Mason University
The Impact of Emerging Technologies on Computer Science and Operations Research

Barth, Peter
Max-Planck-Institut fur Informatik, Germany
Logic-Based 0-1 Constraint Programming

Visualization and Optimization

Christopher V. Jones
University of Washington
Seattle, Washington

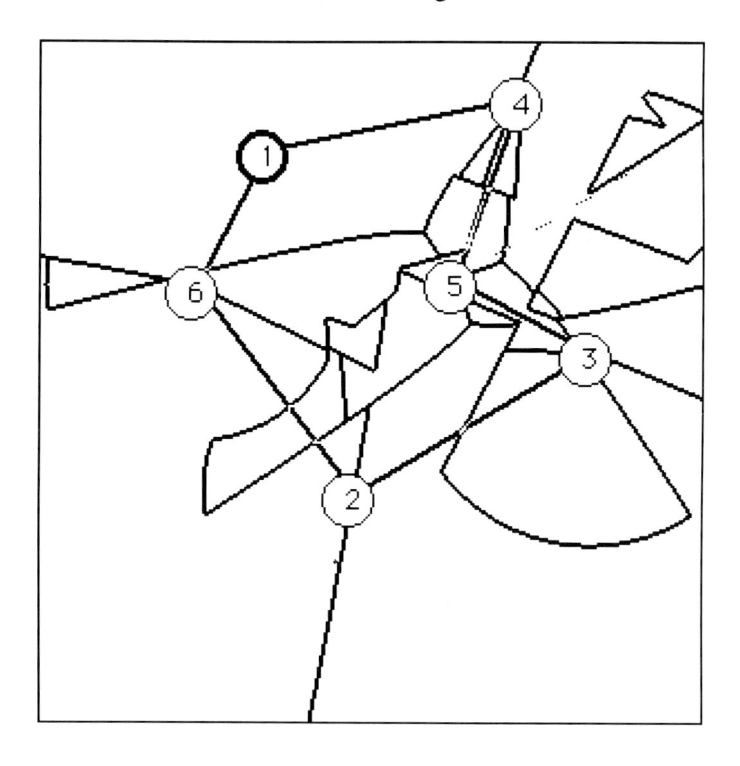

Kluwer Academic Publishers
Boston/Dordrecht/London

Distributors for North America:
Kluwer Academic Publishers
101 Philip Drive
Assinippi Park
Norwell, Massachusetts 02061 USA

Distributors for all other countries:
Kluwer Academic Publishers Group
Distribution Centre
Post Office Box 322
3300 AH Dordrecht, THE NETHERLANDS

Library of Congress Cataloging-in-Publication Data

A C.I.P. Catalogue record for this book is available from the Library of Congress.

Printed on acid-free paper.

Printed in the United States of America

Contents

Preface

This book arose out of an invited feature article on visualization and optimization that appeared in the *ORSA Journal on Computing* in 1994. That article briefly surveyed the current state of the art in visualization as it applied to optimization. In writing the feature article, it became clear that there was much more to say. Apparently others agreed, and thus this book was born.

The book is targeted primarily towards the optimization community rather than the visualization community. Although both optimization and visualization both seek to help people understand complex problems, practitioners in one field are generally unaware of work in the other field. Given the common goals of the respective fields, it seemed fruitful to consider how each can contribute to the other.

One might argue that this book should not be focused specifically on optimization but on decision making in general. Perhaps, but it seems that there is sufficient material to create a book targeted specifically to optimization. Certainly many of the ideas presented in the book are applicable to other areas, including computer simulation, decision theory and stochastic modeling. Another book could discuss the use of visualization in these areas.

The actual writing of the book took two years, but the education and background required for the book took much longer. Tom Baker first interested me in visualization during my Coop term at Exxon in 1978. Bill Maxwell, George Nemhauser, and Don Greenberg at Cornell trusted me enough to allow me to pursue my Ph. D. at the intersection of operations research and computer graphics. Art Geoffrion, Harvey Greenberg and Steve Kimbrough provided ongoing intellectual and spiritual inspiration. Simon Fraser University provided a delightful place for conducting my research. Ramesh Sharda was kind enough to recommend this book to the publisher. Gary Folven provided patient support during a much longer than expected writing process. Mark Quinn kept the rest of my life sane.

Many people and organizations graciously supplied images for use in the book. My heartfelt thanks goes out to all of them.

This research was supported in part by a grant from the Natural Science and Engineering Research Council of Canada.

C. V. J.

Bellingham, Washington

Chapter 1

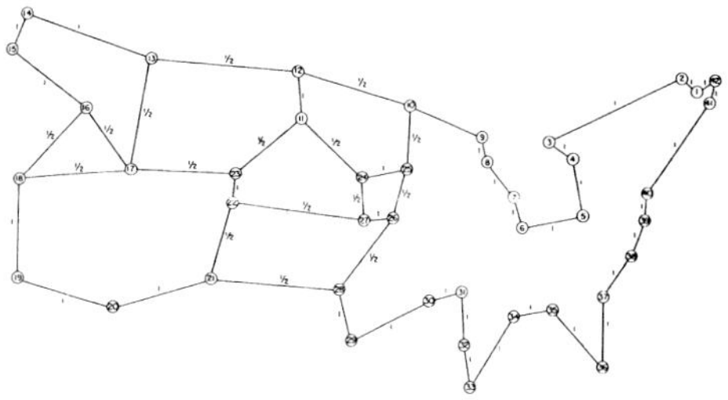

Fig. 18. A fractional solution x satisfying all loop conditions with $\Sigma\, d_{IJ}\, x_{IJ} = 698$.

Introduction

> Imagination or visualization, and in particular the use of diagrams, has a crucial part to play in scientific investigation.
> *René Descartes, 1637*, as quoted in [1].

1.1 Preliminaries

In order to understand the complexity of the real world, people—whether they realize it or not—build abstract descriptions or *models*. By selectively and carefully omitting details and including just the relevant factors, a model can provide a useful tool for understanding a particular problem. People construct models because the real problem of interest cannot be studied easily if at all by itself. Humans are extremely adept at filtering out detail in order to produce a simplified, and therefore, accessible representation of reality. People construct models of just about everything, from airplanes to zebras, from quarks to galaxies. Most models, though, are rather mundane. Every sentient person has some sort of mental model of the streets surrounding their home. They know the route from home to the grocery store, to work, to the post office, to the airport, to the train station. Residents of a large city might maintain a mental map of the subway or underground railway system.

Models, however, only have value when then can provide insight on some situation, that is, to answer questions. For example, if a tourist visiting London needs to take the subway from Piccadilly Circus to Heathrow Airport, a Londoner's mental model of the geography of the underground can be analyzed to generate directions that might even be comprehensible.

In order to answer a question, though, a model must be queried, probed, tested, solved, or *analyzed*. When giving directions, for example, the provider must concentrate to develop an appropriate route. In that act of concentrating, the provider is analyzing her mental model of the London subway system. Systematized techniques for analyzing models are called *algorithms*.

Because they are systematized, algorithms can be implemented on a computer. Computer implementations are valuable since a computer's great calculation speed can analyze many problems heretofore practically intractable.

Algorithms and models are not the same, however. The model is a description of a situation of interest. An algorithm analyzes that description in order to produce some useful information. Several different algorithms, perhaps from vastly different paradigms, can be used to analyze the same model, each often producing using "answers." Many people, though, when they speak of a "model" often lump together both the description of the problem and the technique used to solve it. Possessing a map of a subway system—a model—is not the same as using that map to find a route between two stations—an algorithm.

This book is concerned, in particular, with *optimization* models and algorithms. Optimization seeks the provably best solution to a particular problem, according to some objective criteria. Criteria often used include minimum cost, maximum profit or minimum risk. Since the origins of optimization in the late 1940's, particularly with the development of fast computers and efficient algorithms, optimization has been applied successfully to a variety of real world problems. These include blending gasoline, routing vehicles, assigning employees to tasks, scheduling factories, saving billions of dollars. Returning to the London tourist, he might want to know the fastest route between from Piccadilly Circus to Heathrow.

One should never, forget, however, that the "optimal" solutions produced by optimization techniques are only as good as the underlying model and data. The "fastest" route produced by an optimization model could be quite wrong if the information on traffic congestion were inaccurate as it often is. The model represents an abstraction of the actual situation, and the data fed into the model almost certainly is inaccurate or likely to change. In short, the purpose of optimization modeling, like all modeling, is to provide useful information, not necessarily the one, true, "optimal" solution. As Geoffrion[2] noted, paraphrasing Hamming's [3]'s characterization of computing, the purpose of optimization is "insight, not numbers."

Many wonderful, even award-winning, books have been written describing optimization algorithms and models. This book does not. Rather this book concentrates on a more difficult issue. In particular, the models constructed for solution by optimization algorithms and the answers they produce must be understandable by real people. Although a native may know the proper train to take to the airport, he may not be able to describe the route clearly. In order to make use of optimization (or most modeling techniques, for that matter), one must consider how the models will be constructed and how the optimal solutions can be understood by real people.

In short, models and solutions must be represented in a way that is clear and informative. Unfortunately, humans and computers usually disagree about which representations are easily comprehensible. Representations suitable for a computer algorithm are often impenetrable to humans and vice versa. If we are to employ optimization successfully, we must consider not only model and solution representations that computers can understand and exploit but also (probably different) representations that humans can understand and exploit. If a decision maker cannot understand a model nor the solution produced by an algorithm, then the model and algorithm are useless.

In some ways, with the major improvements in optimization algorithms and computer technology, the difficulty of understanding models, algorithms and solutions has grown worse. Optimization problems having hundreds of thousands of decision variables and constraints are now routinely solved. The solution to such a large problem would require many pages of output, if it were all printed. A decision maker may be interested in the behavior of a solution over a range of input values. The algorithm may experience numerical instabilities that will have to be understood to be overcome. As Harvey Greenberg [4] has stated, "we can solve far larger problems than we can understand."

In other ways, however, with the proliferation of interactive computer graphics, multi-media, graphical user interfaces, virtual reality, and other techniques of the day, the suite of representations available for optimization modeling has expanded greatly. The study of this variety of representations has been called *visualization* [5].

The use of visualization for modeling crystallized in a report from a task force from the computer graphics community written in 1987 [5]. The computer graphics community had achieved notable success in developing techniques for synthesizing extraordinarily realistic images. Their techniques were increasingly adopted by computer-aided design, film and video producers, in computer games. By 1987 computer graphics was increasingly being applied to supporting scientific exploration. But most work in visualization has concentrated on the physical sciences—physics, chemistry, biology, geology, medicine. More recently has seen increasing application to mathematical modeling, including optimization, hence the motivation for this book. Some call this emerging style of visualization, *information visualization*, to distinguish it from scientific visualization [6].

People have always drawn pictures to help them understand problems. A recent study of expert modelers [7] confirmed that most draw pictures or doodle while they develop a model. Paper and pencil have proven to be effective visualization tools, in the proper hands. With the advent of computer graphics, however, sophisticated pictures are now easier to draw. For example, most anyone can now create professional looking presentation graphics such as line charts and bar charts. With the improvement in our tools, our ability to use (and misuse) visualization has increased.

Visualizing phenomena in the physical sciences is arguably easier than visualizing the abstract worlds considered in mathematics and optimization. The physical sciences, by definition, study with problems that have some form of physical presence. Certainly some of the types of phenomena studied by the physical sciences are difficult and challenging to understand because they do not correspond to everyday reality, for example, quantum mechanics or relativity. Yet many of the problems studied have a palpable physical presence. Physicians need to visualize the human body; meteorologists need to visualize the movement of clouds over the surface of the Earth. Many optimization problems, however, do not have such a physical presence. What is the appropriate visual representation of a production schedule or financial plan?

Note that the purpose of visualization is much the same as the the purpose of optimization modeling: to provide insight on complex problems. Optimization uses sophisticated computer algorithms to uncover optimal solutions to complicated problems. Visualization uses carefully designed representations to help people understand complicated problems.

An example (from Woolsey [8]) illustrates how visualization can sometimes be used to solve "optimization" problems. Due to growth in population, a city fire department needed to build a new fire station. The question was where. The fire station could be located in a variety of areas, and within each area, many sites were possible. How should they decide?

Determining the best location for a facility is a classic optimization problem. With enough time and effort, a very sophisticated optimization model could probably be developed. However, such an analysis might easily ignore an extremely important part of the problem: the firefighters. Firefighters are generally not well versed in mathematics, and certainly not optimization. Yet, the firefighters will be the ones who will have to live with whatever decision is made. If the decision is made incorrectly, the firefighters will be delayed in reaching a fire, giving the fire a chance to grow. A larger fire would increase the risk to the firefighters, who already work in a risky job. Although one could almost certainly develop a wonderfully sophisticated mathematical model of this problem, if the firefighters cannot understand the reasons behind the recommendation of the model, it will probably not be accepted. The challenge, therefore, is not only to build a useful model, but to be able to make its results transparent to non-mathematicians.

In an actual case for the Denver fire department, described by Woolsey [8], no sophisticated mathematical or computer techniques were used. A map of Denver was pasted on the wall. Pushpins were inserted into the map, midpoints determined, and city-owned land drawn onto the map. When the simple technique was demonstrated to the fire chief, he said, "Well (expletive deleted) I can do that..." As the fire chief wrote, the technique "simplified an otherwise cumbersome methodology and made for a more efficient and expedient manner in the decision-making process..." Although an optimization algorithm might solve this problem more accurately than this physical technique, it is unclear whether the solution provided would have been accepted. One might have obtained a theoretically better answer that no one believed. Using simple, non-computerized techniques, the non-technical fire chief could *visualize* the solution, and thereby understand the problem more effectively.

Displaying an intricate visualization on a super whiz-bang graphics computer can cause as much confusion as it is attempting to relieve, however. Similarly, using the zippiest new optimization algorithm on the fastest new computer will not guarantee success either. The success of a modeling endeavor depends on many issues, including the accuracy of the model, the quality of the data provided, and also the level of communication among the participants in the problem-solving project. Computerized optimization or visualization techniques are not even always necessary as shown by the firehouse location example.

This book takes the point of view that techniques from visualization, when applied appropriately, can improve our ability to solve people's problems using optimization, but there is no silver bullet. Visualization, like optimization can easily be misused. One of the goals for this book is to provide some guidelines to promote better visualizations.

In short, this book discusses how visualization can be used to support building and understanding optimization models, algorithms and their solutions. Its intended audience consists of both practitioners attempting to apply optimization techniques as well as researchers attempting to develop new optimization techniques. In addition, optimization algorithms are increasingly being employed to generate visualizations; this book will discuss that work as well.

Those looking for the level of mathematical certainty provided in typical books on optimization will be disappointed. Of course it is comforting to be able prove that a particular algorithm will produce the optimal solution to a given problem in a predictable amount of time. A provably optimal solution may be wasted if it is not understood, however. It is far more difficult, unfortunately, to prove that a given visualization is the "most understandable" for a particular situation. This should not be surprising. Optimization algorithms inhabit the precise, well-defined world of mathematics. Visualization, by definition, inhabits a much less well-defined world—the world of human beings. There *is* some science that will prove useful, but most of that science is not mathematical.

1.2 Some History

The use of visualization to support mathematical modeling is certainly not new. The quote from Descartes that begins this chapter illustrates the point. That Descartes made such a statement should not be surprising since Cartesian coordinates were named in his honor, although such a coordinate system had been used by the Chinese in the 12th century [1]. Even though Descartes advocated the use of visualization, graphical representations for numerical information were not widely used until the end of the 18th century [1].

As for the history of visualization for optimization, optimization is generally considered a part of Operations Research (OR) and Management Science (MS), although some would claim that it more properly is part of Mathematics or Computer Science. At any rate, OR grew out of World War II. Operations researchers applied mathematics (there were no real computers then) to solve operational problems. Today, OR is known primarily for its use of mathematics, including optimization to solve problems. Yet visualization was also used right at the birth of the field as well.

For example, in the spring of 1940, with Hitler's armies advancing on France, Winston Churchill faced a difficult decision: Should Britain deploy 10 more squadrons of fighter aircraft to France, or should the squadrons stay in Britain [9]? (The author is indebted to Peter Bell [10] for this anecdote.) Churchill felt a strong loyalty to France, but the head of Fighter Command, Hugh Dowding, was convinced that it would be a mistake since the loss of airplanes would leave Britain weakly defended. Dowding asked for an analysis of the situation from the Operational Research Section of Fighter Command (this group coined the term operational research). The presentation of that analysis persuaded Churchill that the Germans would quickly destroy the British squadrons, exposing Britain to great danger. The analysts took the time to *plot* the results because they believed they would be more convincing. As Larnder [9] wrote,

> Knowing from experience that people can be persuaded through their eyes when it is impossible to do so through their ears,

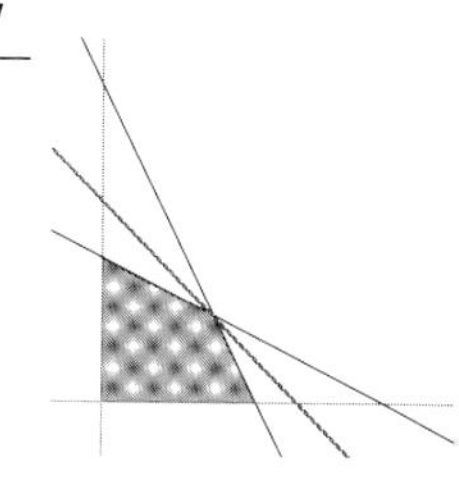

> he [Dowding] laid his graphs in front of the Prime Minister. In Dowding's considered view, "That did the trick." ::: Perhaps the real value of this OR study was not so much in presenting a Commander-in-Chief with facts concerning his own forces—facts of which he himself had shrewd knowledge—but, rather that by presenting them in graphical form OR had provided him with the means to oppose successfully what he knew would have been a fatal decision.

Not only were the 10 squadrons retained in Britain, most of the British squadrons deployed in France were withdrawn. Given the near defeat by Germany during the Battle of Britain (which started just two months later), it was clearly the correct decision. It was not just the mathematics that changed history; it was also the visualization.

In a later example, Dantzig, Fulkerson, and Johnson in 1954 [11] exploited a variety of representations in their seminal research on the use of cutting plane techniques to solve an instance of the traveling salesman problem. They sought to solve a 49-city traveling salesman problem, visiting one city in each of the lower 48 states in the US plus Washington, DC, using linear programming techniques. The original solutions contained subtours (Figure 1.1), so additional constraints—cutting planes—had to be added in order to find the optimal solution. Although this early work was essentially anecdotal, the ideas formed the basis of cutting plane techniques that have now been used to solve much larger traveling salesman problems [12].

Figure 1.1 comes from Dantzig, Fulkerson and Johnson's research [11] on cutting-plane algorithms for the traveling salesman problem. Displayed is the solution to a linear programming relaxation to the problem of finding the tour connecting the capitals of the lower 48 United States and Washington, DC. The optimal integer solution produces a tour, whereas, the linear programming relaxation allows fractional decision variables. By examining visualizations such as this, the authors were able to introduce a few additional constraints (cuts) that did not exclude any feasible tours. With the constraints added, the optimal linear programming solution contained no fractional decision variables meaning that it was the optimal tour.

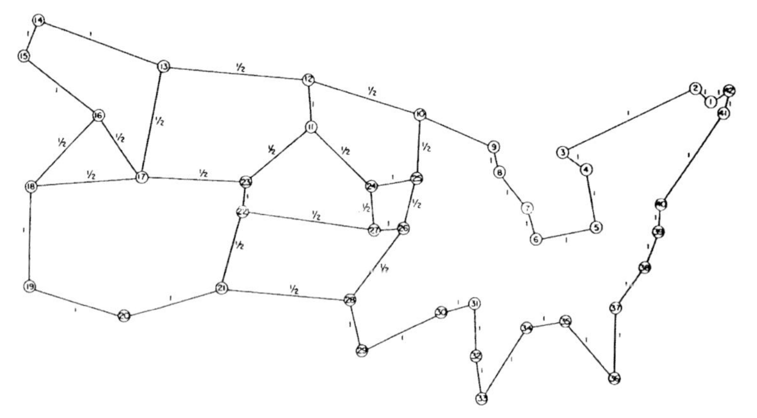

FIG. 18. A fractional solution x satisfying all loop conditions with $\Sigma\, d_{IJ}\, x_{IJ} = 698$.

Figure 1.1

Later on in the 1950's, Ford and Fulkerson [13], expanding on earlier work by Hitchcock [14] and Koopmans [15], developed algorithms for the minimum cost network flow problem (Figure 1.2), which has application to problems in transportation and inventory planning, among many others. Minimum cost network flow problems have a natural graphical representation.

From Koopmans' [15] pioneering research on minimum cost network flow algorithms, the figure below shows the optimal solution to a transportation problem, a special case of the minimum cost network flow algorithm.

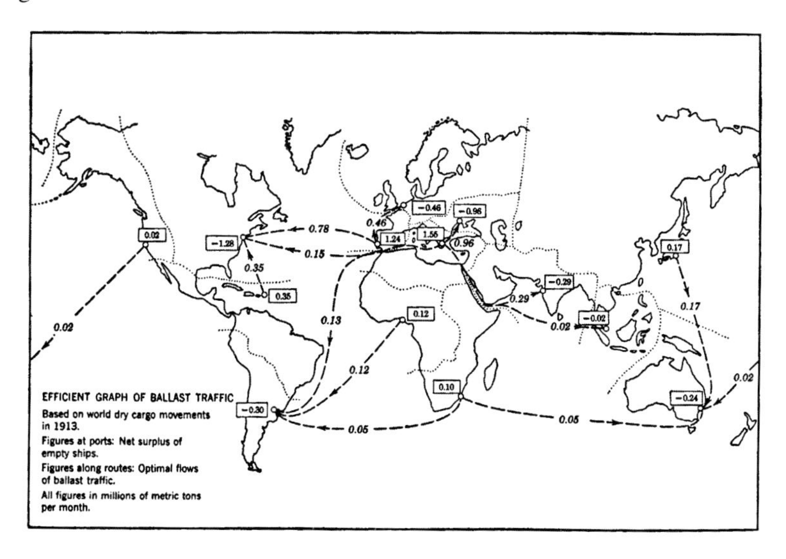

Figure 1.2

Around the same time as Ford and Fulkerson project management techniques such as Project Evaluation and Review Technique (PERT) and the Critical Path Method (CPM) [16], [17], [18] were developed to manage large projects consisting of thousands of interrelated tasks (Figure 1.3). Although specialized algorithms are usually employed to analyze these problems, they can easily be formulated as linear programs. Again, the problem is conveniently represented graphically, which is part of its appeal. Many companies now sell PC-based project management systems that allow project managers to draw network representations of their projects. They can then analyze the criticality of each task to the timely completion of the entire project.

Displayed below is figure 5 from [16], which shows a simple PERT network. Each circle represents a milestone in a project. The arrows represent tasks that must be completed.

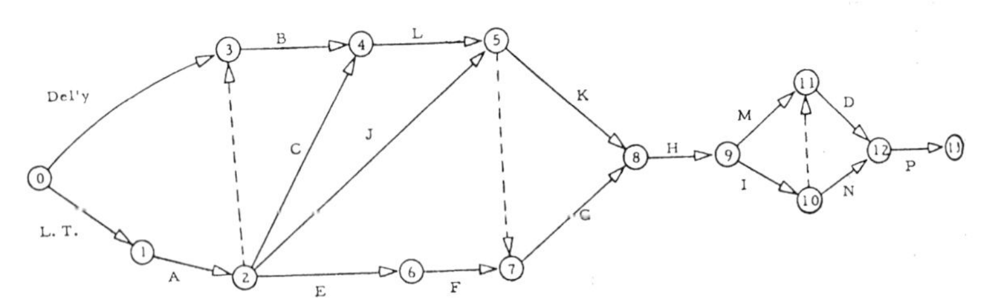

Figure 1.3

In the 1970's, vehicle routing algorithms began to be used widely in industry. The solutions to these problems are frequently represented graphically as routes of vehicles on a map, for example, see Cullen, Jarvis, and Ratliff [19], Babin, *et al.* [20], and Savelsbergh [21]. Users can change the routes interactively, with an evaluation provided of the change. In work that won the Edelman Prize in 1983, Bell, *et al.* [22] developed novel optimization techniques for vehicle routing and relied on an interactive computer graphic implementation (see [23]).

Today, the development of interactive vehicle routing systems continues [24], spurred on in part by the emergence of a *Geographic Information System* (GIS) [25]. As [24] state, GISs "support the capture, management, manipulation, analysis, modeling and display of spatially referenced data." GISs have allowed for much more realistic problems to be solved, since they provide detailed, accurate data concerning travel times and routes between destinations.

In the 1980's, in another Edelman Prize winning application, Lembersky and Chi [26] applied dynamic programming and three-dimensional interactive computer graphics to train loggers how to cut raw logs into lumber to yield higher profit. Loggers see a three-dimensional representation of a particular log to be cut. The loggers can then, by hand, specify how the log should be cut as well as compare their cutting pattern to the optimal pattern. Through this interaction, loggers are trained to cut their logs to yield higher profits.

1.3 Summary

In conclusion, visualization has been used throughout the history of optimization. The difference today is that the tools available to promote visualization have significantly improved. We can easily create multi-dimensional, multi-colored animated displays to help modelers and decision makers understand their problems. This book attempts to:

- Present basic background material from cognitive psychology and other fields relevant to visualization.
- Provide guidance for those seeking to visualize their problems, their models, and their algorithms.
- Survey existing research and practice using visualization to support optimization analysis.
- Suggest directions for academic researchers interested in pursuing research on visualization applied to optimization.

- Survey research that has applied optimization to creating visualizations.

1.4 Outline of the book

The book is divided into three parts. Part I presents relevant background material from cognitive psychology, computer graphics, visual design and other areas. It discusses optimization very little; rather it attempts to present important background material. Part II discusses how visualization has been or might productively be used to support the distinct phases of the modeling life-cycle. Part III discusses how different representation formats have been or could be used to support optimization.

1.5 Flip Chart

On each odd numbered page is a single image in the upper right corner. These images produce an animation when the pages of the chapter are quickly flipped. The animation beginning at the start of this chapter shows a common representation used to teach linear programming. The picture shows the following linear program:

$$\begin{array}{rrcl} \text{max} & x+y & & \\ \text{subject to:} & & & \\ & x+2y & \leq & 12 \\ & 2x+y & \leq & 12 \\ & x,y & \geq & 0 \end{array}$$

as the right-hand side of the first constraint is decreased from 12 to 4.

Bibliography

[1] Collins BM. Data visualization—has it all been seen before? In Earnshaw RA, Watson D, editors, *Animation and Scientific Visualization: Tools and Applications*. London:Academic Press, 1993.

[2] Geoffrion AM. The purpose of mathematical programming is insight, not numbers. *Interfaces*, 1976;7(1):81–92.

[3] Hamming RW. *Numerical Methods for Scientists and Engineers*. New York:McGraw-Hill, 1962.

[4] Greenberg HJ. A summary of remaining goals and strategies for the development of an intelligent mathematical programming system. Technical

report, Denver, Colorado:Department of Mathematics, University of Colorado, 1988.

[5] McCormick BH, DeFanti TA, Brown MD. Visualization in scientific computing. *Computer Graphics*, 1987;21(6):1–14.

[6] Mackinlay JD, Robertson GG, Card SK. The perspective wall: Detail and context smoothly integrated. In Robertson SP, Olson GM, Olson JS, editors, *Proc. ACM Computer-Human Interaction '91 Conference on Human Factors in Computing Systems*, pages 173–179, New York:1991. ACM Press.

[7] Willemain TR. Insights on modeling from a dozen experts. *Operations Research*, 1994;42(2):213–222.

[8] Woolsey G. Where were we, where are we, where are we going, and who cares? *Interfaces*, 1993;23(5):40–46.

[9] Larnder HL. The origin of operational research. In Haley KB, editor, *Operational Research '78*. Amsterdam:North-Holland, 1979.

[10] Bell PC. Visual aid: Interactive graphics come of age in the OR/MS community. *OR/MS Today*, 1992;19(4):24–27.

[11] Dantzig G, Fulkerson R, Johnson S. Solution of a large-scale traveling-salesman problem. *Operations Research*, 1954;2.

[12] Grötschel M, Padberg MW. Polyhedral theory. In Lawler EL, Lenstra JK, Kan AHGR, Shmoys DB, editors, *The Traveling Salesman Problem: A Guided Tour of Combinatorial Optimization*. Chicester, United Kingdom:John Wiley & Sons, 1985.

[13] Ford LR, Fulkerson DR. *Flows in Networks*. Princeton (NJ):Princeton University Press, 1962.

[14] Hitchcock FL. Distribution of a product from several sources to numerous localities. *Journal of Mathematical Physics*, 1941;20:224–230.

[15] Koopmans TC. Optimum utilization of the transportation system. *Econometrica*, 1949;17:3–4.

[16] Kelley J. Critical path planning and scheduling: Mathematical basis. *Operations Research*, 1961;9(3):296–321.

[17] Malcolm DG, Roseboom JH, Clark CE, Fazar W. Applications of a technique for R and D program evaluation (PERT). *Operations Research*, 1959;7(5):646–669.

[18] Moder JJ, Phillips CR, Davis EW. *Project Management with CPM, PERT and Precedence Diagramming*. New York:Van Nostrand Reinhold Company, 3rd edition, 1983.

[19] Cullen FH, Jarvis JJ, Ratliff HD. Set partitioning based heuristics for interactive routing. *Networks*, 1981;pages 125–143.

[20] Babin A, Florian M, James-Lefebvre M, Spiess H. EMME/2: An interactive graphic method for road and transit planning. Technical Report 204, Montréal, Canada:Centre de Recherche sur les Transports, Université de Montréal, 1982.

[21] Savelsbergh MWP. Computer aided routing. Technical report, Amsterdam, The Netherlands:Centrum voor Wiskunde en Informatica, 1988.

[22] Bell WJ, Dalberto LM, Fisher ML, Greenfield AJ. Improving the distribution of industrial gases with an on-line computerized routing and scheduling optimizer. *Interfaces*, 1983;13(6):4–23.

[23] Fisher ML, Greenfield A, Jaikumar R. Vergin: A decision support system for vehicle scheduling. Technical Report 82-06-02, Philadelphia (PA):Department of Decision Sciences, The Wharton School, University of Pennsylvania, 1982.

[24] Bodin L, Fagan G, Levy L. Vehicle routing and scheduling problems over street networks. Technical report, College Park (MD):College of Business and Management, University of Maryland, 1993.

[25] Laurini R, Thompson D. *Fundamentals of Spatial Information Systems.* London:Academic Press, 1992.

[26] Lembersky MR, Chi UH. Decision simulators speed implementation and improve operations. *Interfaces*, 1984;14(4):1–15.

Part I

A Framework for Visualization and Optimization

The five chapters of this part provide a whirlwind tour of some basic theories, frameworks, and ideas from a variety of fields including cognitive psychology, human-computer interaction, computer graphics, computer science, art and design. Clearly those fields are far too broad to be covered in three chapters, let alone a single book. The focus is on some of the more important ideas and results. In the second part of the book, we build on the basics presented here, emphasizing their application to visualization and optimization.

In order to organize our discussion of visualization and optimization, we present a simple framework (Figure 1.4). The framework is:

> The success of visualization for a problem-solving project depends on the *tasks* to be accomplished, the *people* involved, and the representations *formats* used for the task and the people.

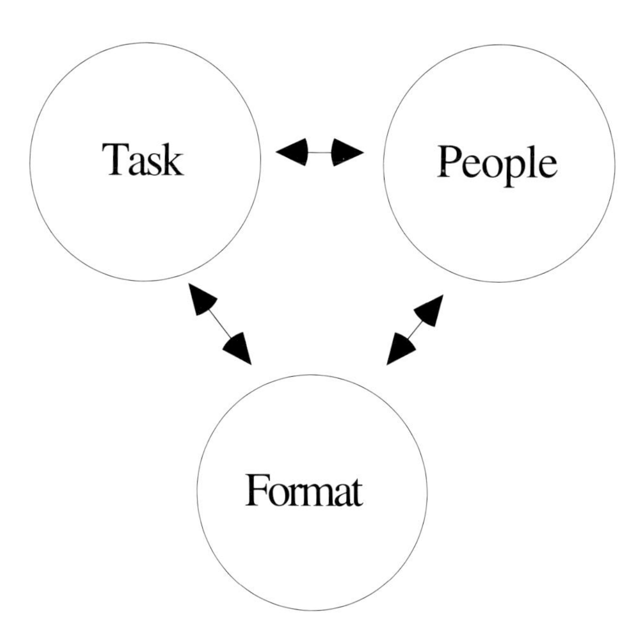

Figure 1.4: A framework for visualization and optimization.

In other words, different people may benefit from different representation formats for different tasks.

Although this framework may seem obvious, it has not always been apparent to many. For example, research comparing the effectiveness of traditional two-dimensional plots to tabular representations (known as the "graphs-vs-tables" question) is widely acknowledged to be "a mess" [1]. The research results to date have been almost completely equivocal. Some studies found no difference in performance, others have found that tables produced better performance, and others found that graphs produced better performance.

The explanation, in the abstract at least, is quite simple. Most of the studies used different tasks, representations, and subjects. Within each study, only variations in representation format were made. Each individual experiment therefore was valid only for the particular task and subject population studied. The real question is not graphs versus tables, but rather, what characteristics of tasks and subjects favor graphs over tables (and vice versa)?

Another example on the importance of representation to problem solving comes from Larkin and Simon [2]. Consider the following game:

> Each of the digits 1 through 9 is written on a separate piece of paper. Two players draw digits alternately. As soon as either player gets any three digits that sum to 15, that player wins. If all nine digits are drawn without a win, then the game is a draw.

One might be surprised to learn that this game is equivalent to tic-tac-toe (sometimes called "noughts and crosses"). If one places the digits 1 through 9 in the following table, however, the correspondence is clear.

6	1	8
7	5	3
2	9	4

Any win in tic-tac-toe corresponds to a selection of digits, three of which sum to 15. In this case, a visual representation seems paramount.

However, another example, from Adams (via Glass and Holyoak[3]) shows the importance of textual representations over visual representations:

> Consider a sheet of paper 1/100th of an inch thick. Fold the paper in half, and then in half again. Repeat this process 50 times in total. Of course, it is not physically possible to perform this task, but imagine if you can. How thick is the resulting folded paper?

In this case, the visual representation gives few clues, but using standard mathematical notation–text–the answer can easily be found, $1/100 \times 2^{50}$ inches.

The importance of matching the characteristics of the task, the viewer and the format has been called *cognitive fit* by Vessey [4], [5]. Vessey [4] defines cognitive fit as

> a cost-benefit characteristic that suggests that, for most effective and efficient problem solving to occur, the problem representation and any tools or aids employed should all support the strategies (methods or processes) required to perform that task.

In a review of the graphs versus tables literature [5], based on the cognitive fit hypothesis, Vessey was able to offer a cogent explanation as to why some studies found graphs superior to tables and others found tables superior to graphs. Each study considered different tasks and different representation formats. Whatever format was best matched to the task yielded superior performance.

The cognitive fit hypothesis implies that representation format matters; it affects human performance in solving problems. By carefully matching the representation format to the task at hand, performance can be enhanced. One of the goals of this book is to explore appropriate representation formats for the many tasks involved in optimization modeling.

1.6 Tasks

Numerous authors have proposed frameworks for classifying the tasks involved in an optimization project. These frameworks assume that optimization projects involve a series of stages or steps, that is, a modeling *life-cycle*.

1.6.1 The Modeling Life Cycle

The life-cycle idea is embodied in Simon and Newell's [6], [7] model of decision making, Intelligence-Design-Choice. According to his framework, in the Intelligence phase of problem solving, information is gathered in an attempt to understand the nature of the problem—the boundaries of the problem, those affected by the problem, the costs and benefits of the problem. In the Design phase, alternative solutions to the problem are constructed. In the Choice phase, one or more of the alternatives are selected. Many problems concentrate more on one of these phases than on another.

For optimization modeling, many different authors have described the stages in the modeling life-cycle, with varying levels of detail. Table 1.1

Authors		
Simon and Newell [6],[7]	*Churchman, Ackoff and Arnoff* [8]	*This Book*
Intelligence	Formulating the problem	Model Development
Design	Constructing a mathematical model to represent the system under study	Algorithm Development
	Deriving a solution from the model	
	Testing the model and the solution derived from it	Solution Analysis
Choice	Establishing controls over the solution	Results Presentation
Implementation	Putting the solution to work: implementation	Implementation

Table 1.1: Comparison of different versions of the modeling life-cycle.

lists three of the different frameworks, attempting to match steps from one framework with those from another.

For purposes of this text, we shall simplify the modeling life-cycle into five stages:

1. *Model Development.* This includes identifying the underlying problem, collecting and analyzing data, formulating a mathematical model.

2. *Algorithm Development.* Depending on the problem identified in the previous stage, algorithm development may be trivial, if an existing piece of software can be used to analyze the model. However, for many optimization problems, the algorithm must be carefully constructed.

3. *Solution Analysis.* The "solutions" produced by algorithms must be tested, probed and debugged. The data could (will?) have errors, the model could have problems, or the algorithm could fail to converge. Furthermore, even with a debugged model, data and algorithm, one must then attempt to understand their behavior. When the price of a set of inputs rises, how does that affect the solution? If one relaxes some constraints, how does the solution change? In other words, in this phase, one also attempts to relate the information provided by the algorithm back to the real problem being attacked.

4. *Results Presentation.* The results generated by the previous phase must be presented to the people in charge, the people paying for the study, the people actually responsible for solving the problem. They will probably be skeptical; they must be convinced that the recommendations being made make sense. Usually they will not really understand the sophisticated mathematical and computer techniques used to solve the problem. But, they need to be convinced that the recommendations are sound.

5. *Implementation* Given the results of the analysis, the choice must be implemented. It must be communicated to those responsible for effecting the change, and the process of change must be continually monitored.

Although this framework contains fewer steps than many of the proposed modeling life-cycles, it arguably captures the essence of how models evolve over time, starting with identifying the problem and ending with final results.

One can easily be lulled by the clean picture presented by any such life-cycle model. In actual modeling projects, progress does not follow cleanly and surely from one stage to another. Often problems that are undetected in previous stages are only uncovered in later stages, forcing a retreat back to an earlier stage to fix the undetected problem. In a recent study [9], [10] expert modelers were found to move quite often among the different tasks in the modeling life-cycle.

1.6.2 Visualization and the Modeling Life Cycle

From the viewpoint of visualization, the modeling life-cycle transforms vague, poorly defined representations of a problem into an understandable, convincing solution to the problem. During the modeling life-cycle, representations such as mathematical formulations, inputs to algorithms, program listings, algorithm outputs, among others can be of use. The field of optimization has concentrated most of its energy on merely one phase of the modeling life-cycle: algorithm development. The modeling life-cycle, on the other hand, involves many other activities. Representation is a common thread underlying all the phases of the modeling life-cycle. Therefore, the study of those representations should help improve the chances for success in a modeling project.

Different representations are required at different phases of the modeling life-cycle (Table 1.2). In the next chapter, for each phase of the modeling life-cycle, we discuss appropriate representations. In the following chapter, we discuss how the different representations can be used to support various aspects of optimization.

1.6.3 Summary

This section has presented several versions of the modeling life-cycle. Although they differ, they all basically follow the same idea: modeling consists of a variety of different tasks. Although actual modeling projects are usually far messier than the clean picture presented by these life-cycle models, the tasks listed in the life-cycle models do in fact occur.

The chapters in the rest of this part of the book (Chapters 2-6) discuss the other two components of our basic framework, people and format. Chapter 2 discusses models of human perception and cognition, as well as some of the differences among people that have been identified by cognitive psychologists. Chapters 3-6 discuss in detail theories, frameworks and recommendations for the variety of different representation formats that can be used to represent complex information.

Part II of the book considers how visualization can support each phase of the modeling life-cycle. Part III of the book discusses how different representation formats can be used to support optimization.

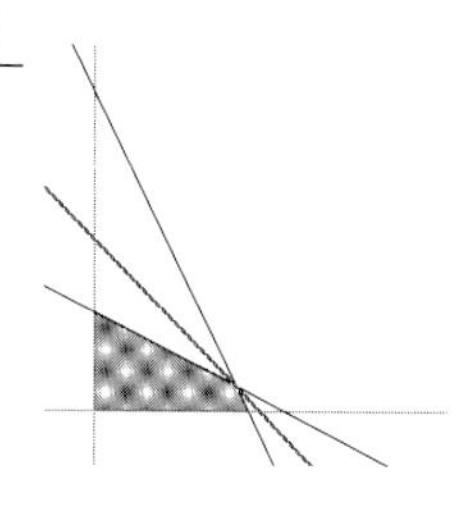

	Model Development	*Algorithm Development*	*Solution Analysis*	*Results Presentation*
Text	Algebraic Languages	Programming Languages	Standard Output	Narrative
Tables	Spreadsheets; Block Structured	Matrix Images	Matrix Images	Summary Tables
Static Graphics	Graph-Based	Visual Languages	Presentation Graphics	Presentation Graphics
Animated or Interactive Graphics	Modeling by Example	Algorithm Animation	Animated Sensitivity Analysis	Algorithm Animation; Animated Sensitivity Analysis
Sound	?	?	?	?
Touch	?	?	?	?

Table 1.2: The different types of representations that can be useful in different phases of an optimization project

Bibliography

[1] Coll RA, Coll JH, Thakur G. Graphs and tables: A four-factor experiment. *Communications of the ACM*, 1994;37(4).

[2] Larkin JH, Simon HA. Why a diagram is (sometimes) worth ten thousand words. *Cognitive Science*, 1987;11:65–99.

[3] Glass AL, Holyoak KJ. *Cognition*. New York:Random House, 2nd edition, 1986.

[4] Vessey I. Cognitive fit: A theory-based analysis of the graphs versus tables literature. *Decision Sciences*, 1991;22:219–241.

[5] Vessey I, Galletta D. Cognitive fit: An empirical study of information acquisition. *Information Systems Research*, 1991;2(1):63–86.

[6] Newell A, Simon HA. *Human Problem Solving*. Englewood Cliffs (NJ):Prentice-Hall, Inc., 1972.

[7] Simon HA. *The New Science of Management Decision*. New York:Harper and Row, 1960.

[8] Churchman CW, Ackoff RL, Arnoff EL. *Introduction to Operations Research*. New York:John Wiley and Sons, 1957.

[9] Willemain TR. Insights on modeling from a dozen experts. *Operations Research*, 1994;42(2):213–222.

[10] Willemain TR. Model formulation: What experts think about and when. *Operations Research*, 1994;page forthcoming.

Chapter 2

People

The information processing capabilities of computers are reasonably well understood, at least when compared to the information processing capabilities of human beings. With computers, the underlying architecture is at least known, and much theory has been developed. With respect to human information processing capabilities, the underlying architecture remains less clearly understood, and the theories that do exist are only provable in the empirical sense. Yet there are nuggets of information about human information processing that are relevant to user interfaces. A general review of cognitive psychology can be obtained from texts such as Lindsay and Norman [1], Bailey [2] or Monsell [3]. Both Allen [4] and Card, Moran and Newell [5] give a more thorough discussion of cognitive issues related to user interfaces than can be presented here.

Card, Moran and Newell [5] proposed perhaps the most widely adopted model of human information process. According to their model, the human information-processor is comprised of three different systems:

The Perceptual System External stimuli are detected by the senses and transmitted to the cognitive and motor system for further processing or action.

The Motor System Physical actions are produced through the actions of muscles moving parts of the body.

The Cognitive System The perceptual stimuli, stored knowledge, and stored procedures for responding to the stimuli based on the knowledge are processed to produce appropriate actions, either extra stored knowledge or procedures, and physical actions (effected through the motor system).

Each one of these processing systems has particular limitations.

2.1 The Perceptual System

Each component of the perceptual system, vision, hearing, taste, smell, and touch, has limitations and strengths. For visualization, vision clearly is of most interest, but the other senses can be exploited to enhance understanding. Some [6] have recommended that a better term than visualization is *perceptualization*.

For vision, for example, the human retina has a certain minimum response time, that is, the time required for a spot on the retina to register a change in the light impinging on it. For human beings, this time is approximately 60-100 msec. This phenomenon, known as *persistence of vision*, underlies any moving image system such as film and television. To create the appearance of a moving image, the image must be changed at least once every 100msec, which implies drawing at least 10 frames per second. Although a crude approximation, it does provide an accurate lower bound on commercial television refresh rates (25-30 frames per second) and motion picture refresh rates (24 frames per second).

2.2 The Motor System

Arms, fingers and legs can move only with limited speed and limited accuracy. The minimum reaction time is approximately 30 msec (ranging to a high of 70 msec) [5]. This would imply that the fastest typing speed would be approximately 1 keystroke every 140 msec, the down stroke and upstroke each taking 70 msec. This corresponds reasonably well with measured results with actual typists, that are in the range of 100 to 300 msec per keystroke, though there is great variability depending on the type of text being typed and the skill level of the typist.

Another limitation of motor skill has applications to the design of screen layouts. Fitt's law [5] allows one to estimate the time in msec T required for one's hand to be move a distance D to a target located at position of size S:

$$T = K \log_2(2D/S)$$

where K is a constant approximately in the range of 27 to 122 msec. Examining the equation, movement time increases only as the logarithm of the distance to be traveled, and decreases as the logarithm of the size of the target increases. Or alternatively, movement time depends only on the ratio of the distance to be moved and the size of the target. For example, Card, Moran and Newell [5] give an example of the improvement in time possible if a gold "shift" key on a calculator is moved closer to more commonly used keys on the calculator.

2.3 The Cognitive System

Many authors organize human memory roughly into short and long-term memory. Using common terms from computer architecture, short-term memory can be thought of as a small set of registers that store readily accessible information. Long-term memory can be roughly thought of as main memory or secondary storage such as disks or tapes and stores much larger quantities of information than short-term memory. Long-term memory takes much longer to access ("to bring to consciousness") than short-term memory.

2.3.1 Short-Term Memory

Much research has been devoted to refining models of the architecture of these memories. Miller [7], in a classic paper, suggested that the size of short-term memory was seven plus or minus two items (e.g., registers), or chunks of information. Other authors have suggested somewhat smaller sizes for short-term memory, e.g., Simon [8]. Although capacity in terms of the ill-defined chunk appears small, the concept of a chunk is variable. In other words, although the capacity of short-term memory is limited, the information contained in each chunk expands as people learn. This implies that as users become familiar with a system, they usually want to replace a sequence of commonly used commands or actions by a single command or action.

2.3.2 Long-Term Memory

Much research has been devoted to the idea that as facts are assimilated, more complicated structures are assembled. These structures facilitate the creation of more detailed chunks, as discussed in the previous section. Yet, the more details stored in the memory, the greater the chance for interference effects. For example, since different computer programs provide different mechanisms for halting (e.g., "end", "quit", "halt"), the more computer programs that one learns, the more confusion as to which command is appropriate in a particular circumstance. One of the advantages of user interfaces such as on the Apple Macintosh and Microsoft Windows is that commonly used functions are performed in the same fashion, regardless of the particular application program.

2.3.3 Anchoring

People seem to prefer to work with their problems using the representations they already know. Ghani [9] found, for example, that people preferred to use the representation format they originally used when given a choice between graphs and tables. In an example cited by Larkin and Simon [10], the oral form of Croatian differs from Serbian about as much as the oral form of American English differs from British English. Since Croatian is written using the Roman alphabet and Serbian is written using the Cyrillic alphabet, Croats and Serbs find it difficult to read each other's literature. Each is tightly anchored to a particular alphabet.

We call this phenomenon *anchoring*. Anchoring, of course, has been widely studied in the decision analysis community [11], wherein people often seem to base decisions on an initial starting value rather than on a "rational" exploration of the entire range of alternatives.

The user interface community seems well aware of anchoring; one of the principal commandments of user interface design is: "know the user." The now common "desktop metaphor" for graphical user interfaces (GUI) as found in Microsoft Windows, OS/2 and the Apple Macintosh has arguably achieved success because it attempts to simulate a physical desktop.

Similarly, the optimization community seems implicitly aware of the phenomenon. Algebraic languages including GAMS [12], AMPL [13], MPL [14], LPL [15] and MODLER [16], [17] attempt to duplicate the commonly used algebraic notation in a computer-readable fashion. Rather than having to translate from a familiar representation style (algebraic syntax) into an unfamiliar style (matrix generator) (see [18]), modelers can work with a widely used representation. Correspondingly, advocates of the use of spreadsheets for optimization [19], [20] argue that millions of spreadsheet users can now easily become optimizers with little extra effort, though others have cautioned about the dangers of allowing unsophisticated users access to powerful tools [21], [22].

Anchoring also seems to be at the heart of *Visual Interactive Modeling* (VIM) [23], [24], [25]. More a philosophy than a specific technique, VIM argues that modeling should begin first with an elicitation of the decision maker's problem in graphical form, not with the development of a mathematical model. Through this process the actual underlying problem can be revealed. By discovering understandable representations early, one hopes to do a better job of conveying the results of the mathematical models, once they are developed.

2.4 Different Types of Users

Different people, however, may be anchored to different representations. *Analysts*, experts in optimization technology after years of training, are (usually) quite comfortable with algebraic formulations. *Decision mak-*

ers, experts in the problem that needs to be solved, often find algebraic representations hopelessly obtuse. Decision makers are usually quite familiar with their problem. They have their own representations of aspects of the problem. Unfortunately, that preferred representation format is usually quite different from algebraic notation. Analysts and decision makers use distinct vocabulary, terms, and representations in their job.

One of the principal tasks of the analyst, therefore, is to understand the milieu of the decision maker. The decision maker, although she may of necessity need to learn about the analyst's bag of tricks, should not have to. The analyst must translate from the decision maker's representations into the representations needed for optimization and back.

Of course, one could make much finer distinctions among the types of people involved in an optimization project. Decision makers could be divided into different levels, for example, those responsible to day-to-day interaction with the optimization project, and higher level management to whom final results are to be presented. Similarly, analysts could consist of computer programmers and optimization specialists, among others. The greatest divide, however, usually occurs between those who understand the technical details of the mathematics and computer science and those who understand the problem to be solved. In an ideal problem-solving project, these two groups should converge towards a unified, valid solution. The challenge is to facilitate communication between these groups.

This level of difference fits roughly into a branch of cognitive psychology that has studied the differences between experts and novices. In a classic experiment in 1973, Chase and Simon [26] tested the difference between master and novice chess players. Both the novice and master players were shown chess boards with chess pieces arranged on the boards for a short time. They were then asked to reconstruct the positions of the pieces on the boards. When the pieces were arranged as they might be found in a typical chess game, master players were far better than novices at reconstructing the positions of the pieces. However, when the pieces were placed at random, there was no significant difference between expert and novice performance. One explanation for the difference lies in the differing abilities of expert and novice players to organize information. In this hypothesis, the master players are better able to group together arrangements of players on a board, as long as the board represents a realistic game. They do not possess any extra grouping ability over novice players when the arrangement does not correspond to his/her area of expertise. This experiment has been conducted in many other contexts with similar results.

Orlikowski and Dhar [27] conducted similar experiments in the realm of building mathematical programming models. They found that novice LP modelers created models out of chunks such as variables and constraints, whereas expert LP modelers used chunks consisting of larger model pieces.

2.5 Visualization Ability

Researchers in cognitive psychology have developed a construct called *visualization ability* that measures "the ability to manipulate or transform the image of spatial patterns into other arrangements" [28]. People with high visualization ability have been shown in a variety of contexts to learn more easily and have higher performance than those with lower abilities [29]. Furthermore, those with poor visualization ability are helped by specifically tailoring materials to augment their poor visualization ability. Good visualizers are also helped by better visualization tools [29]. This result provides support for the use of visualization tools. First, those who are poor visualizers can theoretically be helped by using visualization tools. Second, those who are good visualizers are also helped by appropriate visualization tools.

2.6 Left Brain, Right Brain

The brain is divided into two halves or *hemispheres*, one on the left and one on the right. During the past 150 years, a variety of observations and formal experiments have explored the differing cognitive abilities of each hemisphere [30]. Damage to the left hemisphere can result in a loss of ability to talk (*aphasia*), for example, whereas damage to the right hemisphere can result in a loss of ability to navigate in unfamiliar surroundings. Much more has been learned during the past 150 years about the differences in the two hemispheres. and we cannot possibly do justice to that work here. A very thorough survey can be found in [30].

A great deal of that research has relied on so-called *split-brain* subjects. For a particular medical reason, for example, epilepsy, the communications pathways (consisting of over 200 million nerve fibers in the corpus callosum) connecting the two hemispheres of the brain are sometimes cut. By presenting stimuli to the eyes or hands on one side of the body of these split-brain subjects, experimenters can communicate directly and exclusively with one hemisphere of their brains. They can then assess the differing cognitive abilities of the two hemispheres. Note that, in general, the left hemisphere of the brain actually controls the right side of the body, and the right hemisphere controls the left.

Consider the following simple experiment described in [30]. A picture of an object, a cup, say, is shown to the left brain (through the right eye) of a split-brain subject. When the subject is asked to name what they saw, she verbally identifies the cup. In contrast, if another object, a spoon, say, is shown to the right brain (through the left eye), the subject states that she has seen nothing. However, when asked to select the object by using her left hand to feel a collection of objects hidden from view, she picks up the spoon. When asked to name the object, she calls it a pencil.

Based on many similar experiments, in general, the left hemisphere specializes in verbal, analytical, logical processing whereas the right hemisphere specializes in visual, holistic processing. The following table summarizes some of the differences between the left and right hemispheres of the brain.

Left Hemisphere	*Right Hemisphere*
Verbal	Nonverbal, visuo-spatial
Sequential, temporal, digital	Simultaneous, spatial, analogical
Logical, analytical	Gestalt, synthetic
Rational	Intuitive
Western thought	Eastern thought

Figure 2.1

This left brain, right brain dichotomy has captured a place in popular consciousness, with claims often made that lack little scientific support. Some, as noted in [30], have even "equated the left hemipshere with the evils of modern society." In fact, with respect to the table in Figure 2.1, Springer and Deutsch cautioned,

> The descriptions near the top of the list [Figure 2.1] seem to be based on experimental evidence; the other designations appear more speculative.

Although the difference in cognitive abilities of each hemisphere is real, the exact differences are at best characterized by rather general descriptions (what are the exact characteristics of an analytical, visual, logical or temporal cognitive process?). Moreover, exactly what effects are caused by the differences remains not well established. All the world's problems cannot be attributed to differences in cognitive abilities of the left and right hemispheres of the brain.

One can only hypothesize about how left and right brain differences relate to visualization and optimization. Optimization is highly analytical, with many details to be mastered. Furthermore, optimization models are typically expressed in textual or numeric form. One could then hypothesize that optimization modeling uses a relatively larger measure of the left brain, since the left brain specializes in verbal and analytical skills. On the other hand, one also could hypothesize that scientific visualization involves a large measure of right hemisphere processing, since the right hemisphere specializes in visual skills. If both of these hypotheses are true, then one might propose that using both visualization and optimization (and other analytical tools) might lead to a better understanding of a particular problem, since literally more brainpower would be attacking the problem. Actual experiments to confirm this hypothesis have not, to our knowledge, been performed.

2.7 Summary

This chapter has discussed a variety of issues related to the people involved in a problem-solving endeavor. Limits of the human perceptual, cognitive and motor systems limit our performance. By understanding these limits, one can hope to avoid them in designing visualizations. Furthermore, humans demonstrate a variety of biases in their ability to learn and adapt to new technologies. Anchoring for example, limits people's ability to understand unfamiliar representations no matter how elaborate or technologically sophisticated. In the next three chapters, we discuss issues relating to different representation formats that can be used for visualization. Issues discussed include results from cognitive and perceptual psychology, computer science, graphic design, among others.

Bibliography

[1] Lindsay P, Norman DA. *Human Information Processing.* New York:Academic Press, 1977.

[2] Bailey RW. *Human Performance Engineering: A Guide for System Designers.* Englewood Cliffs (NJ):Prentice-Hall, Inc., 1982.

[3] Monsell S. Representation, processes, memory mechanisms: The basic components of cognition. *Journal of the American Society of Information Science*, 1981;32.

[4] Allen RB. Cognitive factors in human interaction with computers. In Badre A, Shneiderman B, editors, *Directions in Human/Computer Interaction.* Norwood (NJ):Ablex Publishing Corporation, 1982.

[5] Card S, Moran TP, Newell A. *The Psychology of Human-Computer Interaction.* Hillsdale (NJ):Lawrence Erlbaum Associates, 1983.

[6] Erickson T. Artificial realities as data visualization environments: Problems and prospects. In Wexelblat A, editor, *Virtual Reality: Applications and Explorations.* Boston:Academic Press, 1993.

[7] Miller AR. *The ABC's of AUTOCAD.* San Francisco:Sybex, 1988.

[8] Simon HA. How big is a chunk? *Science*, 1974;183:482–488.

[9] Ghani J. *The Effects of Information Representation and Modification on Decision Performance.* [dissertation], Philadelphia (PA):The Wharton School, The University of Pennsylvania, 1981.

[10] Larkin JH, Simon HA. Why a diagram is (sometimes) worth ten thousand words. *Cognitive Science*, 1987;11:65–99.

[11] Kahneman D, Slovic P, Tversky A, editors. *Judgement Under Uncertainty: Heuristics and Biases.* Cambridge, United Kingdom:Cambridge University Press, 1982.

[12] Brooke A, Kendrick D, Meeraus A. *GAMS: A User's Guide, Release 2.25.* South San Francisco (CA):The Scientific Press, 1992.

[13] Fourer R, Gay DM, Kernighan BW. A mathematical programming language. *Management Science*, 1990;36(5):519–554.

[14] Maximal Software, Inc. *MPL Modeling System.* Arlington (VA):Maximal Software, Inc., 1994.

[15] Hürliman T. LPL: A structured language for linear programming modeling. *OR Spectrum*, 1988;10:53–63.

[16] Greenberg HJ. MODLER: Modeling by object-driven linear elemental relations. *Annals of OR*, 1992;38:239–280.

[17] Greenberg HJ. *Modeling by Object-Driven Linear Elemental Relations: A User's Guide for MODLER*. Boston (MA):Kluwer, 1993.

[18] Fourer R. Modeling languages versus matrix generators for linear programming. *ACM Transactions on Mathematical Software*, 1983;9:143–183.

[19] Savage S. Lotus Improv as a modeling language. In *ORSA/TIMS Joint National Meeting*, San Francisco (CA):1992.

[20] Vazsonyi A. Where we ought to be going: The potential of spreadsheets. *Interfaces*, 1993;23(5):26–39.

[21] Gass SI. Model world: Danger, beware the user as modeler. *Interfaces*, 1990;20(3):60–64.

[22] Wagner H. The next decade in operations research: Comments on the CONDOR report. *Operations Research*, 1989;37(4):664–672.

[23] Bell PC. Visual interactive modeling as an OR technique. *Interfaces*, 1985;15:26–33.

[24] Bell PC. Visual interactive modeling in operational research: Successes and opportunities. *Journal of the Operational Research Society*, 1985;36:975–982.

[25] Bell PC. Visual interactive modeling in 1986. In Belton V, O'Keefe R, editors, *Recent Developments in Operational Research*, pages 1–12. Oxford, United Kingdom:Pergamon Press, 1986.

[26] Chase WG, Simon HA. The mind's eye in chess. In Chase WG, editor, *Visual Information Processing*. New York:Academic Press, 1973.

[27] Orlikowski W, Dhar V. Imposing structure on linear programming problems: an empirical analysis of expert and novice modelers. In *Proceedings of the National Conference on Artificial Intelligence*, volume 1, 1986.

[28] Ekstrom RB, French JW, Harman HH. *Manual for Kit of Factor-Referenced Cognitive Tests*. Princeton (NJ):Educational Testing Service, 1976.

[29] Sein MK, Olfman L, Bostrom RP, Davis SA. Visualization ability as a predictor of user learning success. *International Journal of Man-Machine Studies*, 1993;39:599–620.

[30] Springer SP, Deutsch G. *Left Brain, Right Brain*. New York:W. H. Freeman and Company, 3rd edition, 1989.

Chapter 3

Text and Tables

3.1 Introduction

Information can be represented in many forms, including text, tables, static graphics, animated graphics and sound(Table 3.1). For lack of a suitable graphic representation, touch is omitted from the table, though it also can represent information in a useful manner. People certainly have been using these formats for quite some time. With the advent of computers, however, our ability to use these formats has been greatly enhanced. In this and the following three chapters, we discuss results from cognitive psychology, frameworks and guidelines, and techniques and technologies from computer science to support the use of the different representation formats.

3.2 Introduction to Text and Tables

This chapter in particular discusses the use of text (Section 3.3) and tables to represent quantitative information. In a book on visualization one might be surprised to find a discussion of plain, boring text and tables of text. After all, "a picture is worth a thousand words." Many pictures, however, include textual annotation. Furthermore, text can be a potent visualization tool. Ask any poet. Finally, our ability to work with text on computers has enjoyed significant improvement. Only recently has it been possible to display any font in any size on a computer screen, for example.

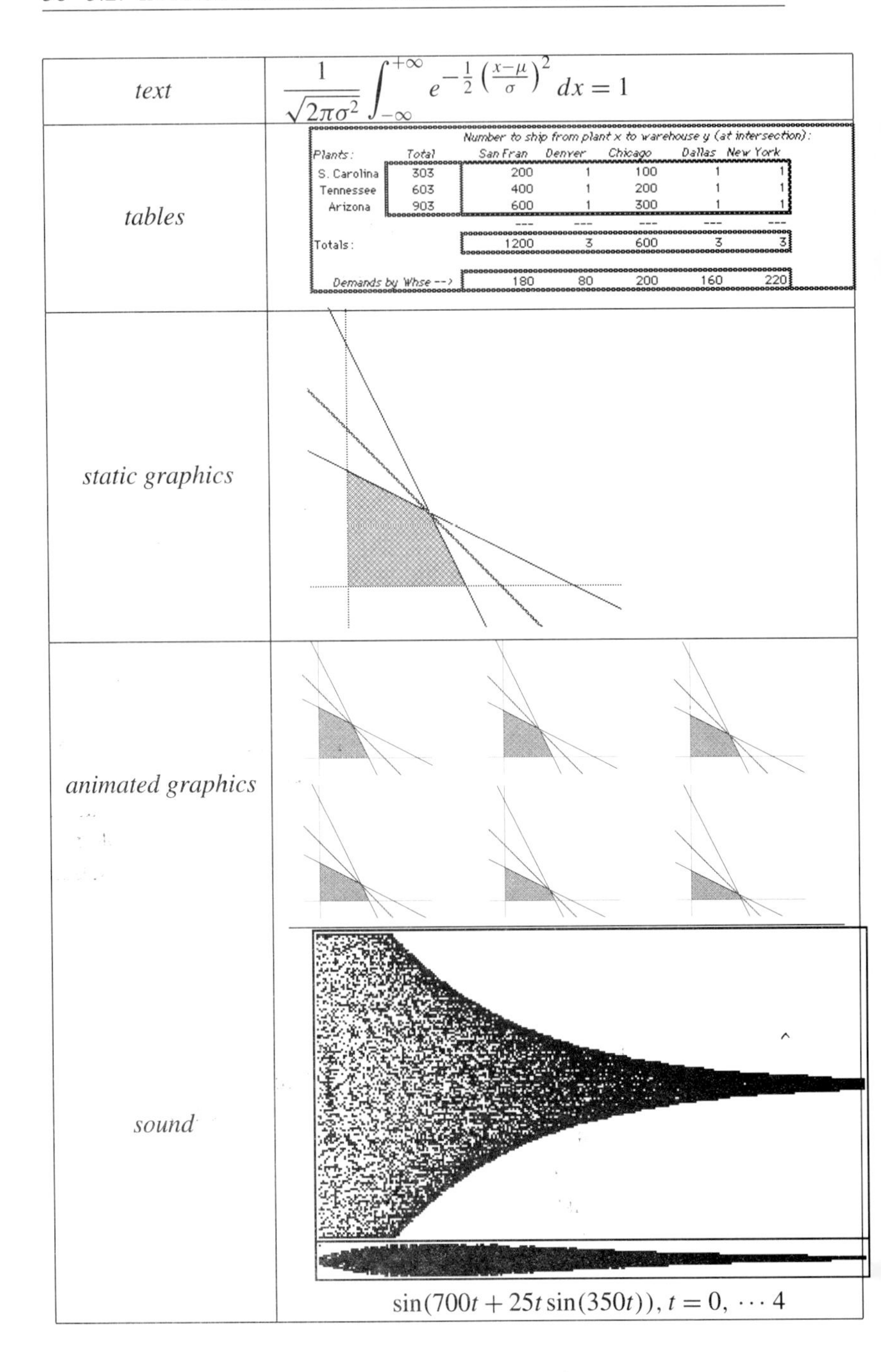

Plants:	Total	San Fran	Denver	Chicago	Dallas	New York
S. Carolina	303	200	1	100	1	1
Tennessee	603	400	1	200	1	1
Arizona	903	600	1	300	1	1
		---	---	---	---	---
Totals:		1200	3	600	3	3
Demands by Whse -->		180	80	200	160	220

Table 3.1: Table of representation formats.

This book will certainly not make recommendations about proper English grammar (just ask the editor). One should be familiar, however, with some facts concerning text and tables if one is interested in visualization.

3.3 Text

Text is "drawn" using a particular set of character shapes, or *font.*

Until recently, creating documents containing different fonts was quite challenging. Early computer printer and display technology could typically support only a single type of font, and one was lucky to have both upper and lower case letters. Those fonts were also generally *fixed width* with every letter, from 'i' to 'X' having the same width. Below are ten i's and ten x's displayed in a fixed width font. Note that both the x's and the i's have the same width.

```
iiiiiiiiii
XXXXXXXXXX
```

Variable width fonts allow letter shapes to have different widths. Below are ten i's and ten x's displayed in a variable width font. The i's are much less wide than the x's.

iiiiiiiiii
XXXXXXXXXX

Moreover, some fonts also support the use of *ligatures*, which combine two or more characters into a single character. For example, compare the two characters, 'fi,' with the ligature 'ﬁ.'

These font characteristics can affect reading speed and require different amounts of space. For example, fonts whose characters all have the same width and are only upper case require on average 30% more space and take 12% longer to read [1].

Using a variety of fonts has become easy today because of the development of *outline* fonts at right. As the name implies, an outline font is defined by its outline or boundary, defined mathematically. This allows a font to be readily scaled to any desired size.

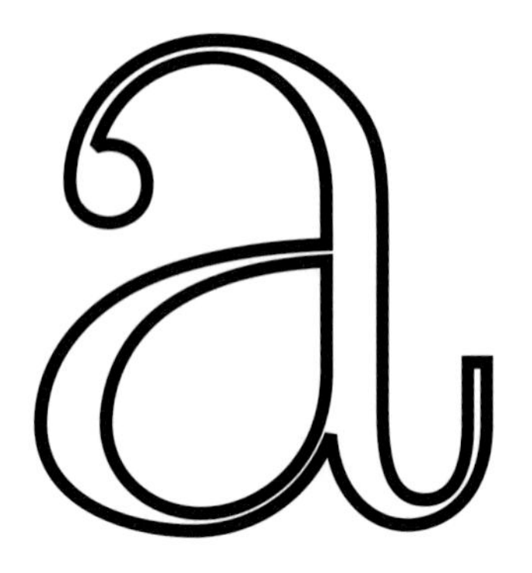

Figure 3.1

In contrast to outline fonts are *bitmap* fonts, whose shapes are defined as a grid of dots. Bitmap fonts cannot be so easily scaled in size, since they are tied to a particular screen or printer resolution. Bitmap fonts, when made larger, usually appear jagged (Figure 3.2).

Figure 3.2

In order to use an outline font, the outline must be filled in with "ink." Since different display devices have different resolutions (printers usually have finer resolution than computer monitors), the algorithm for filling in character shapes must be able to generate readable characters at a variety of resolutions. The two leading standards for outline fonts are Postscript [2], defined by Adobe Systems, and TrueType [3], defined by Apple and Microsoft, respectively.

Fonts have a "natural" size, usually expressed in terms of points or picas (12 points = 1 pica; 1 in = 72.27 points). Although outline fonts allow a font to be scaled to any size, font designers carefully craft different versions of a font for different sizes. For example, Figure 3.3 shows the letterform for the letter 'a'. Both letter-forms have been scaled to approximately the same size. The letterform on the left is the "5 point" version of the font, and the letterform on the right is the "17 point" version of the same font.

Figure 3.3

3.3.1 Formal Languages

In the previous section, we discussed the how individual characters appear on the printed page or on a computer display. It is not primarily the form of the characters that makes text important, however. Characters are combined into appropriate linear sequences in order to express a complicated idea, be it a poem, a short short, a recipe or a computer program. Although computers still have difficulty understanding natural human languages, they are quite effective at translating languages specifically designed for them. One of the major accomplishments of computer science has been the development of fast translators (compilers and interpreters) for textual languages. This section briefly presents some of the major results. This discussion may seem wildly inappropriate for a book discussing visualization. However, formal language theory can provide a solid foundation for developing a theory of visual representation (as discussed in more detail in Chapter 13).

A *language*, be it English, Urdu or FORTRAN, consists of the set of all possible ordered sequences of characters that satisfy properties inherent to the language. The FORTRAN language, for example, consists of the infinite number of possible FORTRAN programs, from those consisting of a few statements to those containing millions or billions of statements. FORTRAN possesses its own particular characteristics and constructs, which are quite different from those of Urdu and English. The great subtlety of natural human languages such as English makes it difficult to define proper English grammar. Computer languages, on the other hand, are generally defined more precisely. We limit ourselves to such languages for the rest of our discussion.

To define a language one must define the valid sequences of characters or *sentences* in the language. Given the vast variety of possible FORTRAN programs, we need a rather powerful mechanism for specifying valid sentences in the language. Once we have a precise definition of a language, we can program a computer so that it can understand the sentences of the language.

We should make a distinction here between the notions of *syntax* and *semantics*. The techniques outlined here for defining languages allows one to specify concisely valid sentences (or programs in the language). But what do we mean by "valid?" It is all too easy to create a FORTRAN program that compiles properly but does not produce the desired result when executed.

It would be great if we could define the FORTRAN programming language to consist of only those programs that perform the correct calculation in any context without the need for any debugging. This lofty goal, however desirable, would at the least require FORTRAN to read the mind of the programmer. The best that FORTRAN can do, really, is to limit the external form or *syntax* of the language. Compilers can detect if a comma is missing, but cannot detect if a computer program contains an infinite loop. In short, the syntax of a language determines only the valid sentences of the language, not what each sentence actually means, that is, the *semantics*. Syntax only specifies that 'X=2+2' is a valid FORTRAN statement, but does not specify what that the statement should accomplish.

The meaning of 'X=2+2' is perhaps so obvious that the distinction between syntax and semantics is unclear. Consider, however, the following valid statements in the programming language C:

```
b = 2;
a++ = ++b;
```

These statements are valid according to the syntax of C. But what are the final values of the variables `a` and `b`? Determining those values requires a precise understanding of the meaning of the sequence of characters, for example, `++b`. In the example, after both statements are executed, `a` will have value `4` and `b` will have value `3`. To keep the discussion tractable, we concentrate here on the relatively simple task of defining syntax, not semantics.

It is now time to give a more precise definition of a language.

Definition 1 *A* language $L = \langle \sum, \sum^*, S_O, G \rangle$ *is a four-tuple where:*

- $\sum$ *is the* alphabet *of* terminal *characters in a sentence of the language,*
- $\sum^*$ *is an alphabet of* non-terminal *symbols where* $\sum \cap \sum^* = \emptyset$,
- $S_0 \in \sum^*$ *is an initial non-terminal that is used as a base from which to construct any sentence in the language,*
- *G is the* grammar *for the language, which describes how to construct valid sentences.*

To define the FORTRAN language, for example, one need only define the alphabet for FORTRAN, the starting (blank) sentence, and a grammar that can be used to construct all valid FORTRAN programs. A grammar G consists of a a set of rules or *productions*, $p \in G$. A production takes a sentence containing terminals and non-terminal characters and modifies it by replacing a non-terminal and replacing it by a sentence consisting of terminals and possibly other non-terminals. Note that we define a sentence containing no characters by the special character ϵ.

More formally, a production $p \in G$ is an ordered pair $p = \langle l, r \rangle$ where l and r are called the left-hand and right-hand sides, respectively, since each is traditionally written on the corresponding side of the page. l is a non-terminal character and r is a string of characters chosen from $\sum \cup \sum^* \cup \{\epsilon\}$. The left-hand side specifies a pattern to match in an existing sequence of terminal and non-terminal characters. The right-hand side replaces the pattern matched by the left-hand side.

Each production is typically written as $l \rightarrow r$. Note that we are defining here a *context-free* grammar because the left-hand side can only consist of a single non-terminal character. More powerful—and computationally more expensive—grammars can be defined that allow multiple terminals and non-terminals on the left-hand side. The terminal characters that appear on the left-hand side provide "context," and those grammars are called *context-sensitive* grammars.

Consider a simple example language, almost universally used to illustrate formal languages, a language for performing simple arithmetic calculations using a calculator. Valid sentences in the language include:

$$\begin{gathered} 1 \\ 2+2 \\ 234 * 345 \\ (-56+37)/(87+(45-34)) \end{gathered}$$

For this language, L^C, we define its four components:

$$L^C = \left\langle \sum^C, \sum^*, S_0^C, G^C \right\rangle$$

where:

$$\sum{}^C = \{0, 1, 2, 3, 4, 5, 6, 7, 8, 9, +, -, *, /, (,)\}$$

$$S_0^C = E$$

$$\sum{}^* = \{E, N\}$$

and the grammar, G^C, consists of the following productions:

$$E \rightarrow -E \tag{3.1}$$

$$E \rightarrow E+E \tag{3.2}$$

$$E \rightarrow E-E \tag{3.3}$$

$$E \rightarrow E*E \tag{3.4}$$

$$E \rightarrow E/E \tag{3.5}$$

$$E \rightarrow (E) \tag{3.6}$$

$$E \rightarrow N \tag{3.7}$$

$$N \rightarrow 0N \tag{3.8}$$
$$N \rightarrow 1N \tag{3.9}$$
$$N \rightarrow 2N \tag{3.10}$$
$$N \rightarrow 3N \tag{3.11}$$
$$N \rightarrow 4N \tag{3.12}$$
$$N \rightarrow 5N \tag{3.13}$$
$$N \rightarrow 6N \tag{3.14}$$
$$N \rightarrow 7N \tag{3.15}$$
$$N \rightarrow 8N \tag{3.16}$$
$$N \rightarrow 9N \tag{3.17}$$
$$N \rightarrow \epsilon \tag{3.18}$$

Note that we chose the non-terminal character E to represent an expression, and N to represent a number.

To build a valid expression for the calculator, one merely *applies* a sequence of productions in any order. Consider how one would construct the expression $2 + 2$ using the productions specified. One starts with S_0, which in this case equals E. If we choose production 3.2, our string becomes:

$$E + E$$

We would then choose production 3.7, which will replace a non-terminal E with non-terminal N. If we apply this production once, one of the E's in $E + E$ will be replaced by an N. Note that the grammar does not specify exactly which of the E's will in fact be replaced. We assume that we shall replace the first match found, so by applying production 3.7, the string becomes:

$$N + E$$

Applying production 3.7 once again, the string becomes:

$$N + N$$

Next, we need to replace each N by a number, so we apply production 3.10, obtaining the following string:

$$2N + N$$

We now apply production 3.18 to remove the first N, obtaining the following string:

$$2 + N$$

since production 3.18 replaces an N by an empty string.

We apply productions 3.10 and 3.18 again and the final expression becomes:

$$2+2$$

At this point, we stop, because our string contains no non-terminals, so no further productions can be applied. We have constructed a valid sentence in the language, $2+2$.

Of particular note is production 3.6. Production 3.6 essentially specifies that any valid expression can be surrounded by parentheses.

In this example, we used the grammar to build a valid sentence in a formal language. Formal languages are more commonly used not to build a sentence in a language, however, but to establish if a given piece of text is in fact a valid sentence for a given language. For example, one can use a text editor to prepare a program in C, but then it must be *compiled* in order to detect syntax errors, as well as to translate it into object code. This process of identification and translation of a sentence is called *parsing*. Parsing, in essence, seeks to identify the sequence of productions that can be applied to S_0 in order to construct the given sentence, if possible.

Parsing is such a well-developed activity that specialized languages have been developed to help in the construction of parsers. Perhaps the most widely known such parser-generator is `yacc` [4], which is a standard part of the unix operating system, though a free version, `Bison`, is available for most other operating systems [5]. By specifying the terminals, non-terminals, and productions, yacc produces a program in C that will parse raw text. There is more to be said about formal languages, parsers, syntax and semantics. We return to the topic of formal languages in Chapter 13 when we apply them not to textual languages, but to pictorial languages.

3.3.2 Concluding Remarks About Text

Although text might seem a rather pedestrian topic in a book on visualization, text will continue to be a key medium for communication. Perhaps the power of text lies in its linear or *sentential* (according to Larkin and Simon [6]) arrangement of information, as opposed to graphics with their two-dimensional information. In text, information is presented in a fixed sequence. An author leads a reader through a particular train of thought, from one idea to another. Graphics, on the other hand, with their two or three dimensional representation does not give such a clear map for the reader. Where does one start to "read" a picture? When one reads text on a page, one can at least be certain that one has perused all the information in the text, though there may be an infinitude of hidden meanings. When viewing a picture, however, can one be certain that all the information contained in the picture has been perused? Of course, many texts can be reread many times, each time the reader uncovering new insights, but at least the basic convention for viewing the text is fixed. Different types of pictures require different strategies for viewing. As such, they may not always be as efficient as plain ordinary text.

Larkin and Simon [6] propose that if the strategies for viewing a graphic representation are known to the viewer then a graphic representation can often be more readily understood. Text allows only sequential search for information, whereas a graphic representation can place related information in close proximity, thereby allowing non-sequential search. Without prior familiarity with the graphic representation format, however, graphics may offer no advantages. In other words, if one is anchored to text, then a graphic representation may not be more effective. Before discussing graphics in detail, however, we discuss two-dimensional textual representations, that is, tables.

3.4 Tables

Although Descartes presented his ideas for Cartesian coordinates in 1637 [7] (although an example of the use of rectilinear coordinates exists from the 10th or 11th Century A.D. [7]), the use of coordinate plots to display information did not become popular until the 19th century [8]. Rather, a "vociferous group of social scientists" [7] promoted the use of tabular representations for the increasing quantities of data collected by the governments of the day. In fact, when data was collected using automated plotting devices, the data was almost always converted into tabular form before it was considered usable. One such device invented by Christopher Wren in the late 1600's recorded some meteorological information graphically. Yet in order to use the information, the Royal Society passed a motion that:

> Mr. Hooke give his directions and assistance to Mr. Hunt to reduce into writing some of the first papers marked by the weather-clock, that thereby the Society might have a specimen of the weather-clock's performances before they proceed to the repairing of it (as quoted in [8]).

In other words, the information that was automatically plotted by the weather-clock was not understandable until it was converted into textual form.

It was not until the 19th century that graphical representations of information, with some exceptions, began to be regularly seen in scholarly literature [7], [8]. In this section, we discuss the use of tables as a representation format.

3.4.1 Spreadsheets and Relational Databases

Tabular representations could just be considered a special form of text. After all, a table is just a rectilinear, almost graphical, arrangement of text. Tables are so widespread, however, that they deserve consideration on their own. Furthermore, with the development of the spreadsheet and relational databases, both fundamentally based on tabular representations, our ability to create, query, calculate with, and manipulate tables has improved tremendously.

Much has been written both about spreadsheets [9], [10], [11], [12] and relational databases [13], [14]. With respect to spreadsheets, many have cited their wide adoption as providing a great medium for delivering optimization (and other mathematical modeling) technology, though some (e.g., [11]) have cautioned about the dangers of putting modeling "elephant guns" into the hands of "children." Relational databases have become the principal standard for storing and manipulating large quantities of data.

The success of both of these ideas arguably arises from anchoring effects, as discussed in Chapter 2. People are comfortable with dealing with information represented in tabular form. Both spreadsheets and relational databases systematize tabular representations and package them in a form accessible to human beings. Spreadsheets are primarily designed for performing calculations, and relational databases are primarily designed for storing and retrieving large quantities of data, but each can perform some of the capabilities of the other.

When relational databases were first proposed in 1970, two other database models existed, neither based on tabular representations. One, the *hierarchical* model, essentially represented data as a tree (or hierarchy); the other, called the *network* model, a generalization of the hierarchical model, represented data as a graph (or network). For a time, a major debate occurred as between advocates for network databases and advocates for relational databases. The advocates for network databases argued that actual network databases had been constructed, whereas relational databases remained academic toys. Advocates for relational databases, in response, railed against the lack of theoretical rigor and conceptual simplicity of network databases. Relational databases have clearly won the battle. Perhaps reason lies in the conceptual simplicity of the representation format upon which relational databases are based: the table.

3.4.2 Multidimensional Tables

Tables are often thought of as purely two-dimensional—rows and columns. Tables, however, can straightforwardly represent multi-dimensional information. At least two schemes are commonly used, the first is seen commercially in relational databases, and the second is seen commercially in certain spreadsheet programs, notably Lotus Improv and Microsoft Excel (version 5). For example, consider the problem of representing a three-dimensional parameter c_{ijt}, which could be the cost to transport a unit of material from location i to location j in time period t.

A typical relational data table would appear as at right.

i	j	t	c_{ijt}
1	1	1	12
1	2	1	23
1	3	1	28
2	1	1	35
2	2	1	84
2	3	1	92
1	1	2	52
1	2	2	16
1	3	2	88
2	1	2	100
2	2	2	37
2	3	2	17

Figure 3.4

More dimensions correspond simply to more columns in the relational database. For large domains for each dimension, the size of a relational-style table can grow quite large, but at least tables can theoretically represent multiple dimensions.

In the second style of multidimensional tables, rows and columns can represent multiple dimensions. For example, another way to represent the data in figure 3.4 is shown at right.

	$t=1$			$t=2$		
	$j=$			$j=$		
i	*1*	*2*	*3*	*1*	*2*	*3*
1	12	23	28	52	16	88
2	35	84	92	100	37	17

Figure 3.5

The allocation of dimensions to the rows and columns in Figure 3.5 is essentiallly arbitrary. Another valid arrangement is shown in at right

	$i=1$		$i=2$	
	$t=$		$t=$	
j	*1*	*2*	*1*	*2*
1	12	52	35	100
2	23	16	84	37
3	28	88	92	17

Figure 3.6

Still another arrangement uses several tables, one for each element of a dimension.

$i = 1$

	t	
j	*1*	2
1	12	52
2	23	16
3	28	88

$i = 2$

	t	
j	*1*	2
1	35	100
2	84	37
3	92	17

Figure 3.7

Commercial spreadsheets like Lotus Improv [15] and pivot tables in Microsoft Excel[16] allow users to change the arrangement of the dimensions of a table quite easily. Consider the following Lotus Improv worksheet for a multi-period transportation problem. Along the left-hand side are listed the sources of supply and some of the problem parameters and decision variables (Unit Cost, Shipments and Total Cost). Across the top are listed the sinks (demand centers) and the time periods.

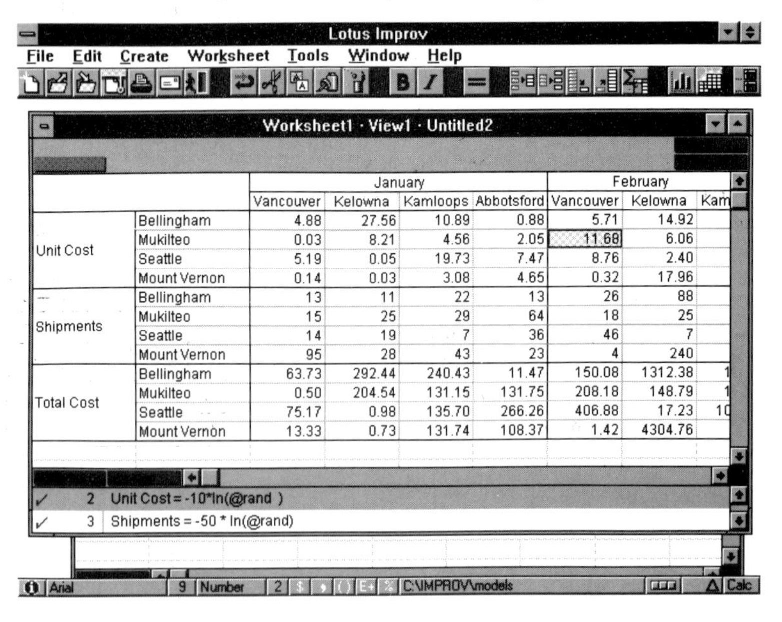

		January				February		
		Vancouver	Kelowna	Kamloops	Abbotsford	Vancouver	Kelowna	Kam
Unit Cost	Bellingham	4.88	27.56	10.89	0.88	5.71	14.92	
	Mukilteo	0.03	8.21	4.56	2.05	11.68	6.06	
	Seattle	5.19	0.05	19.73	7.47	8.76	2.40	
	Mount Vernon	0.14	0.03	3.08	4.65	0.32	17.96	
Shipments	Bellingham	13	11	22	13	26	88	
	Mukilteo	15	25	29	64	18	25	
	Seattle	14	19	7	36	46	7	
	Mount Vernon	95	28	43	23	4	240	
Total Cost	Bellingham	63.73	292.44	240.43	11.47	150.08	1312.38	1
	Mukilteo	0.50	204.54	131.15	131.75	208.18	148.79	1
	Seattle	75.17	0.98	135.70	266.26	406.88	17.23	10
	Mount Vernon	13.33	0.73	131.74	108.37	1.42	4304.76	

Figure 3.8

Users can easily rearrange this table. Note the two rectangles at the bottom, left labeled "Attributes" and "Sources." Similarly note the two rectangles at the upper, right labeled "Periods" and "Sinks." The user rearranges the layout of the table merely by dragging these rectangles into new positions.

For example, simply dragging the "Sources" rectangle to the left of the "Attributes" rectangle rearranges the table into the form shown below

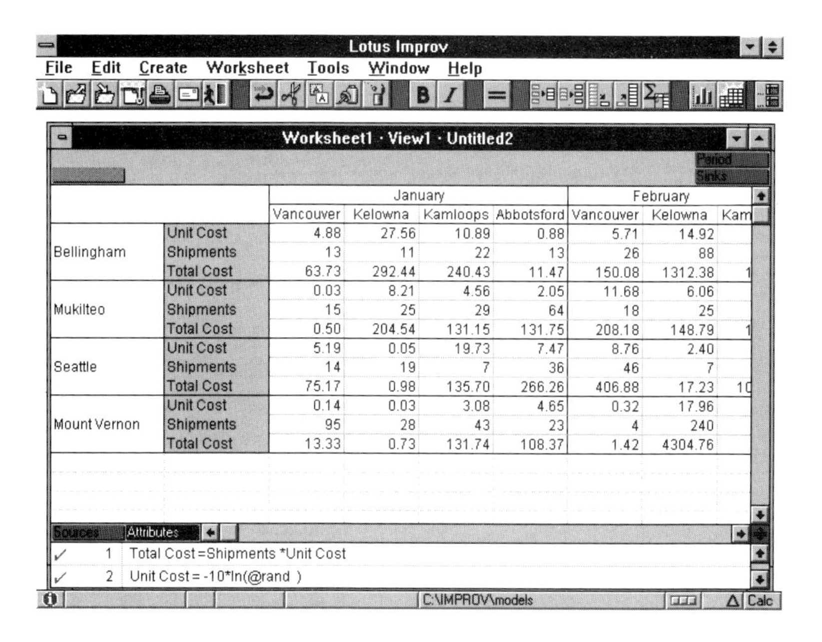

		January				February		
		Vancouver	Kelowna	Kamloops	Abbotsford	Vancouver	Kelowna	Kam
Bellingham	Unit Cost	4.88	27.56	10.89	0.88	5.71	14.92	
	Shipments	13	11	22	13	26	88	
	Total Cost	63.73	292.44	240.43	11.47	150.08	1312.38	1
Mukilteo	Unit Cost	0.03	8.21	4.56	2.05	11.68	6.06	
	Shipments	15	25	29	64	18	25	
	Total Cost	0.50	204.54	131.15	131.75	208.18	148.79	1
Seattle	Unit Cost	5.19	0.05	19.73	7.47	8.76	2.40	
	Shipments	14	19	7	36	46	7	
	Total Cost	75.17	0.98	135.70	266.26	406.88	17.23	1C
Mount Vernon	Unit Cost	0.14	0.03	3.08	4.65	0.32	17.96	
	Shipments	95	28	43	23	4	240	
	Total Cost	13.33	0.73	131.74	108.37	1.42	4304.76	

Figure 3.9

Similarly, simply dragging the "sinks" rectangle from the upper right to the lower left corner and dragging the "sources" rectangle from the lower left to the upper right, the table below results.

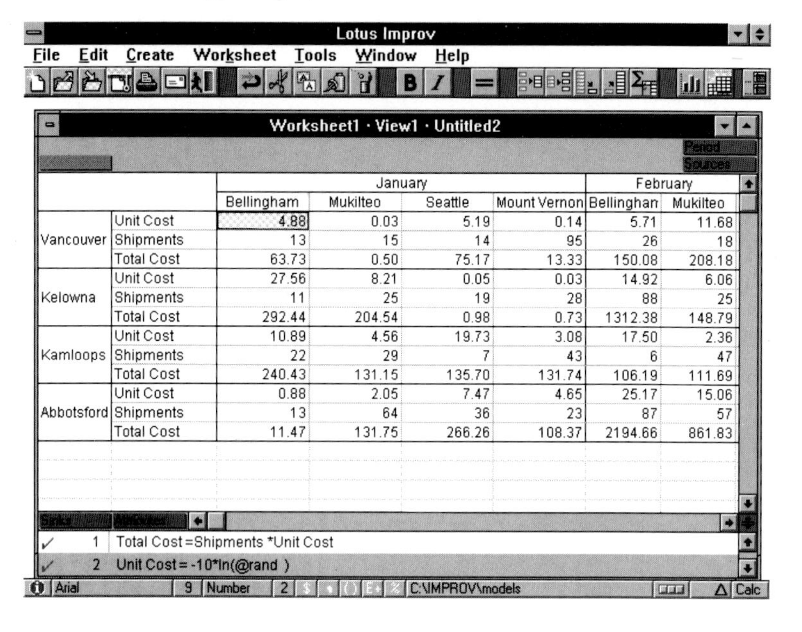

		January				February	
		Bellingham	Mukilteo	Seattle	Mount Vernon	Bellinghan	Mukilteo
Vancouver	Unit Cost	4.88	0.03	5.19	0.14	5.71	11.68
	Shipments	13	15	14	95	26	18
	Total Cost	63.73	0.50	75.17	13.33	150.08	208.18
Kelowna	Unit Cost	27.56	8.21	0.05	0.03	14.92	6.06
	Shipments	11	25	19	28	88	25
	Total Cost	292.44	204.54	0.98	0.73	1312.38	148.79
Kamloops	Unit Cost	10.89	4.56	19.73	3.08	17.50	2.36
	Shipments	22	29	7	43	6	47
	Total Cost	240.43	131.15	135.70	131.74	106.19	111.69
Abbotsford	Unit Cost	0.88	2.05	7.47	4.65	25.17	15.06
	Shipments	13	64	36	23	87	57
	Total Cost	11.47	131.75	266.26	108.37	2194.66	861.83

1 Total Cost=Shipments *Unit Cost

2 Unit Cost=-10*ln(@rand)

Figure 3.10

Note further that Improv allows users to specify formulas that use the names of the table dimensions. For example, the formula to calculate the total cost (shown at the bottom of Figures 3.8–3.10) is written `Total Cost = Unit Cost * Shipments`. Although one can use named ranges in traditional spreadsheets to achieve a similar effect, Improv does this more elegantly.

To conclude, tables represent a convenient, powerful mechanism for representing data. With the advent of relational databases and multi-dimensional spreadsheets, tables can be quite effective at representing information having many dimensions.

3.5 Summary

This chapter has discussed guidelines, recommendations and technology useful for displaying information as text and as tables. Although such a discussion may seem outside the scope of a book on visualization, these representation forms remain dominant and will almost certainly remain so. The pictures, animations, and other representation formats discussed in the next two chapters almost always will also use some form of textual annotation. An understanding of how text and tables can and should be used is therefore an important part of learning about visualization.

Bibliography

[1] Marcus A. *Graphic Design for Electronic Documents and User Interfaces.* New York:ACM Press, 1992.

[2] Adobe Systems. *Postscript Language Reference Manual.* Reading (MA):Addison-Wesley, 1985.

[3] Apple Computer. *Inside Macintosh*, volume 6. Reading (MA):Addison-Wesley, 1991.

[4] Mason T, Brown D. *lex & yacc.* Sebastopol (CA):O'Reilly and Associates, Inc, 1990.

[5] Donnelly C, Stallman R. *Bison: the YACC-compatible Parser Generator.* 675 Massachusetts Avenue, Cambridge (MA) 02139:Free Software Foundation, 1992.

[6] Larkin JH, Simon HA. Why a diagram is (sometimes) worth ten thousand words. *Cognitive Science*, 1987;11:65–99.

[7] Beniger JR, Robyn DL. Quantitative graphics in statistics: A brief history. *The American Statistician*, 1978;32(1):1–9.

[8] Tilling L. Early experimental graphs. *The British Journal for the History of Science*, 1974;8(30):193–213.

[9] Vazsonyi A. Where we ought to be going: The potential of spreadsheets. *Interfaces*, 1993;23(5):26–39.

[10] Bodily S. Spreadsheet modeling as a stepping stone. *Interfaces*, 1986;16(5):34–52.

[11] Gass SI. Model world: Danger, beware the user as modeler. *Interfaces*, 1990;20(3):60–64.

[12] Savage S. Lotus improv as a modeling language. In *ORSA/TIMS Joint National Meeting*, San Francisco (CA):1992.

[13] Codd EF. A relational model for large shared data banks. *Communications of the ACM*, 1970;13:370–387.

[14] Date CJ. *An Introduction to Database Systems*. Reading (MA):Addison-Wesley, 4th edition, 1986.

[15] Lotus Development Corporation. *Improv for Windows Release 2.0 Reference Manual*. Cambridge (MA):Lotus Development Corporation, 1993.

[16] Microsoft Corporation. *Microsoft Excel User's Guide Version 5.0*. Redmond (WA):Microsoft Corporation, 1994.

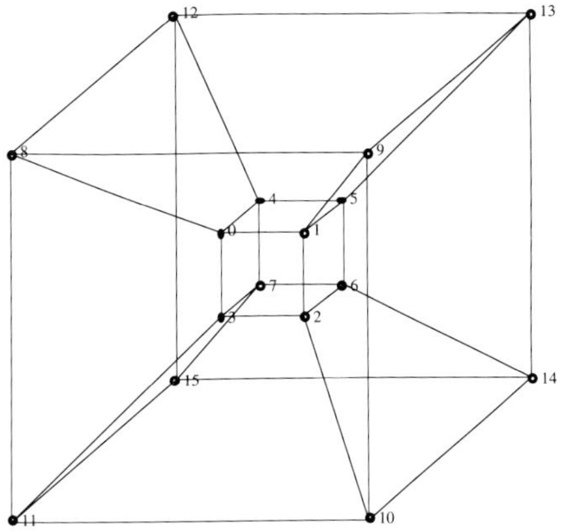

Chapter 4

Graphics and Animation

This chapter discusses both static and moving images. Each section surveys some of the basic results from cognitive psychology related to how one perceives images. Building from these results some of the recommendations made by graphic designers and others on the use of graphics and animation for representing quantitative information are discussed. We devote special attention to issues surrounding the perception and use of color (Section 4.2) and three-dimensional images (Section 4.3). Clearly it will only be possible to provide a brief survey of the relevant results. Sufficient information is provided to give a sense of the difficulty first in understanding how people perceive images, as well as how to design helpful, coherent graphical representations of quantitative information.

4.1 Static Graphics

4.1.1 Results from Cognitive Psychology

Cognitive psychology seeks to develop models of the human visual perception system that account for its remarkable ability to make sense out of the mass of information transmitted to the eyes. Based on these models, researchers in cognitive psychology conduct experiments to validate or invalidate their models. We only give a brief overview of some of the results from cognitive psychology. A detailed survey of the field can be found in [1], [2].

Many of the results in cognitive psychology will probably seem to be very "low level." For example, one question that has been studied by cognitive psychologists is whether color or shape is more effective in distinguishing one object from another in a scene. In order to study this question, experiments are conducted that measure the time people require to identify objects of a particular shape or color in different types of scenes. Generally [3] color is superior to shape with respect to time to search for a given displayed item.

In a planning board or Gantt chart (below, courtesy of Chesapeake Decision Sciences, Inc.), for example, color is typically used to group together related activities. What are the limits of this strategy? How many different colors can be used before people find it difficult to distinguish differently colored activities? In general, it is recommended that only a few colors be used to distinguish items (e.g., less then 10), without additional training [4]. This number can be seen to be an instance of the 7 ± 2 limit on the size of short-term memory (see Section 2.3.1).

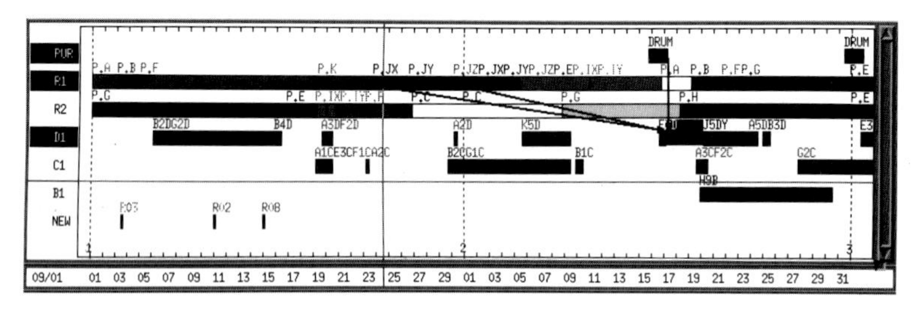

Figure 4.1

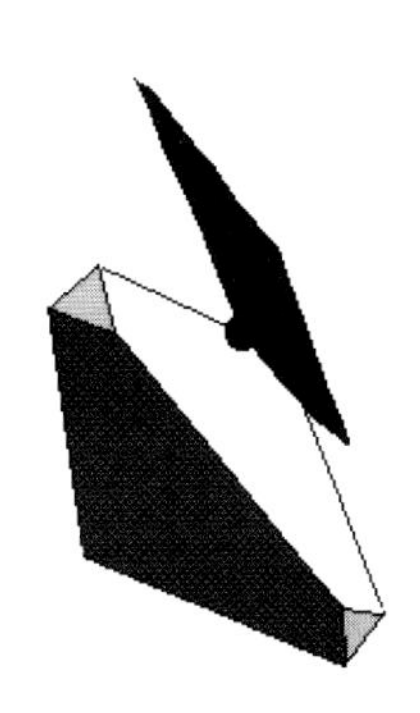

Cognitive psychology can be viewed as primarily a descriptive science, merely trying to generate testable hypotheses about the human perceptual processing system. Analogizing from models from physics however, once one has a robust descriptive model of a phenomenon, that same model can be used to predict behavior. Once prediction is possible, a normative science can emerge.

In this section, we first give a taste of some of the basic ideas developed by cognitive psychology. In the next section, we present some normative frameworks for constructing effective graphic representations of information. Although those frameworks do not explicitly rely on results from cognitive psychology, many of their ideas can be seen to be related to results from cognitive psychology.

4.1.1.1 Early Results

Before the early 1900's, the basic model of visual perception assumed that the human visual system directly and accurately perceives the outside world. That is, light passes through the lens of the eye, is received by the retina as a matrix of stimuli, and that matrix simply and purely represents the outside world to the brain. This model, however, is overly simple. As noted by [5], we easily identify a clock as circular, even though the image on our retina is almost certainly not a perfect circle but an ellipse. Similarly, most squares never appear on the retina as a perfect square, but usually are tilted so that they appear as a trapezoid or parallelogram. In short, the human visual system must perform additional processing on the image sensed by the retina in order to account for the level of perception that humans possess.

4.1.1.2 Gestalt Psychology

As a response to the shortcomings of this simplistic model, *gestalt* psychology emerged in the early 1900's. Gestalt psychologists observed that the image perceived depended not only on the set individual objects that constitute the image, but on interrelationships among the objects. The classic gestalt slogan is "the whole is greater than the sum of its parts." This section provides some illustrations of the slogan. From these and other examples, gestalt psychologists developed a series of principles to begin to describe how we see what we do.

For example, perhaps the most fundamental gestalt principle involves the concept of *foreground* and *background*. This principle refers to the separation that people are able to make between objects in an image. In figure at right, for example, one could perceive the image as a circle (foreground) embedded in a square. One might also perceive the image as a set of four corners.

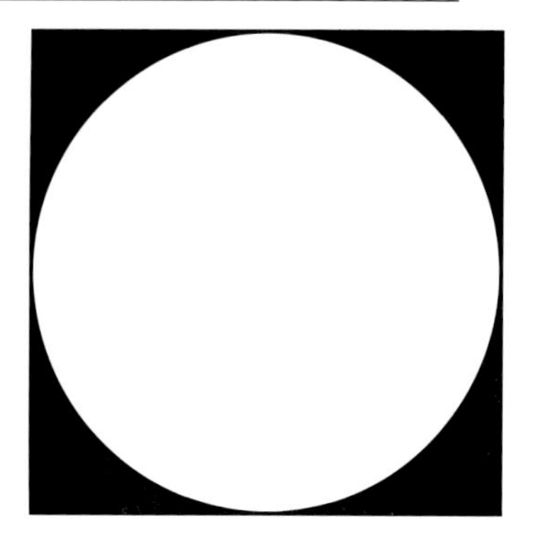

Figure 4.2

In another example, people tend to group together items that have similar visual characteristics. Consider the figure at right, which consists of 36 circles. People tend to describe the image as a series of vertical circles because of the different shading of the circles. To many people, there is apparently no other way to describe the image. But one could also have described it as consisting of six rows of six circles each, wherein each row consists of an alternating series of black and white circles.

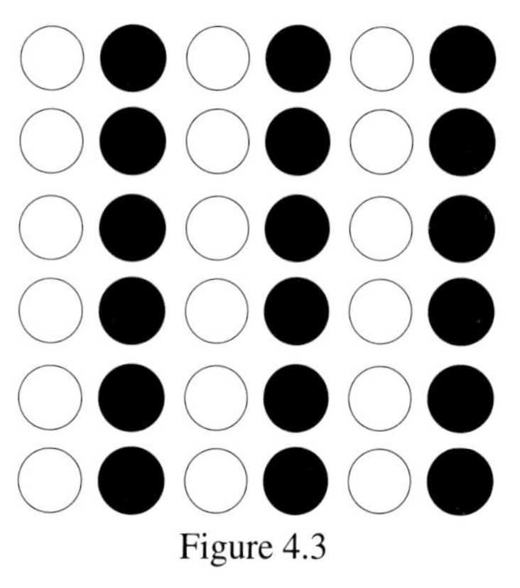

Figure 4.3

If all the circles have the same shade, then the vertical orientation disappears.

Figure 4.4

Another principle proposed by gestalt psychologists is *proximity*. This principle roughly states that people group together objects that are close together in an image. For example, in Figure 4.5 most people group together the circles into horizontal stripes, since the circles are closer horizontally than vertically. If this principle did not exist, one could perceive the image as a set of vertical columns of circles.

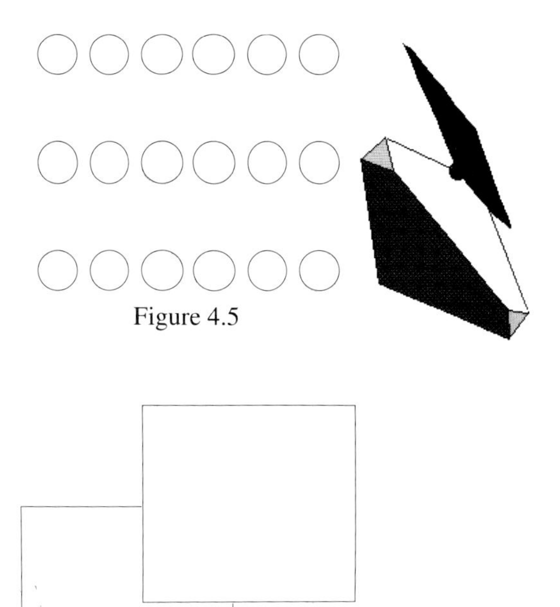

Figure 4.5

The final gestalt principle that we shall discuss is the principle of *continuity*. In this principle, people tend to "complete" objects in an image. For example, consider the figure at right. Most people perceive the left-most shape to be a rectangle—even though it is actually a more complex polygon. This seems to arise because people usually interpret the figure at right to consist of two overlapping rectangles.

Figure 4.6

If the two shapes truly overlap, the right-most "rectangle" could just as easily have the shape shown here.

Figure 4.7

Gestalt psychology made the contribution that the human visual system is not a perfect processor of visual information; rather, it infers structure when convenient. In designing graphical representations of information, one can exploit this "weakness" in the human visual system, for example, by grouping together related objects using proximity or color.

Gestalt psychology rejected the idea that one could explain perception by building from small, independent parts. Rather, gestalt psychologists insisted that one must consider the image as a whole, that individual parts interact to produce non-trivial perceptions.

4.1.1.3 Other Approaches

Other researchers, however, have attempted to explore biological bases for perception and have produced some important results. They have discovered for example that individual neurons in the brain are responsible for detecting particular visual components. For example, an individual neuron is responsible for processing the information generated from a particular part of the retina, or for extracting a particular color [1].

Another approach emphasizes the use of higher level cognition in human visual perception [2]. In this model, visual perception should only be viewed in the context of the goal of the viewer. From an understanding of the goal, one can then develop an appropriate algorithmic model of human perception.

4.1.1.4 Summary

These examples of some of the basic results from cognitive psychology may seem far removed from optimization. After all, how can these "low level" results from cognitive psychology have relevance to the practical problem of creating appropriate representations for complex information? To some extent, cognitive psychology really is concerned with a different question, the seemingly simple question of how people perceive the world. They are primarily interested in creating a model of the human perceptual system that describes how we see what we in fact see. Like optimization modelers, cognitive psychologists build models, but their models are models of the human perceptual and cognitive systems, not factories, vehicle routes, nor budgets. The results from cognitive psychology provide some basic guidance on how to design appropriate representations of information.

Many authors from the art and design community have proposed eloquent, beautifully illustrated books describing frameworks and recommendations for representing complex information graphically. We shall discuss their work in the next section. Some of their recommendations can be seen to trace their origins from some of the phenomena studied by cognitive psychologists. At worst, those interested in visualization should have had at least some exposure to the work of cognitive psychology.

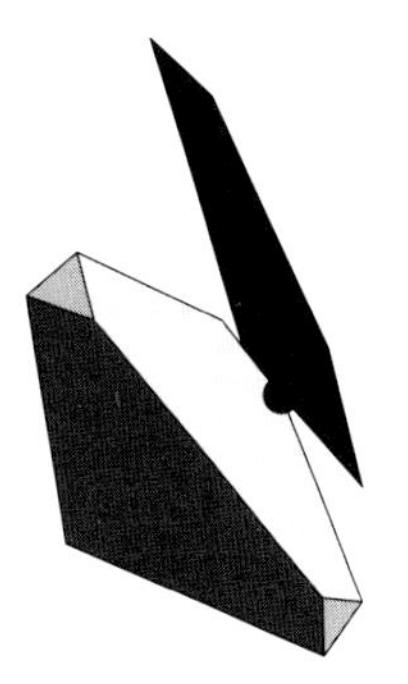

4.1.2 Frameworks for Constructing Graphic Representations of Quantitative Data

This section presents two different authors' frameworks and recommendations for using graphics to represent complex information. The section first describes the work of Jacques Bertin, who in his book *Semiology of Graphics*, attempted to develop a taxonomy for graphics (Section 4.1.2.1). The taxonomy can help in constructing an appropriate graphic representation for a particular problem. The section then presents the eloquent work of Edward Tufte (Section 4.1.2.2) who presents several useful recommendations for representing complicated data graphically.

4.1.2.1 Bertin's Semiology of Graphics

Jacques Bertin attempted to develop a comprehensive framework for graphical representations of information. Although the book is sometimes impenetrable, it does provide some simple foundations for organizing a study of graphic representation.

In Bertin's definition of "graphics," figurative or abstract art is not included. Although art can be beautiful, inspiring, enthralling and entertaining, it generally is not intended to represent complex information. Rather, graphics represents its underlying information as precisely as mathematical notation does. As Bertin says, "On this point, graphics and mathematics are similar." Although graphics and mathematics attempt to be precise, mathematics generally works in the symbolic realm, unlike graphics.

Bertin's framework also separates the graphic from the task to be performed with the graphic, similar to how this book separates task, user and representation format.

Bertin breaks information down into its separate *components*, which are roughly analogous to dimensions found in mathematical programming formulations. One component of information might might be time, another, plant locations, a third, products. Once one has identified the components, one must next identify their type, or as Bertin calls it, their *level*. Appropriate levels include qualitative, ordered, and quantitative. Given the components and levels, one then translates the information into appropriate graphical variables. Bertin identifies eight graphical variables. They are:

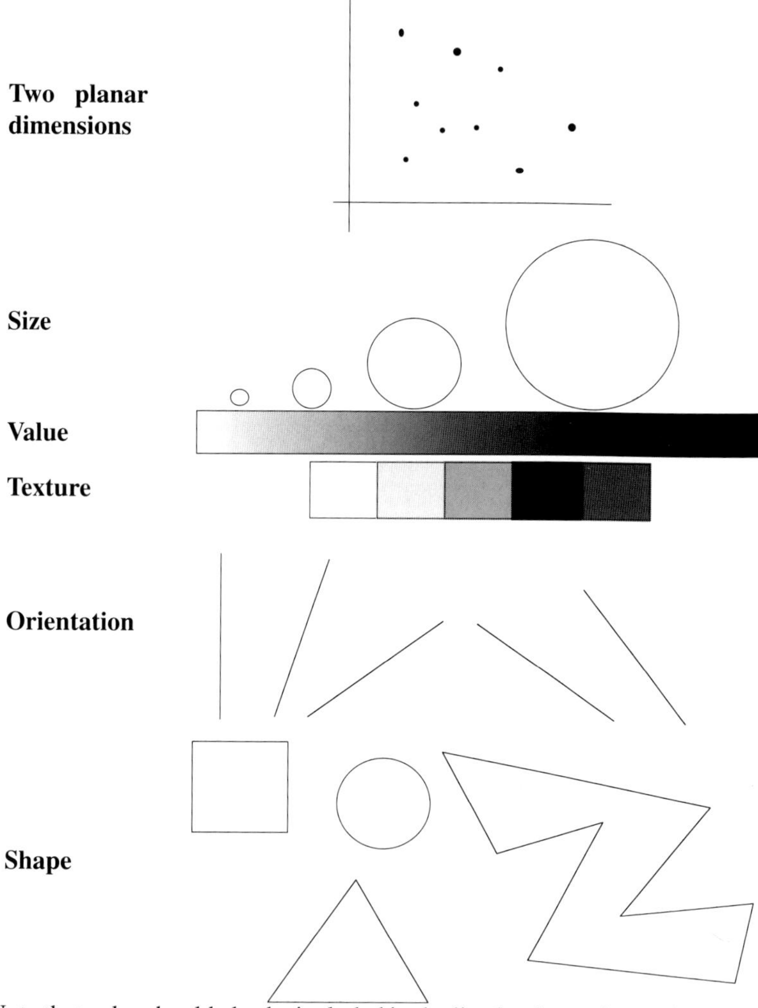

Note that color should also be included in the list, but is not due to the expense of color printing.

Different types of components or levels will require different encodings into graphic variables. He then explores how different types of information can be encoded into these graphic variables, providing hundreds of examples.

Consider a simple problem to represent the the amount of material transported from warehouses to customers. This problem has three components: material, warehouses and customers. Material has a quantitative level, whereas warehouses and customers have a qualitative level, perhaps represented by the name of the warehouses or customer. One possible encoding would map warehouses and customers into the two spatial dimensions. Customers could be distinguished from warehouses by using different shapes, colors, textures or orientations. The movement of material from a warehouse to a customer could be represented by arrows connecting a warehouse to a customer. The amount of material could be represented by the width of the line, by its color or texture.

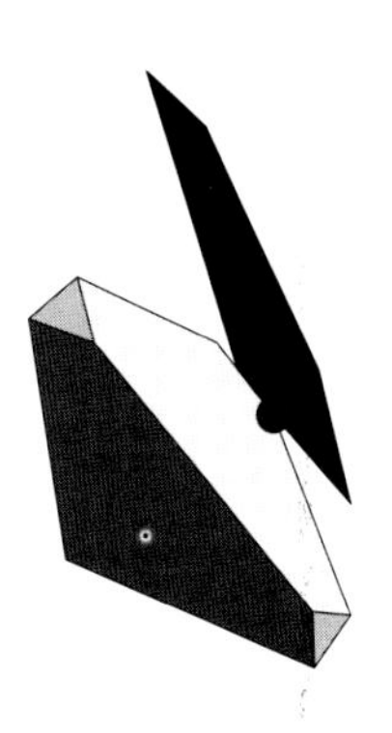

Bertin defines an *efficient* encoding that minimizes the amount of time required to extract needed information from a given graphic. Ideally, one should just need to glance at a graphic in order to obtain the needed information. The information conveyed by such a glance, Bertin calls an *image.* If one needs to study the graphic for some time, that is, require multiple images (in Bertin's sense of the word), then the representation is not as efficient as desired for that particular task.

Bertin's notion of image seems closely related to gestalt psychology. That is, Bertin's 'image' corresponds to the holistic view of a graphic proposed by gestalt psychology. The interaction of various elements in a graphic, as described by gestalt principles, can produce different overall images, in Bertin's sense of the word.

Bertin's framework, at one level is quite simple: merely decompose information into its constituent components and map each of the components into an appropriate graphical form. Exactly how to accomplish the decomposition and mapping operations, to create an effective graphical representation of information is not made terribly clear, unfortunately. In the next section, we discuss some practical guidelines to help in the construction of effective graphical representations of quantitative information.

4.1.2.2 Tufte's Visual Display of Quantitative Information

Edward R. Tufte, in two beautifully composed texts [6], [7], eloquently presents useful guidelines for creating graphics to represent complex information effectively. We attempt to summarize his recommendations here. At the heart of Tufte's recommendations is a sense of economy and simplicity. With the advent of computerized presentation graphics software, it has become extremely easy to construct gaudy, flashy "Las Vegas" style graphics. Often the principal purpose of such graphics is not to inform but to impress or conceal. If one truly wishes to convey the meaning buried in complexity, one should not have to dress up the information in designer clothes. The information should be interesting in and of itself; if it is not, then why is it being presented?

To this goal of simplicity and economy, in his first book [6], Tufte makes several specific recommendations. These include:

- *Maximize the ratio of data to ink.*

Consider a three-dimensional bar chart:

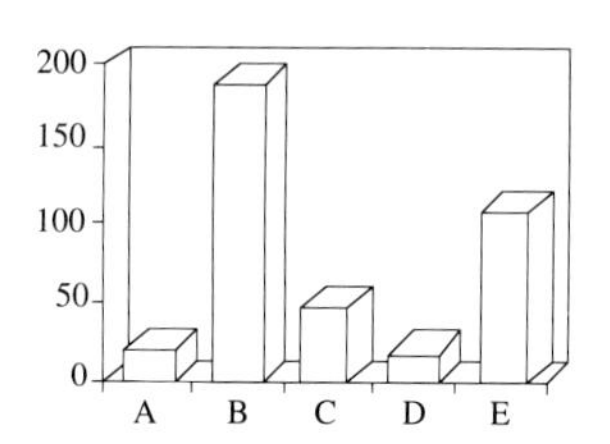

Figure 4.8

What purpose does the three-dimensional effect serve? It certainly does not represent the data. In fact, it arguably distorts the data since one's eye may be drawn to the line at the rear top of each bar, not the line at the front top, which actually represents the data. One might argue that the three-dimensional effect looks more "impressive," and thus has value in drawing attention to the graphic. Tufte believes such attempts at making the graphics more attractive reflect "contempt both for information and for the audience."

A simple solution can sometimes be found merely by erasing as much as possible from the graphic without eliminating any of the information conveyed by the graphic. At right is a first attempt to apply this technique to the graphic in Figure 4.8. This version merely erases the lines producing the three-dimensional effect, and the graph becomes much clearer, without any loss of information. This is the typical sort of bar chart produced by today's presentation graphics software.

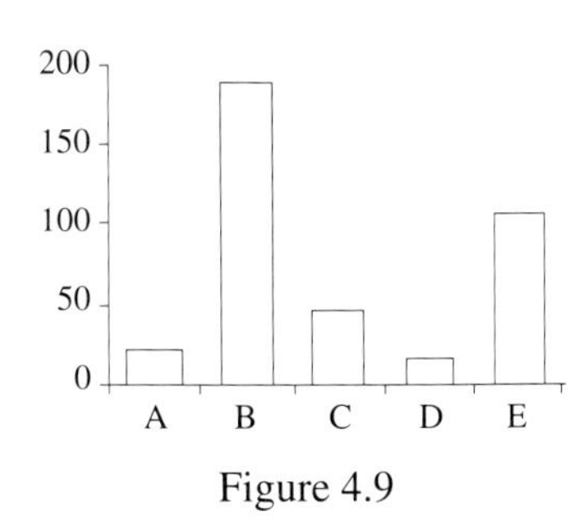

Figure 4.9

One can erase even more, however. Only the tick marks along each axis are needed, not the lines representing the axes themselves.

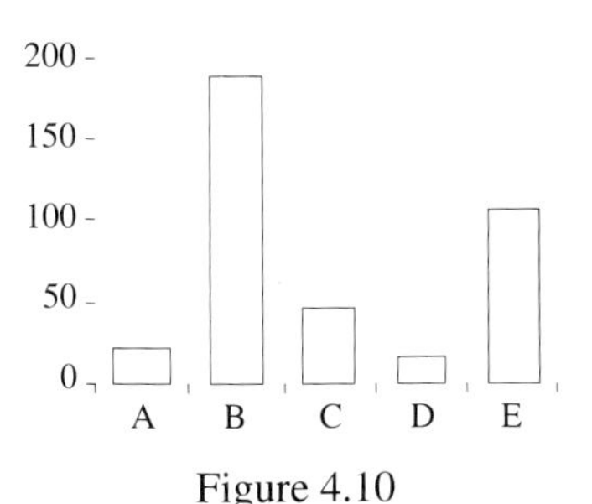

Figure 4.10

Furthermore, each bar needs only a single vertical line to represent the underlying information. This final version represents the same data, but far more economically and simply than the gaudy original (Figure 4.8).

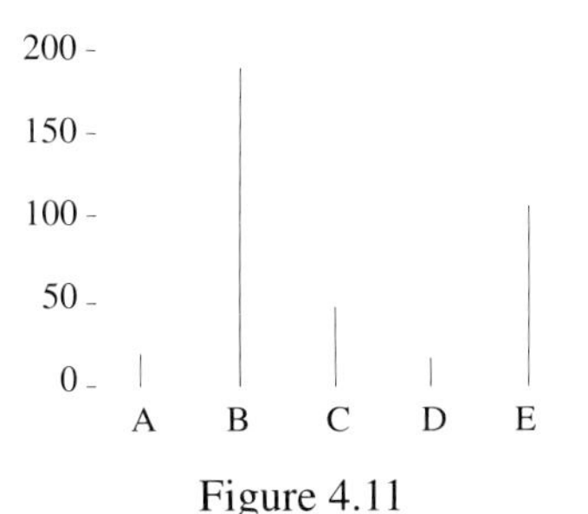

Figure 4.11

- *Use graphics only for large quantities of data.* Again, with the easy ability to create presentation graphics, one sees graphics created for information consisting of only two data points. In fact, if the information contains fewer than seven data points then it probably should not be represented graphically; it would be better represented in a table. The value of seven comes from the capacity of short term memory established by cognitive psychology (see Section 2.3.1). So although Figure 4.11 represents an improvement over the original, it still represents a misuse of graphics, since the underlying data could have been represented as effectively in a table.

- *Integrate text with graphics smoothly.* A picture, by itself, is usually not worth a thousand words. Rather, a picture, combined with text *can* provide effective communication. If text were not valuable, then this book on visualization might contain no text at all.

  Tufte recommends that a graphic be viewed as simply another noun, incorporated into the body of the text. This contravenes many of the traditional stylistic guidelines promoted by book and journal editors, where figures 'float' outside the traditional linear arrangement of text.

- *Show complex information at different levels of detail.*

  With large amounts of data to display graphically, given the small capacity of short-term memory, an effective graphic should be organized to provide both detail and overview simultaneously. Perhaps the best example of such a graphic is a high-quality map. One can see both the overall geography of a region as well as focus in on details of towns, rivers, elevation, among others. In Tufte's second book [7], he calls this *micro* and *macro* readings.

- *Do not lie or distort.* This recommendation may seem obvious, but it is often inadvertently violated, especially when needless adornment is added. The basic rule here is to encode numeric quantities into perceptual quantities in a linear fashion. Given numeric value x encoded into some perceptual quantity $p(x)$ (length, area, for example), then $2x$ should be encoded into a perceptual quantity such that the underlying value is perceived to be twice as large as $p(x)$, that is $p(2x) = 2p(x)$. A classic way that this is violated occurs in using area to encode numeric values. Below are three squares. Square (b) has $\sqrt{2}$ times the width of (a) and hence twice the area of (a). Square (c), in contrast has twice the width and hence four times the area of (a). If square (a) represents a numeric value x, a common error is to encode the numeric value of $2x$ using square (c) rather than square (b).

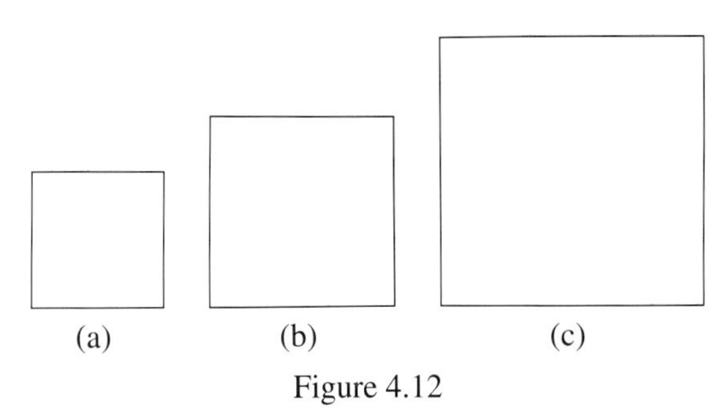

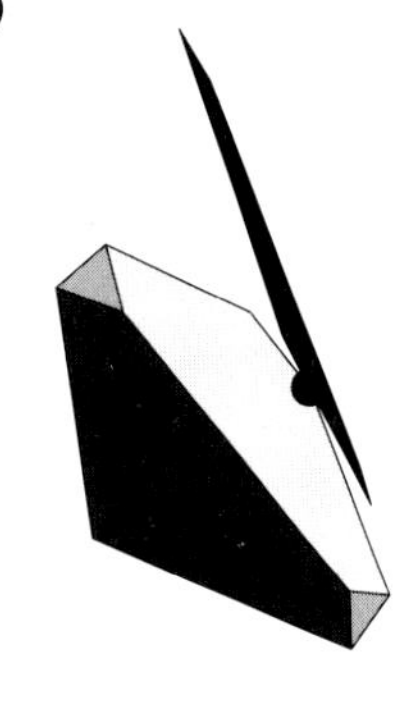

Figure 4.12

In his second book [7], Tufte provides additional recommendations for displaying complex information as discussed on the following pages.

- *Small Multiples.*

  Display several variations of similar data side by side, typically in a tabular format. For example, if one displays the individual frames from the flip chart for this chapter, one can compare the small differences among adjacent frames quite easily. Of course, the animation provided by the flip chart provides a different mechanism for perceiving the differences (animation is discussed in the next section).

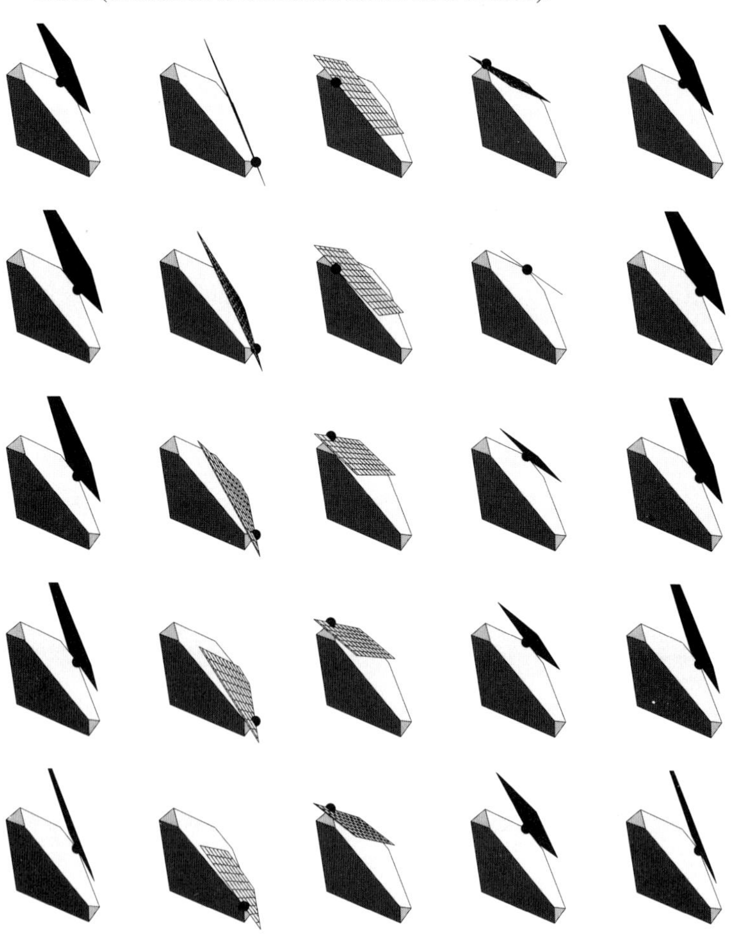

Figure 4.13

- *Layering and Separation.*

  Displaying each element of a graphic with the same color may unintentionally group together, because of gestalt phenomena, elements in unintended ways. Color and texture should be used carefully to distinguish separate graphic elements. In a road map, for example, major highways are typically drawn using one color and secondary roads are drawn using a third color. In spreadsheets, for example, the gridlines identifying cells usually appear as a dim grey, whereas the contents of the cells usually appear as pure black on a white background. The gridlines almost seem to fade into the background; at least they do not distract unduly from the content of the spreadsheet.

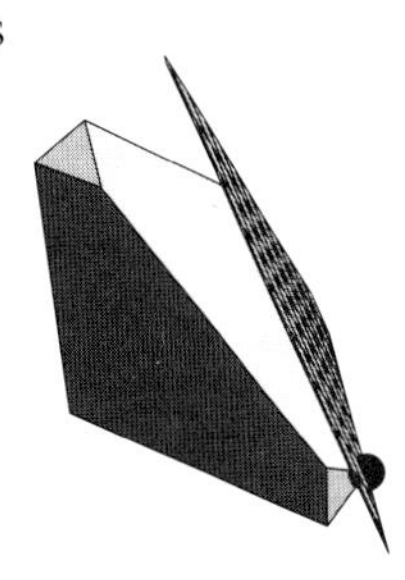

4.1.3 Cleveland and McGill's Experimentally Derived Guidelines

Bertin and Tufte attempted to develop some basic principles that can be *applied* (unlike most results from cognitive psychology) to help develop graphical representations of complicated information. They relied on their own judgement in developing their principles, though, and generally did not rely upon theories from cognitive psychology nor on formal experimentation.

Some formal experimentation has been conducted however to explore the issue of how to represent complicated information. In particular, [8], [9], [10] conducted an ongoing set of research in an attempt to identify which graphic variables are most easily perceived, at least in the case of representing quantitative data. They used a slightly different version of Bertin's graphic variables.

Cleveland and McGill's variables include:

Position on an aligned scale

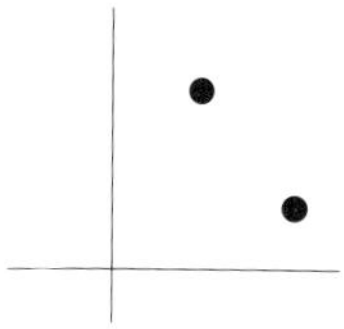

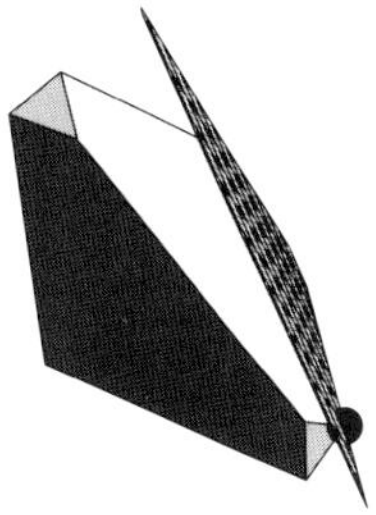

Position on a non-aligned scale

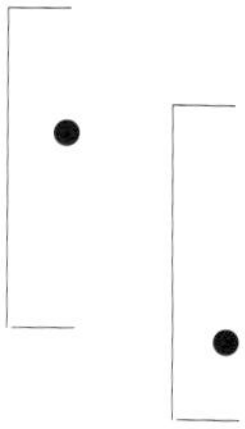

Length

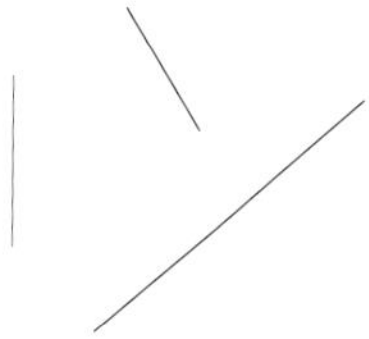

Angle

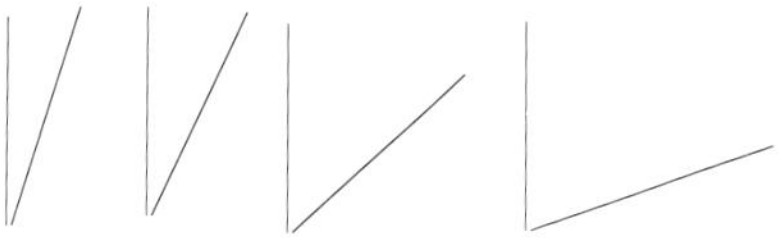

Area

Volume

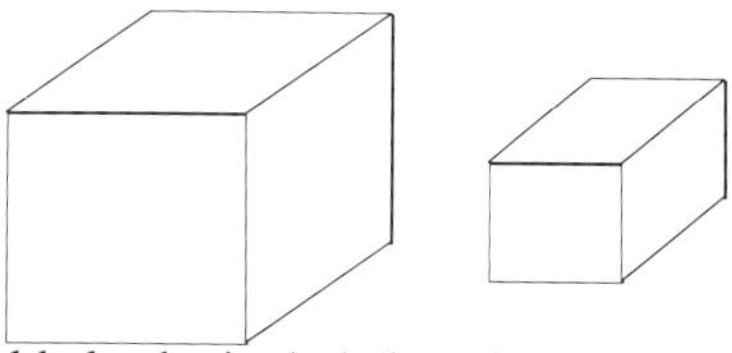

Color (e.g., hue, saturation, lightness) should also be included on the scale but is not shown since this book is being printed in black and white.

Their experiments compared how accurately data encoded in each representation style was perceived. From their experiments, they developed a rank ordering of the different representation styles. The list of Cleveland and McGill's graphic variables as given above is ordered from most accurate to least accurate.

Cleveland and McGill's ordering can be used to guide the selection of graphic format. For example, since data encoded as area or angle is less accurately perceived than data aligned along a common scale, a bar chart representation of data would be more accurately perceived than a pie chart. For example, consider the pie chart at right:

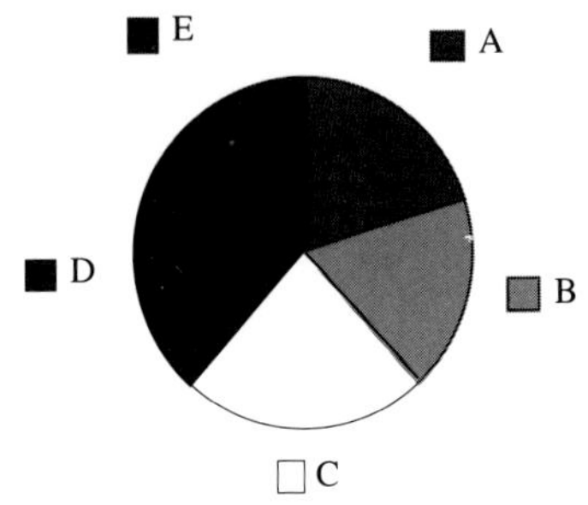

Figure 4.14

The same data can be represented using the bar chart at right.

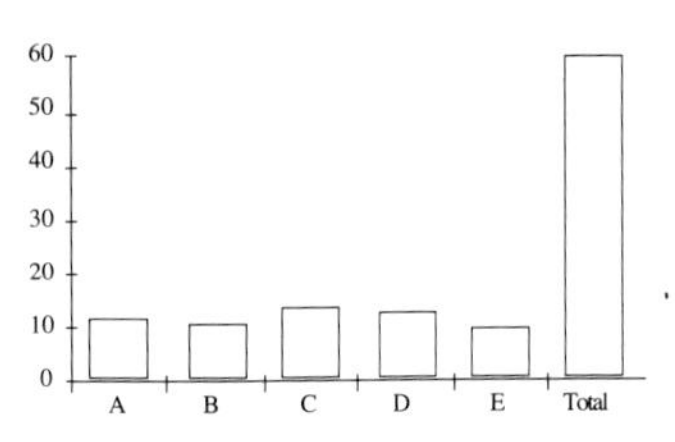

Figure 4.15

Note that percentage of the whole for any item can be gauged using this bar chart because it includes one bar representing the total of all the other bars. Cleveland and McGill would argue that the bar chart is easier to perceive than the pie chart. Is it easier to order the size of the items represented in the pie chart or the bar chart?

Cleveland and McGill's work, providing support to at least one of Tufte's recommendations, also suggests that the three-dimensional bar charts such as shown below that are now ubiquitous should be avoided. Based on Bertin's framework and Cleveland and McGill's results on different levels of perceptual accuracy, Mackinlay [11] developed an expert system for creating presentation graphic representations of data automatically.

Even though pie charts and other common representations are less accurately perceived than bar charts, pie charts will likely still be used. This can be attributed to anchoring effects: if one expects to see certain data as a pie chart, other formats may not be well received by the intended audience.

4.2 Color

Color displays are now the norm. The proper use of color in visualization is still under investigation. A tremendous amount is known about the physicals and perceptual properties of colors. Numerous guidelines on the proper use of color have also been proposed. This section provides a brief overview of some of the more relevant issues concerning color, particularly its use with computers. More on color can be found in [12], [13].

4.2.1 Physics of Color

From the standpoint of physics, color is produced by electromagnetic radiation having a wavelength between 380nm and 770nm. Any light reaching the eye can be characterized by a graph such as the one at right, which plots wavelength versus intensity. If the eye were a perfect receptor, each different spectral curve would be perceived to be a different color. In short, different spectral curves can be perceived by the eye to be exactly the same color.

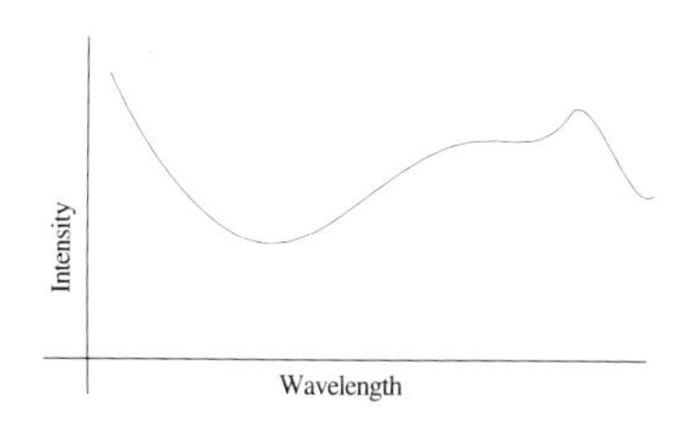

Figure 4.16

This perceptual shortcoming lies at the heart of artificial color reproduction found in photography, printing and computer displays. In each of these systems, mixing three or four different colors produces a wide (but not complete) spectrum of colors as perceived by the human eye.

4.2.2 Color Models

The use of three different primary colors is defined to be a *tri-stimulus* model of color. In tristimulus color systems, three *primary* colors are carefully chosen so that when mixed a wide variety of different colors are produced. For computer displays, the colors are red, green and blue. For printing and photography, the colors chosen are cyan (a light blue), magenta and yellow. A color is specified as a certain percentage of the primary colors. The primary colors define a *color space*.

Tri-stimulus color models are derived from a simple analogy to the human retina. The human retina senses an image using two types of specialized cells, rods and cones. Since the cones come in two varieties, the retina actually uses three different types of receptors. Each different type of receptor has varying sensitivity to different wavelengths of light. Since the retina uses three different receptors, it is argued, using three different primary colors to reproduce visible colors makes sense.

Is there a precise definition for every perceivable color? An international standard developed by the Commission Internationale de l'Eclairage (CIE), defines three primary colors from which all perceivable colors can be created. The CIE primaries are useful mainly for defining colors, but no display or printing device has been constructed based on those primaries. All commercial devices can reproduce only a subset of the colors defined by CIE.

The Red-Green-Blue (RGB) space for computer displays, as well as the Cyan-Magenta-Yellow (CMY) for color printing and photography are not always the most convenient for specifying colors. For example, what percentage of red, green and blue should be mixed in order to create a dark brown? Given the percentages of red, green and blue for a dark brown, what is the appropriate mixture for a light brown? Several proposals have been made that specify color spaces that correspond more carefully to common characteristics of color.

For example, the Hue-Lightness-Saturation (HLS) color space [14] defines one primary (Hue) corresponding to the shade or hue of the color, e.g., red, yellow, green. The Lightness primary defines how bright or dark the color is. The Saturation primary defines how pure the color is. A highly saturated color covers only a very small range of wavelengths as shown in the figure below. A desaturated color covers a wide range of wavelengths. Pastel colors are highly desaturated, whereas "neon" colors are highly saturated. When a color is fully desaturated, it contains no hue at all; in other words, a fully desaturated color is grey. By choosing a particular, hue, lightness and saturation, one can often more easily control the desired characteristics of a color.

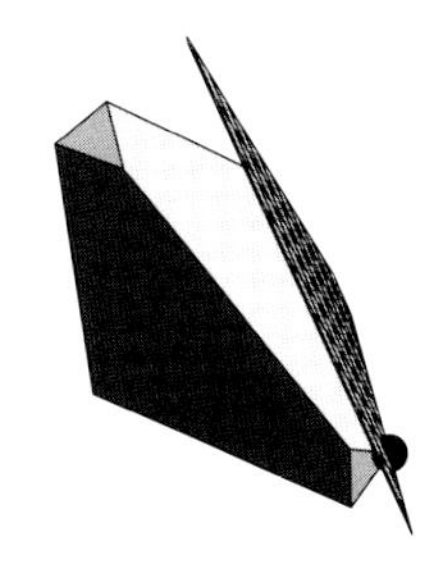

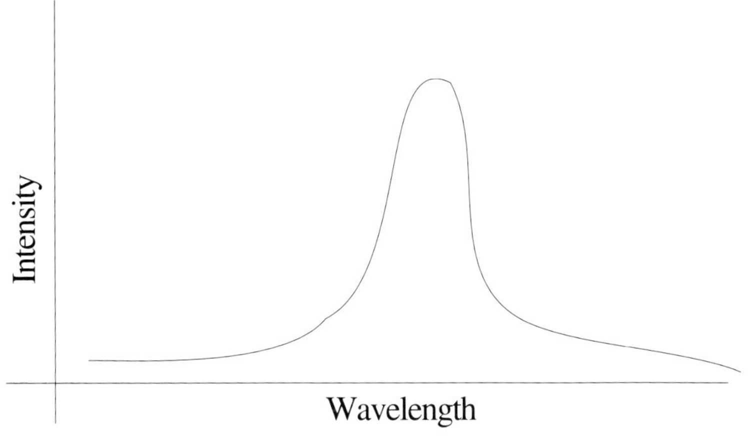

Figure 4.17

The color space that defines color television in North America and Japan (NTSC) defines its own set of primary colors denoted by the abbreviation YIQ. One of the primaries defines the black-and-white signal used by black-and-white televisions. The other two primaries span different ranges of color. They were chosen to maximize the quality of the image, and yet fit into the available bandwidth.

It is possible to convert between any two color spaces. Many of the conversions involve simple linear algebra. In fact, one can draw an analogy between a color space and a vector space. The primary colors correspond to the basis vectors. Different color spaces correspond to different choices of the basis vectors. The conversion, therefore, between two color spaces merely involves a matrix multiplication. For example, to convert from RGB to YIQ space, the transformation is [15]:

$$\begin{bmatrix} Y \\ I \\ Q \end{bmatrix} = \begin{bmatrix} 0.299 & 0.587 & 0.144 \\ 0.596 & -0.275 & -0.321 \\ 0.212 & -0.528 & 0.311 \end{bmatrix} \begin{bmatrix} R \\ G \\ B \end{bmatrix} \tag{4.1}$$

Converting from YIQ to RGB merely involves computing the inverse of the above matrix.

Unfortunately, neither computer displays nor color printers can reproduce all the colors that the eye can see. For example, using only cyan, magenta and yellow inks only a darkish green "black" can be produced. Most color printers add a fourth primary ink—black—to ensure high quality blacks are produced. Moreover, the color space of a computer display is not the same as the color space of a color printer, nor is it the same as commercial television. For example, "pure" red on a computer display (100% Red, 0% Blue, 0% Green), when converted to YIQ using equation 4.1 produces a Y value of 113%. Even worse, as printers and displays age, the colors produced by the device can change. Reproducing identically perceived colors on different devices is an ongoing challenge.

Reproducing colors accurately on different media requires careful color *calibration* among the media. One way to solve the color calibration problem is to buy colors. The Pantone color matching system [16], [17], for example, is a set of colors that the Pantone company has certified for use on different media. The apple-green that you see on the computer screen will match precisely the apple-green on the printer, or on the paint on the wall, or the dye on the fabric.

4.2.3 Color Guidelines

The retina has relatively few receptors of blue and green, especially in the center of the retina where detail is perceived [3]. This information would suggest that small detail and text not be colored blue. Similarly, the lens of the human eye is not color corrected. It is not possible for the human eye to focus simultaneously on both a red or blue background. The lack of color correction combined with the poor suitability of blue for details makes blue a superb background color.

Colors interact [18]. Even different shades of gray interact. For example, compare the two images below. The small, light square in (b) appears darker than the large square in (a), yet they are the same shade of gray.

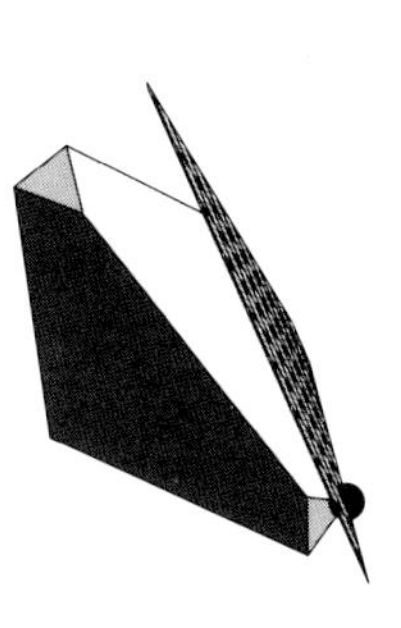

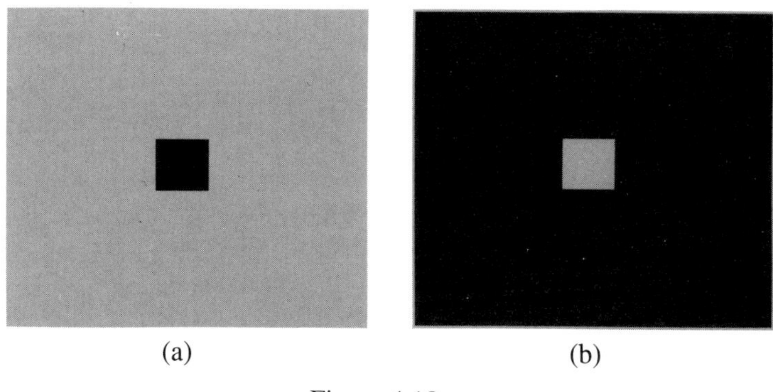

(a) (b)

Figure 4.18

Linear changes in the values of the primary colors used do not necessarily correspond to equivalent perceptual changes. On a computer display, for example, 50% Red, 0% Blue, 0% Green is not necessarily perceived as half as bright as 100% Red, 0% Blue, 0% Green. They eye is not a linear device.

Marcus [13] developed a concise set of recommendations for using color in computer displays. They include:

1. Do not use more than five plus or minus two colors. More colors than that can saturate a user's short term memory. A corollary to this recommendation is to design initially using just black and white, and then add color carefully.

2. Do not use bright, saturated colors for backgrounds; rather use desaturated colors.

3. Use bright saturated colors for small areas, particularly to attract attention.

4. Use color consistently, not only on computer displays, but also in accompanying documents such as user's manuals.

5. Consider cultural interpretations of color. Red can mean danger or stop. Green can mean go, vegetation, among others. White is a symbol of death in some cultures and a symbol of peace in others.

6. Use color to group similar items. Watch out for guideline number 1, however. To support color deficient users (approximately 8% of the population has some trouble seeing a full-range of colors) [3], double code the information contained in color using shape or some other encoding scheme.

4.3 Three-Dimensions

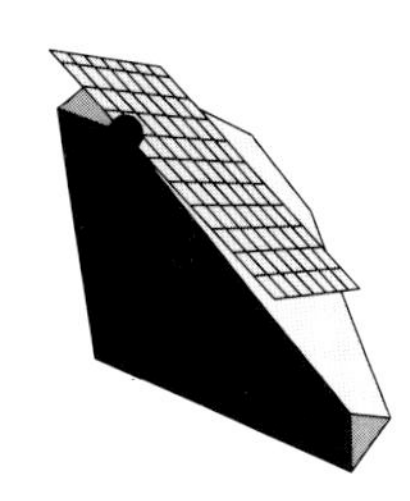

Except for expensive display devices, every computer graphic image is inherently two-dimensional, since it is composed merely of glowing phosphors in a glass surface, chemical pigments on photographic paper, color dyes on printed paper, or light shining through a plane of colored filters in an LCD display. Yet we often "see" three-dimensional (3D) images on our two dimensional (2D) media.

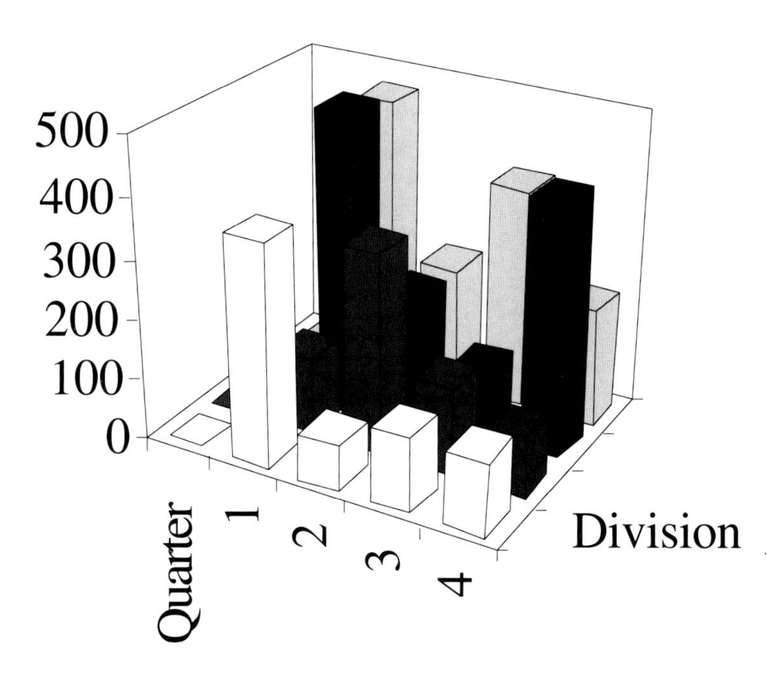

Figure 4.19

4.3.1 History

This illusion is called *perspective*. It was essentially unknown before the 15th century when the Florentine artist Brunelleschi proposed and the mathematician Alberti formalized the idea of perspective [19]. Before the advent of perspective, objects in a scene were drawn larger than others only if they were more important than the others; in Christian paintings, therefore, Jesus generally was the largest person in the scene. With the advent of perspective, objects were only drawn larger if they were closer than others in the modeled world.

Creating perspective drawings involves the mathematical operation of *projection*. To create a perspective drawing, one is essentially projecting a three-dimensional scene into two dimensions. The projection operation is not invertible since the projection operation causes information to be lost. It is a tribute to the human visual system, however, that we are able able to infer with reasonable accuracy the original three-dimensional scene from the two dimensional projection. Projecting from worlds having four or higher dimensions into two, however, taxes the human visual system. Adding a third dimension to an image may produce no net benefit. For production scheduling, for example, we proposed a three-dimensional Gantt chart [20] that is quite similar to a three-dimensional bar chart. Although this type of Gantt chart is an interesting piece of conceptual art, it has no practical application.

4.3.2 Computing Perspective

One of the fundamental accomplishments of the computer graphics field since its origins in the 1960's is the development of efficient algorithms and computer hardware for calculating perspective projections. Rotation, scaling and translating objects in a scene can be accomplished using simple linear transformations expressed as matrix multiplications. Perspective projection (see [14] for details) requires a final division operation, essentially dividing the horizontal and vertical coordinates of an object by its depth coordinate. In short, every point in a modeled world must be multiplied by a matrix transforming that point into an appropriate orientation. Every transformed point is then divided by an appropriate value in order to determine its proper location on a two-dimensional plane.

The principal idea behind modern algorithms for performing perspective projections involves delaying the relatively costly division operation to as late as possible. This involves accumulating the perspective divisor through every other transformation of points in the modeled world, and then, only at the end of the cycle, performing the final division. The divisor is accumulated by adding a fourth dimension to every point, and using four by four transformation matrices rather than three-dimensional transformation matrices.

4.3.3 Perceiving Three Dimensional Perspective

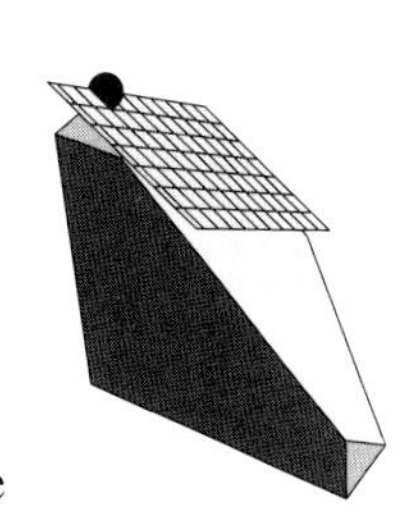

The human visual system perceives three-dimensional depth partially because we see using both our eyes. Each of our eyes sees a slightly different version of a scene. This *binocular disparity* [1] provides one mechanism allowing us to "see" three dimensional information in the real-world. Binocular disparity, however, does not account for how people who have only one good eye navigate through a three dimensional world. One source of three-dimensional perception involves *motion parallax*. Motion parallax occurs when objects in a three-dimensional scene move. Objects at different depths demonstrate different relative motions, and the human visual system is particularly adept at inferring the appropriate depth relationships based on such differences.

4.3.4 Three-Dimensional Displays

It is possible to exploit this ability using artificial displays. If slightly different images of a scene are conveyed to each eye, the human vision system will perceive three-dimensional depth even when none is there. This idea was first exploited in the 1800's [1]. In 1838 Charles Wheatstone developed a stereoscope which used a set of mirrors to project two separate images into the eyes. David Brewster, in 1849, developed a lens based device to produce the same effect, and it was his invention that saw widespread use. The "3D" color movies hyped in the 1950's and which are still produced sporadically require the viewer to wear special glasses containing polarized lenses. Two images are simultaneously projected onto the silver screen using polarized light. Each image is projected with opposite polarity. Each lens in the set of glasses worn by the viewer filters out one image, thereby transmitting a different image to each eye.

Computer displays that can show three-dimensional images also exist. In one type of device, the viewer wears glasses whose "lenses" consist of fast shutters synchronized to the computer display [21]. The shutters, using liquid crystal technology, alternately admit light into each eye, switching at a very fast rate. They are synchronized with the computer display which alternately displays different images for each eye. Note that this requires a very fast computer display. This type of three-dimensional display is sometimes called a "fish-tank," since one sees three-dimensional images inside the computer display—the fish-tank.

Another approach uses a different type of "glasses"—more a helmet than glasses—called a *head-mounted display* [22], [23]. The head-mounted display contains two displays, one for each eye, coupled with a set of carefully designed lenses to compensate for the relatively short distance from the display to each eye. As opposed to fish-tank three-dimensional display, such glasses provide an *immersive* visual environment, since the image fills the user's field of view. Coupled with devices to track the position and orientation of the user's head (see Section 5.2.2), the image can be adjusted as the user moves, thereby creating an artificial or virtual reality. We discuss virtual reality more in Chapters 6 and 16.

Yet another approach to generating three-dimensional images mounts two small computer displays on a counter-weighted mechanical boom [24]. The user presses her eyes to lens in front of the displays, thereby providing binocular vision. The user can move the display; as it moves, its position and orientation are tracked by measuring the angles of the joints on the boom. Through this tracking, the image displayed can be updated so that the user can see an artificial world.

Note that none of these stereoscopic display devices perfectly recreates normal three-dimensional vision. In all of these cases, the eyes must stay focused on a fixed two-dimensional plane, no matter the relative distance of the image being perceived. In a three-dimensional movie, for example, if an object is perceived to move close to the viewer, the eyes still must remain focused on the far distant screen in order to see the object clearly. Similarly, in three-dimensional movies and fish-tank displays, since the image does not fill the viewer's entire field of view, the three-dimensional imagery stays confined within a frame. Although it is possible to induce the perception of small objects extending from the screen or display into the real world, the outside frame can easily cue the visual system into seeing the object only as existing inside the screen or display.

Lacking a head-mounted display or liquid crystal shutters, one can exploit motion parallax to help with three-dimensional depth perception of a two-dimensional projection. Merely rotating an object slightly can greatly help one perceive three-dimensional depth in a two-dimensional projection [1].

Holography represents another solution to the problem of producing the illusion of three-dimensional depth. A hologram stores the complete pattern of light intersecting the image plane. A hologram stores not a single image, but in essence a series of images, one for each possible viewing direction. When light is shone through or on a hologram, the recorded images are transmitted to the viewer. Clearly a hologram stores an incredible quantity of information. To create a hologram requires very fine grain film; although computer-generated holography is currently the subject of research [25], it is not yet become practical.

Mechanical devices have also been proposed to create the illusion of depth without the need for special glasses. These devices typically use a vibrating or rotating plane of a light generating matrix. As the plane moves, the lights are turned off and on at appropriate times to generate an object at that location in space [26].

4.3.5 Volume Visualization

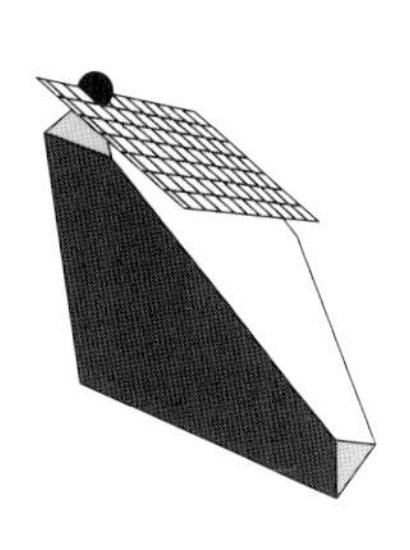

Much three-dimensional imagery is composed of two-dimensional surfaces such as polygons. The resulting three-dimensional objects are essentially hollow. In contrast, there are many applications (e.g., medicine or CAD) which require solid or dense three-dimensional objects. Visualizing such information has become known as *volume visualization* [15]. In volume visualization, the point $p = (x, y, z)$ and its associated value w_p is called a *voxel*, analogous to a pixel for two-dimensional display.

Perhaps the most common application of volume visualization has been in medicine with the explosion in medical imagery such as CT, MRI and PET scans. In their raw form, such scanning devices produce a set of cross-sectional images. Volume visualization allows the images to be "stacked up" to construct a solid, three-dimensional visualization of the patient. In order to create a useful volume visualization, voxels are allowed to have different levels of transparency, so that voxels inside the scene can be visible. Otherwise, if all the voxels were opaque, one would only see the voxels on the boundary. Below is a typical volume visualization of some MRI data [27]. The image consists of a stereo pair of images. Note the diaphanous nature of the resulting images. (Images copyright ©1994-1995 Vox-L Inc., A Lateiner Dataspace Company, Woburn, MA. Reprinted with permission)

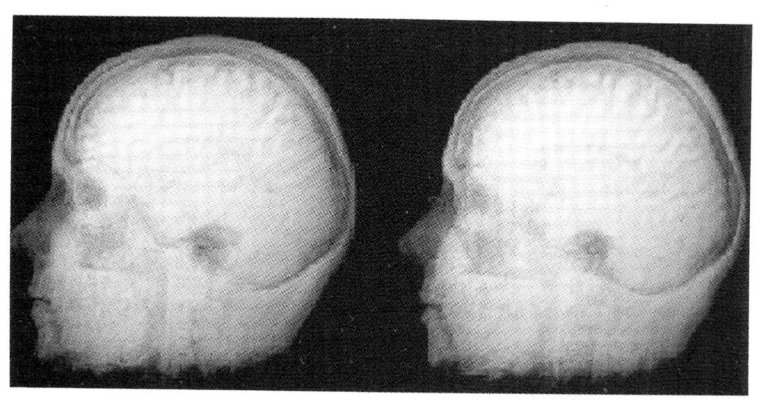

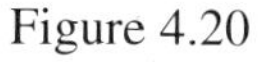

Figure 4.20

4.3.6 Summary

Displaying information in three-dimensions is now easy. Much of what is labeled "virtual reality" is simply the display of three-dimensional scenes (see Chapter 6). Whether or not one should display information in three-dimensions is another matter. Moreover, most optimization problems involve thousands of dimensions, not just three. We discuss how to represent information having more than three dimensions in Chapter 14.

4.4 Animation

Leland Stanford, railroad tycoon, first governor of California, and eventual founder of Stanford University wanted to understand a complicated system: the motion of a trotting horse. He actually just wanted to settle a long-standing question that tended to float around saloons and taverns: are the hooves of a trotting horse all off the ground at the same time, or does at least one hoof always stay on the ground? Although this question is far from momentous, the technology that emerged from this study was.

Stanford turned to a photographer acquaintance, Eadweard Muybridge for help. (The unusual spelling of Muybridge's name came from the place of his birth, Kingston in Great Britain, where several early English kings were crowned, two of which were named Eadweard) [28]. Using a crude apparatus, Muybridge photographed trotting horses running along a track, taking several images in succession.

Muybridge discovered that when the sequence of photographs was projected in rapid succession, the motion of the horse was reproduced. In addition to answering the original question (all four hooves of a trotting horse hooves *are* in the air at the same time), this was one of the first examples of motion pictures. Moreover, it was one of the first uses of moving pictures to study a complicated system.

Another example illustrates the importance of motion in our ability to perceive the world. Consider a person in complete darkness to whom lights have been attached at various parts of the body. If the person does not move, the lights will appear to be arranged arbitrarily. However, as soon as the person moves, immediately (within 1/10th second), one recognizes that the lights are attached to a person. Moreover, solely from the motion of the lights, it is possible to identify the gender of the person [1].

In short, pictures need not be static; they can move. Motion can convey a great deal of information. In fact, some cognitive psychologists, notably Gibson [29], [30], [31] maintain that humans never see static images, only moving images, since our eyes are in constant motion.

Animation is a special type of moving picture, wherein each individual image or *frame* is carefully drawn, i.e., not photographed. Given that smooth motion requires anywhere from 10 to 30 images a second [32], animation has traditionally been quite expensive, and hence rarely used outside of Hollywood. With the advent of computers, however, the cost to produce animation has significantly declined, making animation available even to optimization.

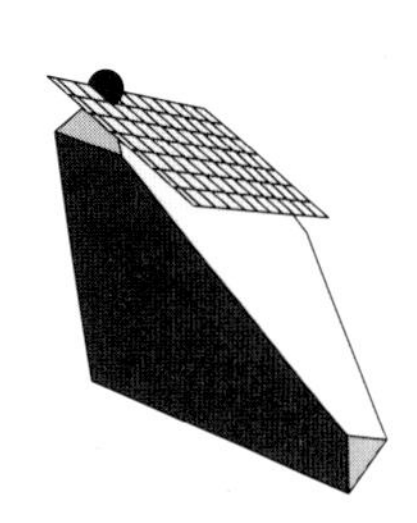

For which tasks and which users is animation helpful? How can animations be created? Unfortunately, our understanding of the proper use of animation is even less well-developed than our understanding of the proper use of static graphics. There is some useful knowledge, however.

4.4.1 Results from Cognitive Psychology

How does the human vision system perceive motion? This is more complicated than it may seem at first. In particular, the changes in the light reaching our eyes are not always sufficient to produce the perception of movement. For example, when a train moves [1], the light coming into our eyes changes, and from those changes, we infer that the train is in fact moving. However, we can also move our heads or our bodies and as we do so the light entering our eyes also changes as *we* move. Even if we run madly around a stationary train we still perceive the train as stationary. Somehow, we are able to distinguish objects that move from the movement induced from our own eyes, heads and bodies.

More remarkably, the human vision system is able to make these deductions extremely efficiently, far more efficiently than is currently possible with current computer technology. Yet the human vision system's motion detection apparatus is not perfect. For example, if a passenger is sitting in a stationary train next to another train that is just starting to move, a common illusion can occur [1]; one can be fooled into believing for a moment that the passenger's (stationary) train is moving and the moving train is actually stationary. Eventually other clues (the stationary train does not vibrate or rumble) allow the passenger to realize what actually happened. Another illusion occurs when clouds pass in front of the moon [1]. People often perceive the clouds to be stationary and the moon to be moving.

As stated earlier, some cognitive psychologists maintain that all vision involves motion, since the eyes are constantly in motion. Rather than analyzing single static images, they believe the human visual system analyzes the data flowing into the eyes over time. This is not the same as analyzing a series of static images, one for each successive instant in time. Rather, under this model, vision involves analyzing a continuous *optic flow* of information through time[1] and [33].

From the optic flow, the human visual system exploits invariant relations among the perceived objects. When one tilts one's head, for example, stationary objects will "move" relative to one another; objects that are nearer will both reveal and occlude portions of objects that are farther away. According to the optic flow hypothesis, the human visual system possesses well developed capabilities to extract reasonably accurate interpretations of the outside world. One of the fundamental principles they postulate involves *invariant* relations among the objects perceived. As described in [33], invariant relations arise from simple projective geometry, which is easily modeled mathematically, as briefly discussed in Section 4.3.2. Apparently, the human visual system has an algorithm for extracting appropriate information based on these mathematical relationships.

4.4.2 Film and Computer Animation

All of the above forms of motion do not cover the type of motion used in film or computer animation. In those forms of animation, the object that appears to move is actually not moving at all. Rather, the viewer is shown a series of fixed, static images or *frames* (remember Stanford's horses). If those images are shown sufficiently rapidly (around 30 frames per second), then the *illusion* of motion is produced even when nothing is actually moving. This type of motion is called *apparent* motion.

The proper *frame rate* has a high degree of variability. If the objects in the scene are not moving at all, theoretically only a single animation frame is ever required. If the frame rate is too slow, the human vision system begins to see the individual static images, and the illusion of motion is lost. This problem is often described as *flicker* (explaining the etymology of the word "flicks" used to describe the movies). Flicker can actually occur at frame rates of around 30 frames per second.

In commercial films, the standard frame rate is 24 frames per second, a rate that was standardized at the introduction of sound films in the late 1920's. Many silent films are properly seen at a slower frame rate; when they are projected at 24 frames per second, the motion is more rapid than was intended. Even though film is projected at 24 frames per second, each individual frame is displayed twice within each 1/24th of a second; a shutter opens and closes the light beam twice for each frame of film.

For television and computer displays, the frame rate is often referred to as the *refresh* rate. Television in North America (NTSC), for example, uses a refresh rate of 30 frames per second, whereas, the European derived systems (PAL and SECAM), use a refresh rate of 25 frames per second [34]. In both these systems, however, a technique called *interlacing* is used to increase the apparent refresh rate. A television frame is composed of a series of horizontal lines. In a typical television set, the lines are drawn using a cathode ray tube, which consists of an electron beam of varying intensity impinging on phosphor causing it to glow. In an interlaced device, the individual frame is drawn in two passes. In the first pass, the odd numbered lines are drawn, and in the next pass the even numbered lines are drawn, with the whole frame drawn in 1/30th of a second (for NTSC). In other words, a low resolution version of each frame is refreshed every 1/60th of a second. Interlacing is used because it reduces the cost of the television set, while producing a sufficiently high quality image.

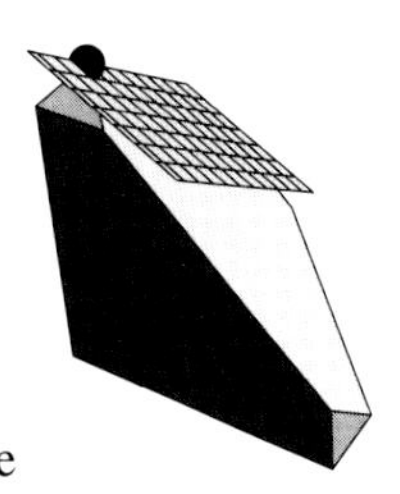

Interlaced computer displays, although increasingly rare, illustrate some of the problems with 1/30th of a second frame rates. A single horizontal line on an interlaced monitor will appear to flicker. Non-interlaced displays, which cost more than interlaced displays, draw each entire frame in 1/60th of a second and eliminate the problem of flicker.

In computer animation, one must distinguish between the refresh rate and the *update* rate. The refresh rate is solely a property of the computer display, and refers to how fast the electron beam (for cathode-ray tube displays, for example) can draw out a single image stored in digital form. The digital image is stored in a *frame buffer*, which is a specialized memory that allows very fast access. The update rate refers to how fast the computer can change the image stored in the frame buffer. For non-interlaced monitors, the refresh rate is generally around 60 frames per second. The update rate for a computer animation depends on how fast the frame buffer can be changed. That speed depends on the complexity of the animation being produced. Techniques for producing such animation are discussed next.

4.4.3 Basic Computer Animation Techniques

The limits of the human vision system place great demands for computer animation. A single frame on a computer monitor with a resolution of 1000 by 1000 *pixels*, each pixel consisting of 24 bits (or more) of information, requires 3 megabytes of storage. If one is to maintain an update rate of 30 frames per second, then 90 megabytes must be updated every second. Luckily, most computer-based animation does not require such resources. For instance, most computer animations do not need to update the entire screen with each frame. Rather, only a relatively small portion of the screen actually changes. If one merely needs to move an object on the screen, the background does not need to be updated. Even for this simple type of animation, several different specialized techniques have been developed. This section discusses them.

The discussion assumes that one needs to move an object on the screen in a smooth, continuous fashion. The principal challenge in moving an object on the screen is *not* moving the object. The real challenge is making certain that the rest of the image is maintained properly. After all, one could merely repeatedly draw the object in each of its new positions. The problem is that the object must be erased in its old position, and the background restored.

4.4.3.1 XOR Animation

A very efficient, easy-to-program mechanism for moving an object on the screen is called *XOR* animation, where XOR refers to the boolean exclusive or function (see table at right).

a	*b*	*a* XOR *b*
0	0	0
1	0	1
0	1	1
1	1	0

Figure 4.21

For the sake of simplicity, assume a black and white screen, that is, each pixel in the frame buffer consists of only a single bit. If the bit is 1, the image seen is black, if it is 0, the image seen is white. In XOR animation, one draws the object to be moved by computing the boolean XOR of the object with the current image in the frame buffer. Call this use of XOR for drawing, *XOR mode*. For example, given a background image (a), the object to be moved (b), the image computed when the object is XOR'd into the background is shown in (c). Note that the object inverts black pixels in the background wherever the moving object is drawn.

a. Background

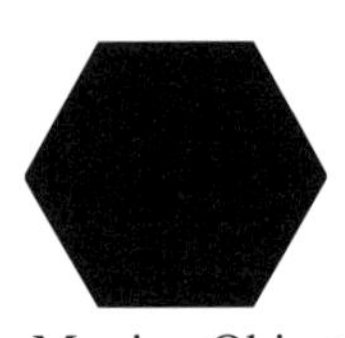

b. Moving Object

c Background XOR Moving Objec

Figure 4.22

The moving object can be erased and the background restored simply by drawing the object in the same location using XOR mode. To move an object using XOR animation, therefore, the object is drawn twice in each position. The first time it is drawn, the object appears. The second time, the object is erased and the background restored. XOR animation is used extensively in the Apple Macintosh and Microsoft Windows, for example, when windows are resized.

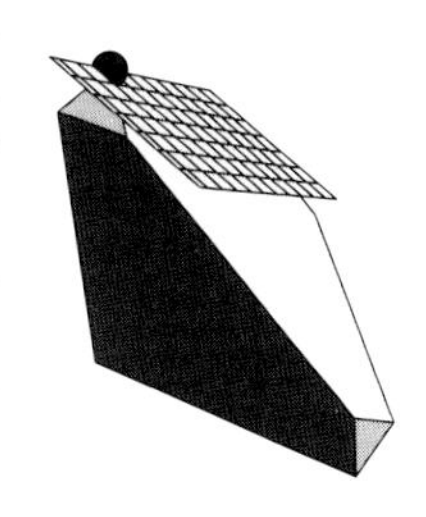

XOR animation is easy to program and does not require expensive computer hardware. It does not, however, produce as smooth an animation as one might desire. First, the drawing and erasing of the moving object may not occur quickly enough, so that flicker is produced. Second, the moving object is not drawn perfectly, but rather, is inverted over the background. When the background is in color, the colors produced can be quite strange. Other techniques, described below, overcome some of the limitations of XOR animation, but at some cost.

4.4.3.2 Color Map Animation

As stated previous, each pixel on the screen is stored as a binary number in the frame buffer. But how is that binary number translated into a color on the screen? Many frame buffers store 24 bits for each pixel. Of those 24 bits, 8 bits usually define the level of each of the red, green and blue phosphors that make up the pixel color. A *color map* is a table of colors. In a color-mapped frame buffer, the bits in the pixel define the entry in the color map table, which then defines the color that will appear on the screen. A color-mapped frame buffer is particularly useful for less expensive frame buffers which use fewer than 24 bits per pixel. If a frame buffer stores only 8 bits per pixel, then at most $2^8 = 256$ different colors can be displayed simultaneously. A color map, however, allows far greater flexibility in choosing those colors. If each color map entry consists of 24 bits, then a wide variety of different colors can be chosen, but still only 256 can be displayed simultaneously.

A color map can be used to support animation. In particular, the color map can be changed independently of the image displayed in a frame buffer, and since the color map is relatively small, it can be changed quite quickly. Merely by changing the colors in the color map, the image will change, even though no objects are actually moving on the screen.

Color maps, however, can be used in a more sophisticated way to produce animation. Assume for now that each pixel in the frame buffer can store at most 8 bits, meaning that at most 256 colors can be used, numbered from 0 to 255. For normal drawing, we will restrict ourselves only to colors 0 to 127. In the color map, we can set entries 0 to 127 to whatever values we wish. For normal drawing, in other words, we will only use 7 of the 8 bits in the pixel. The eighth bit will be used when we are moving an object. For colors 128 to 255 in the color map, we set the color to the same shade, say red. When an object is moved, we will draw it so that only the 8th bit of a pixel in the color map is changed, which is an operation available in most color-mapped frame buffers. When the eighth bit is set to 1, the pixel will be colored red. When the eighth bit is set to 0, its original color will be restored. By drawing the moving object so that only the eighth bit is set, the object essentially "floats over" the background. Although the moving object can only appear in one color in this example, using more bits for the moving object allows it to be displayed in more colors. Note however that each additional bit used for the moving object decreases the number of colors available to the background by a factor of 2.

This style of animation has the advantage that the moving object is unaffected by the underlying background, unlike in XOR animation, but reduces the number of colors available for normal drawing.

4.4.3.3 Double Buffering

Another style of animation corrects both of the problems of XOR and color map animation, but at the expense of special hardware or more computer memory. In this style of animation, often known as *double buffering*, the current image is stored in a frame buffer, just as before. The next frame in the animated sequence, however, is drawn off-screen, in a separate frame buffer. Call the original frame buffer, frame buffer A, and the (currently) off-screen frame buffer, frame buffer B. When the next frame has been drawn in frame buffer B, frame buffer B is channeled to the computer display, and frame buffer A is no longer seen. The subsequent frame can then be drawn in frame buffer B. In short, in a double buffered system, two frame buffers are used; the animation is produced by alternating between the two frame buffers; when one frame buffer is displayed on the computer screen, the next frame is drawn on the other frame buffer.

Double buffering eliminates the flicker produced by XOR animation, since all drawing is hidden from the user. The user does not see each individual line, polygon and circle drawn, but only the completed image when the frame buffers are swapped. It does not consume colors as does color map animation, since one maintains two frame buffers. It does require more resources, since one needs double the frame buffer memory.

A form of double buffering is used on personal computers, even though they do not typically maintain two hardware frame buffers. The image that would be stored in the offscreen frame buffer is stored in main memory. This offscreen image is referred to commonly as an *offscreen bitmap* (for monochrome images) or an *offscreen pixmap* (for color images). Once the image in the offscreen pixmap is constructed, a single instruction, generically known as a *bit-block transfer* or *bit-blit*, is used to move the contents of the offscreen pixmap into the frame buffer for display.

Frame buffer animation still is limited by the speed at which each individual frame can be drawn. If drawing a frame requires more than 1/30th of a second, then the animation will not be smooth, however, the viewer will not see the sequence of drawing operations required to draw each frame. The animation will appear to jump from one frame to another.

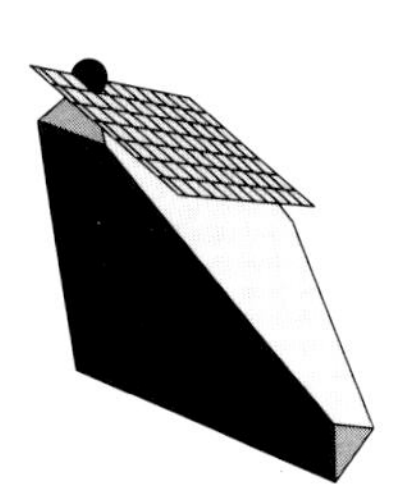

4.4.3.4 Film-Strip Animation

When the time to draw an individual frame is long, one can create the individual frames off-line, store them, and then play them back, producing an animation. This is often called *film-strip* animation or *digital movies*. Although this does not allow the user to interact with the animation, sometimes it is the only alternative.

Given the potential amount of space required to store an individual frame, however, almost always some form of *compression* scheme is used to reduce the amount of space required (see [35], [36] for a survey). Many such schemes have been proposed.

Data compression schemes such as Huffman encoding, run-length encoding, and Lempel-Ziv are currently used to reduce the storage required for standard computer files. Compression schemes specifically tailored to the needs of audio and video are an area of active investigation. Key characteristics of encoding schemes include:

Quality Some compression schemes are tailored specifically for high quality images and sound, others for lower quality images and sound.

Bandwidth Bandwidth refers to the rate of data transmission. Some compression schemes are tailored exclusively for low bandwidth transmission media, e.g., 64Kbit/sec, some for medium bandwidth transmission media, e.g., 1.5Mbit/sec, and some for very high bandwidth transmission capability, e.g., 10Mbit/sec and higher.

Scalability Some compression schemes allow for varying bandwidths (with a corresponding change in quality), others are custom designed for particular bandwidths.

Symmetry Some encoding schemes are more efficient than their corresponding decoding schemes and vice versa. If a source material need only be encoded once, e.g., a commercial film, then an encoding scheme biased towards speed of decoding would be preferable.

Single/Multiple Some schemes are tailored for single images, others for multiple images.

Synchronization Some standards explicitly address the need to synchronize sound and image; others ignore this issue.

Loss Many of the proposed compression schemes for audio and video involve some form of loss, in that the decompressed signal is not identical to the original. Such compression schemes can be acceptable because of some of the limitations of the human perceptual system.

Temporality Some compression schemes have seen principal application in the compression of a single image, others compress multiple images, in order to support full-motion imagery better.

Many more subtle characteristics of compression/decompression schemes have been identified (see, e.g., Lippman [37]).

To give an example, a commonly used compression scheme, the *Discrete Cosine Transformation* (see Rao and Yip [38]), a variation on the Fourier transform, transforms the input image from the spatial domain into the frequency domain. Since the human vision system perceives low frequencies better than high frequencies, the higher frequencies can be less accurately stored, thereby saving storage space, with some loss. A different encoding scheme proposed by Intel, *digital video interactive* (DVI) (in greatly simplified form), breaks down an image into a sequence of primitives, e.g., a square in the upper left corner, a triangle in the lower left, and merely sends the codes for the primitives. Since the codes for the primitives require much less space than the actual image, a savings can result. Current research attempts to extract even more information about the underlying constructs that constitute a scene. Such *model-based* encoding schemes are more similar to a computer graphics rendering program than a traditional coder (Lippman [37]). They generally require specific knowledge about the particular image being encoded, trading more restricted application for greatly increased efficiency.

Two of the more important compression standards are JPEG [39] is MPEG [40], which stand for the Joint Photographic Experts Group and Motion Picture Experts Group, respectively. As an international standards, they enjoys the advantage that it is not tied to any particular private organization. JPEG was originally designed only to compress static images. It has been adapted, however, to be used to compress digital movies. MPEG not only compresses individual frames, but also compresses adjacent frames, since often adjacent frames are quite similar. Currently, filmstrips stored in MPEG format require special hardware in order to be played back in real-time. Another standard for filmstrip animation is QuickTime [41], developed by Apple Computer. Available not only for the Apple Macintosh but also for MS Windows and Unix, QuickTime is not a single compression system, but allows each individual digital movie to use a different compression scheme, e.g., MPEG. Microsoft has proposed Video for Windows [42] for MS Windows. Once an animation is stored in one of these standard formats, it is relatively straightforward to move the animation from one application to another, or from one computer to another. These standards are making the exchange of animation as easy as the exchange of text and graphics has become.

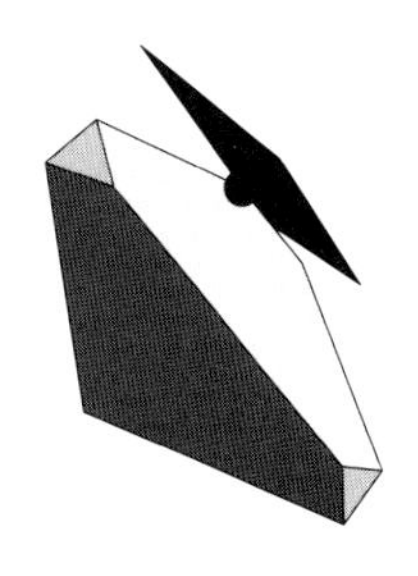

4.4.4 Frameworks for Animation

Fundamentally, animation illustrates changes in state. When the position of an object changes, for example, the object could merely disappear at its old position, and then instantly appear at its new position. In fact, in Microsoft Windows, when one maximizes a window, the window instantly leaves its old position (and size), and transforms into its new position (and size). As recommended by [43], a less jarring transition would involve animating the window as it changes position and size. The user can more easily track how the object makes a transition from one state to another. Animation, as discussed in Section 4.3 is also useful in providing the illusion of three-dimensional perception on a two-dimensional screen.

The most well-developed body of knowledge about animation probably comes from Disney animators [44]. Although designed primarily for frothy entertainment, many of the guidelines that they developed have application to the perhaps more mundane world of scientific visualization [43]. In their lavishly illustrated book, Thomas and Johnston outline a few basic principles for Disney animation.

- *Squash and Stretch.* Objects do not keep fixed shape when moving. They change shape (except for rigid objects).

No squash and stretch.

With squash and stretch

- *Anticipation.* "People in the audience watching an animated scene will not be able to understand the events on the screen unless there is a planned sequence of actions that leads them clearly from one activity to the next. They must be prepared for the next movement and expect it before it actually occurs." This is achieved by preceding each major action with a specific move that anticipates for the audience what is above to happen. The figure below shows a series of frames from an animation of a simple ball starting to move (the left-most ball), and then moving to the right. When the ball starts to move, it vibrates in anticipation of the move.

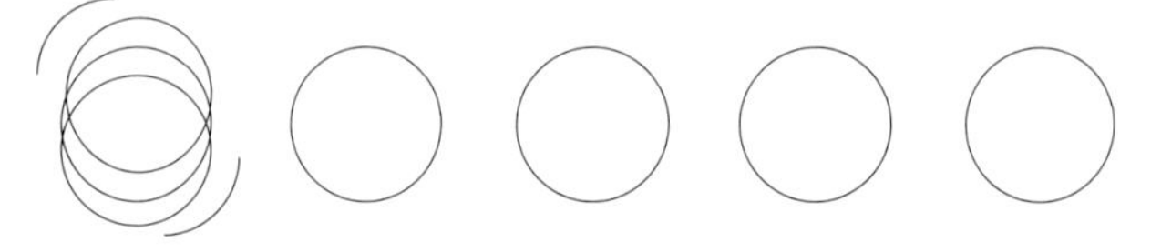

- *Slow In and Slow Out.* When objects begin to move, they move slowly. During the primary portion of the move, objects move relatively quickly, and when objects stop, they again move relatively slowly. The pictures below show frames from animations where a ball is moving from left to right. In the top sequence, the ball starts moving at a constant velocity until it stops on the right. In the bottom sequence, from the left the ball accelerates to a constant velocity and then decelerates to stop on the right.

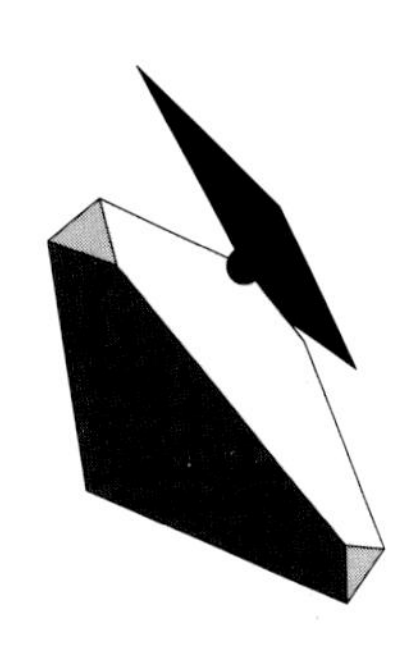

No slow in or slow out

Slow in and slow out.

- *Follow Through and Overlapping Action.* Not everything stops or starts moving all at once. The body may stop moving, but the arms continue. Chang and Ungar [43] used this idea to augment a traditional multi-window user interface. When a window stops moving, it wiggles before it finally comes to rest. In the following picture, again an animation of a ball moving from left to right, as the ball stops on the right, it vibrates slightly to emphasize that it is coming to a stop.

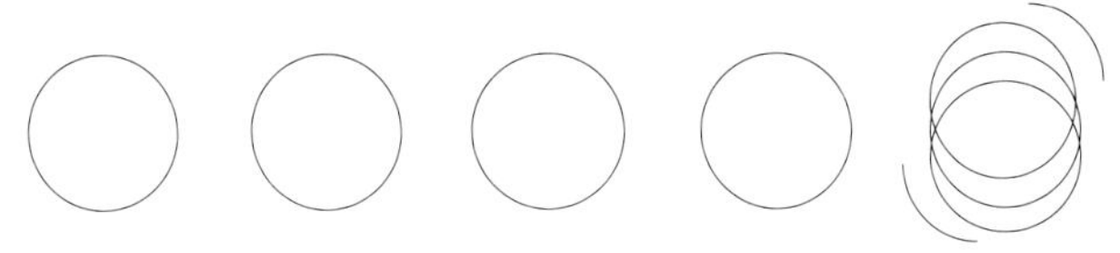

- *Arcs*. Natural objects move in arcs, not straight lines. Straight lines are easy for the programmer to implement, but arcs are more commonly seen in the real world. The pictures below show frames from animations of a bounding ball. The upper sequence shows a ball traveling in straight lines before and after impact. The lower sequence shows a ball traveling in arcs before and an after impact. Arcs reflect natural movement better than linear motion.

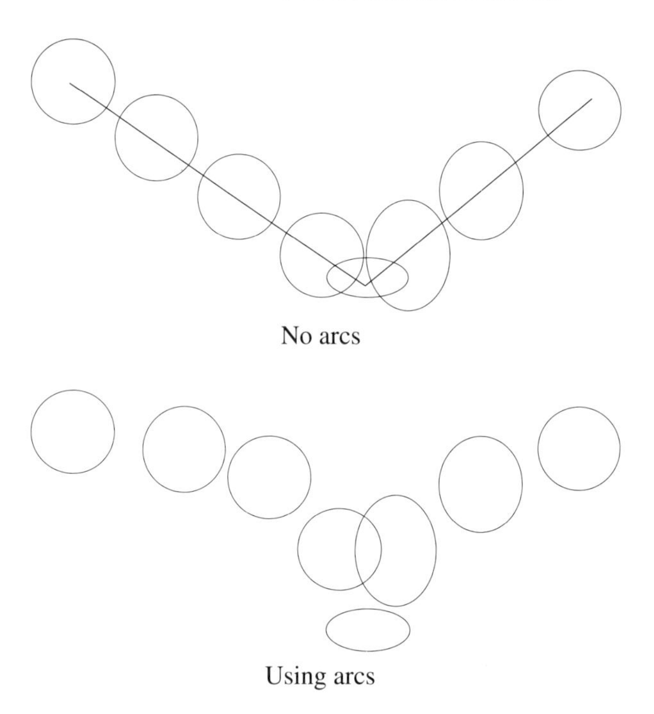

- *Timing*. How fast should an object move? Objects that move quickly require special treatment. In particular, a fast moving object will appear as a blur to the viewer. Therefore, a single frame showing a fast moving object will not show a clear image of the object, but rather will show a blurred image of the object. If this blurring is not provided, then the motion will not appear smooth. Rather, the object will appear to jump suddenly across the screen. For a computer programmer trying to produce an animation, generating such blurred images is non-trivial; it is far easier simply to draw the object in each of its new positions.

- *Solidity*. Objects have a physical presence. They do not usually appear from thin air, nor do they disappear suddenly. Even when objects are moving, even when they squash and stretch, they still retain a solid presence.

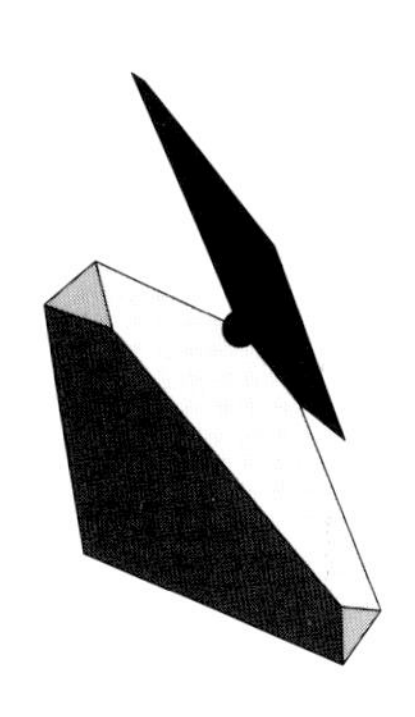

Not all of these are probably applicable to animations produced to support modeling activities. However, some do have relevance. An exaggerated motion may help users pick out an important phenomenon. Slow in and slow out is useful linking together different poses within a scene. Constant velocity, linear motion with abrupt starts and stops is much easier to program than motion following the guidelines of slow in, slow out, arcs, anticipation and follow through. Animation with the latter generally appears more natural. It may not reflect the underlying data. Perhaps one has to "distort" the underlying data in order to make its meaning clear. Such distortions, however, introduce great danger. They should only be introduced when they do not significantly affect the underlying information.

4.4.5 Experimental Evidence

Very little formal experimentation has explored the effectiveness of animation for different subjects and different tasks. Some of the results support the use of animation, others provide support the use of animation in some contexts, and others provide little support for the use of animation.

Rieber [45], [46], [47], [48] explored the use of animation for education. In a series of papers, he considered both children and adults learning Newton's laws of motion by interacting with an animated "starship." Animation increased the scores of children on a post-test, whereas for adults, the animation did not. However, the adults exposed to the animation finished the post-test faster than the other groups of adults, so animation did help adults remember what they learned more easily. Rieber [45], in conclusion, recommended that interactive animations, not just those which are passively viewed, may represent the most effective use of animation for instruction. He further recommended that animation cannot necessarily be presented in isolation, but requires appropriate additional information to cue the learner as to what phenomena to look for and what each phenomenon means.

[49] and [50] studied the effectiveness of animation in teaching people how to use interactive computer systems. Some subjects were shown animated demonstrations of how to use a computer system, others received just a paper-based user's manual, and others received both a a manual and animation. The authors found that user performance was enhanced by animation. Animation, however, was not as effective as manuals in helping people retain knowledge. When tested one week after the experiment, subjects who had only paper-based training remembered more.

Stasko [51] studied the use of animation for teaching computer science students about algorithms. This type of animation, called *algorithm animation*, has been under intensive investigation for over a decade. It is discussed in more detail in Chapter 4. In Stasko's experiment, subjects were asked to learn an algorithm that was unfamiliar to them. Subjects were given either an algorithm animation or merely a paper-based description of the algorithm. After study, the subjects were then tested about their understanding of the algorithm. Stasko found no statistically significant difference between the group using algorithm animation and the group not using algorithm animation in terms of their performance on the test. This result runs counter to his and others' experience in using algorithm animation to teach algorithms. Stasko hypothesized that this result was caused by the lack of guidance to the subjects using animation as to what to look for. In classroom uses of algorithm animation, the instructor points out particularly interesting phenomena that characterize the algorithm. In the experiment, subjects were essentially left on their own to explore how the algorithm behaved. Stasko therefore hypothesized that algorithm animation coupled with some explanatory capability would be more helpful than just a textual description of the algorithm. A follow-on experiment [52], found that students learned most when they created the algorithm animation rather than merely viewing an existing algorithm animation.

4.5 Summary

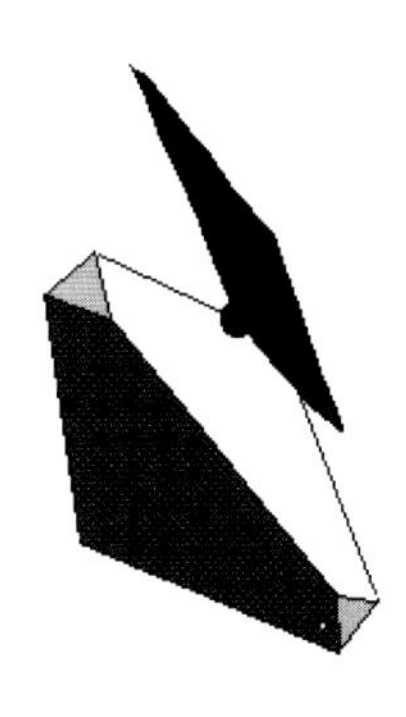

This chapter has presented a whirlwind tour of perceptual psychology, color theory, animation, graphic design, three-dimensional perspective. We have tried to combine underlying psychological theories with both practical and experimentally derived guidelines to provide some general direction to those interested in developing helpful, informative graphical visualizations of complicated systems. In the next chapter, we discuss ways of presenting information using non-visual senses including sound and touch.

Flip Chart

The flipchart that begins at the start of this chapter corresponds to the following linear programming problem:

$$
\begin{array}{lrcrcrcl}
\text{max} & ax & + & by & + & cz & & \\
\text{subject to:} & & & & & & & \\
 & x & + & y & + & z & \leq & 25 \\
 & x & & & & & \leq & 20 \\
 & & & y & & & \leq & 20 \\
 & & & & & z & \leq & 20 \\
 & & & x, y, z & & & \geq & 0
\end{array} \tag{4.2}
$$

where:

$$a = \frac{1}{2}\left\{1 - r\cos\theta + r\sqrt{2}\sin\theta\right\} \tag{4.3}$$

$$b = \frac{1}{\sqrt{2}}\{1 + r\cos\theta\} \tag{4.4}$$

$$c = \frac{1}{2}\left\{1 - r\cos\theta - r\sqrt{2}\sin\theta\right\} \tag{4.5}$$

$$\tag{4.6}$$

and

$$\theta = 0, \pi/10, 2\pi/10, \ldots, 2\pi \tag{4.7}$$

If we consider the objective function $ax + by + cz$ as the dot product of two vectors $\mathbf{d} = (a, b, c)$ and $\mathbf{w} = (x, y, z)$, then equations 4.3 through 4.7 rotate the objective function vector $\mathbf{d}$ around the vector $(1, 1, 1)$ so that the angle between $\mathbf{d}$ and $(1, 1, 1)$ is equal to $\cos^{-1} \frac{1}{\sqrt{1+r^2}}$. In other words,

$$\frac{\mathbf{d} \cdot (1, 1, 1)}{|\,\mathbf{d}\,|\,\sqrt{3}} = \frac{1}{\sqrt{1 + r^2}}$$

For the figures, r was set equal to 0.3.

This is illustrated in the figure below. Each dot represents the value of the objective function coefficients used for one frame of the flip chart.

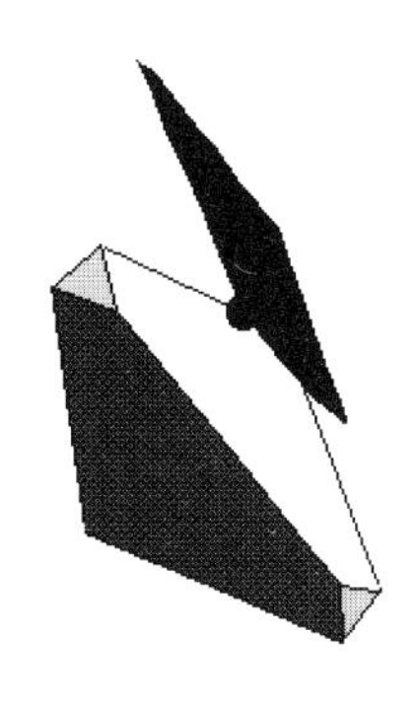

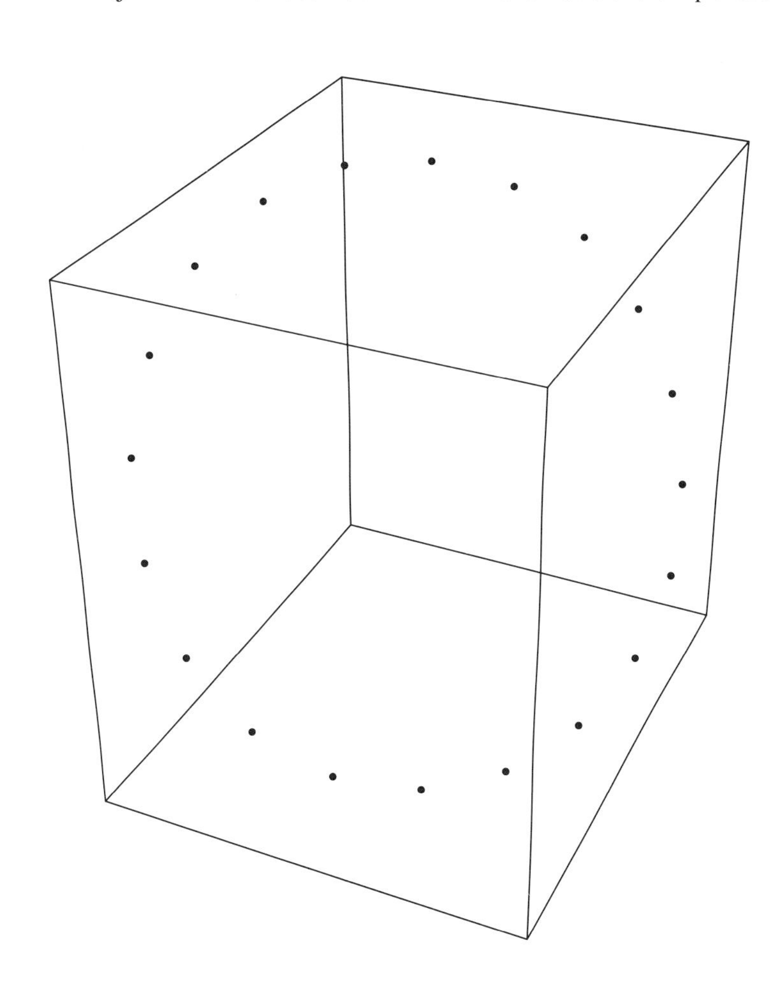

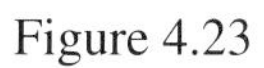

Figure 4.23

Bibliography

[1] Wade NJ, Swanston M. *Visual Perception: An Introduction.* London:Routledge, 1991.

[2] Marr D. *Vision.* San Francisco:Freeman, 1982.

[3] Silverstein LD. *Human Factors for Color DIsplay Systems: Concepts, Methods, and Research*, pages 27–61. Boston (MA):Academic Press, 1987.

[4] Shneiderman B. *Designing the User Interface.* Reading (MA):Addison-Wesley, 2nd edition, 1992.

[5] Glass AL, Holyoak KJ. *Cognition.* New York:Random House, 2nd edition, 1986.

[6] Tufte ER. *The Visual Display of Quantitative Information.* Cheshire, Connecticut:Graphics Press, 1983.

[7] Tufte ER. *Envisioning Information.* Cheshire, Connecticut:Graphics Press, 1990.

[8] Cleveland WS. Research in statistical graphics. *J. of the American Statistical Association*, 1987;82(398):419–423.

[9] Cleveland WS, McGill R. Graphical perception and graphical methods for analyzing scientific data. *Science*, 1985;229:828–833.

[10] Cleveland WS, McGill R. Graphical perception: Theory, experimentation, and application to the development of graphical methods. *Journal of the American Statistical Association*, 1984;79(387):531–554.

[11] Mackinlay J. Automating the design of graphical presentations of relational information. *ACM Transactions on Graphics*, 1986;5(2):110–141.

[12] Hall R. *Illumination and Color in Computer Generated Imagery.* New York:Springer-Verlag, 1989.

[13] Marcus A. *Graphic Design for Electronic Documents and User Interfaces.* New York:ACM Press, 1992.

[14] Foley JD, Dam AV, Feiner SK, Hughes JF. *Computer Graphics: Principles and Practice.* Reading (MA):Addison-Wesley, 2nd edition, 1990.

[15] Wolff RS, Yaeger L. *Visualization of Natural Phenomena.* Santa Clara (CA):TELOS, 1993.

[16] Judd DB, Wyszecki G. *Color in Business, Science and Industry.* New York:John Wiley and Sons, 1963.

[17] Billmeyer FW, Saltzman M. *Principles of Color Technology.* New York:John Wiley and Sons, 2nd edition, 1981.

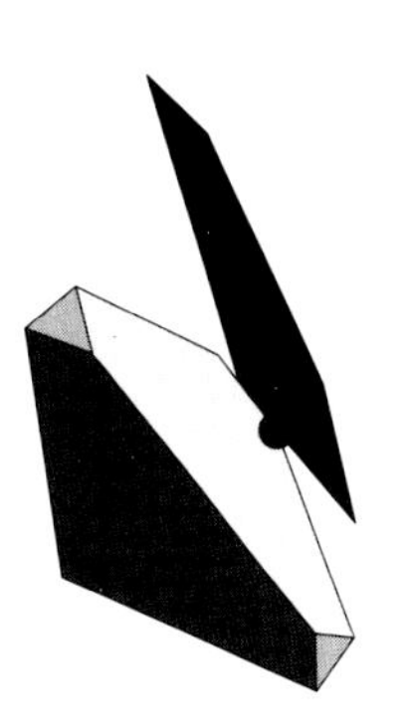

[18] Albers J. *Interaction of Color.* New Haven (CT):Yale University Press, 1975.

[19] Boyer CB, Merzbach UC. *A History of Mathematics.* New York:John Wiley and Sons, 2nd edition, 1991.

[20] Jones CV. The three-dimensional Gantt chart. *Operations Research,* 1988;36(6):891–903.

[21] Tektronix, Inc. *3D Stereoscopic Color Graphics Workstation (TEK 4126 Product Literature).* Beaverton (OR):Tektronix, Inc., 1987.

[22] Sutherland IE. A head-mounted three-dimensional display. In *Fall Joint Computer Conference, AFIPS Conference Proceedings,* volume I, pages 757–764, Washington (DC):1968. Thompson Book Company.

[23] Robinett W, Rolland JP. A computational model for the stereoscopic optics of a head-mounted display. *Presence,* 1992;1(1):45–62.

[24] MacDowall I, Bolas M, Pieper S, Fisher S, Humphries J. Implementation and integration of a counterbalanced CRT-based stereoscopic display for interactive viewpoint control in virtual environment applications. In *Proceedings of the 1990 SPIE Conference on Stereoscopic Displays and Applications,* Santa Clara (CA):1990.

[25] Hilaire PS, Benton SA, Lucente M. Synthetic aperture holography: a novel approach to three-dimensional displays. *Journal of the Optical Society of America,* 1992;9(11):1969–1977.

[26] Fuchs H, Pizer SM, Heinz ER, Tsai LC, Bloomberg SH. Adding a true 3D display to a raster graphics system. *IEEE Computer Graphics and Applications,* 1982;2(7).

[27] Lateiner Dataspace. Stereoscopic volume visualization. Technical report, Woburn (MA):Lateiner Dataspace, 1995. WWW page.
URL: *http://www.dataspace.com/WWW/documents/stereoscopic.html*

[28] Hendricks G. *Eadweard Muybridge: The Father of the Motion Picture.* New York:Grossman, 1975.

[29] Gibson JJ. *Perception of the Visual World.* Boston:Houghton Mifflin, 1950.

[30] Gibson JJ. *The Senses Considered as Perceptual Systems.* Boston:Houghton Mifflin, 1966.

[31] Gibson JJ. *The Ecological Approach to Visual Perception.* Boston:Houghton Mifflin, 1979.

[32] Card S, Moran TP, Newell A. *The Psychology of Human-Computer Interaction.* Hillsdale (NJ):Lawrence Erlbaum Associates, 1983.

[33] Cutting JE. *Perception with an Eye for Motion.* Cambridge (MA):The MIT Press, 1986.

[34] Kallenberger RH, Cvjetnicanin GD. *Film Into Video: A Guide to Merging the Technologies.* Boston:Focal Press, 1994.

[35] Fox E. Standards and the emergence of digital multimedia systems. *Communications of the ACM*, 1991;pages 26–29.

[36] Fox E. Advances in digital multimedia systems. *IEEE Computer*, 1991;24(10):9–22.

[37] Lippman A. Feature sets for interactive images. *Communications of the ACM*, 1991;34(4).

[38] Rao K, Yip P. *Discrete Cosine Transform—Algorithms, Advantages, Applications.* London:Academic Press, 1990.

[39] Wallace GK. The JPEG still picture compression standard. *Communications of the ACM*, 1991;24(4):30–44.

[40] LeGall D. MPEG: A video compression standard for multimedia applications. *Communications of the ACM*, 1991;34(4):46–58.

[41] Apple Computer. *QuickTime Developer's Kit, Version 1.0.* Cupertino (CA):Apple Computer, 1991.

[42] Ozer J. *Video Compression for Multimedia.* Boston:Academic Press, 1995.

[43] Chang BW, Ungar D. Animation: From cartoons to the user interface. In *Proceedings of the ACM Symposium on User Interface Software and Technology*, pages 45–55, New York:1993. Association for Computing Machinery.

[44] Thomas F, Johnston O. *Disney Animation: The Illusion of Life.* New York:Abbeville Press, 1984.

[45] Rieber LP. Animation in computer-based instruction. *Educational Technology Research and Development*, 1990;38(1):77–86.

[46] Rieber LP. Using computer animated graphics in science instruction with children. *Journal of Educational Psychology*, 1990;82(1):135–140.

[47] Reiber LP, Boyce MJ, Assad C. The effects of computer animation on adult learning and retrieval tasks. *Journal of Computer-Based Instruction*, 1990;17(2):46–52.

[48] Reiber LP, Kini AS. Theoretical foundations of instructional applications of computer-generated animated visuals. *Journal of Computer-Based Instruction*, 1991;18(3):83–88.

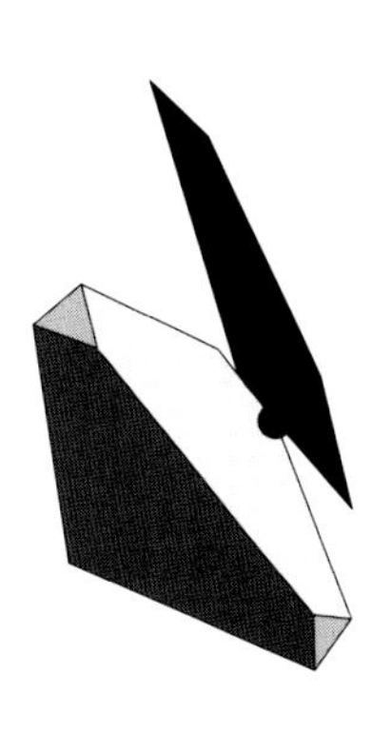

[49] Palmiter S, Elkerton J, Baggett P. Animated demonstrations versus written instructions for learning procedure tasks. Technical Report C4E-ONR-2, Ann Arbor (MI):Center for Ergonomics, Department of Industrial and Operations Engineering, University of Michigan, 1989.

[50] Palmiter S, Elkerton J. Animated demonstrations for learning procedural computer-based tasks. *Human-computer interaction*, 1993;8(3).

[51] Stasko J, Badre A, Lewis C. Do algorithm animations assist learning? an empirical study and analysis. In *Human Factors in Computing Systems, INTERCHI'93 Conference Proceedings, Conference on Human Factors in Computing Systems, INTERACT '93 and CHI '93*, pages 61–66, Amsterdam, The Netherlands:April 1993.

[52] Lawrence A, Badre A, Stasko J. Empirically evaluating the use of animations to teach algorithms. In *Proceedings of the 1994 IEEE Symposium on Visual Languages*, pages 48–54, St. Louis, MO:October 1994.

Chapter 5

Sound and Touch

Graphics, text, tables and animation exploit only the human vision system. Human perception relies on four other senses, however. For presenting complex information, at least some of those senses can also be exploited. This chapter discusses how two of them, sound and touch, can be used to present information.

5.1 Sound

The "silent" films that flourished before the age of the "talkies" were actually never silent. Although the piano might have been out of tune and the piano player might have needed to be shot, "silent" films were always meant to be accompanied by some type of sound. Major films would even have an elaborate musical score played by a full orchestra. Apparently the filmmakers felt that the music enhanced audience's engagement with the film.

Except for video games, most computer graphic imagery is silent. With the proliferation of more capable sound hardware and software, it is now possible to use more and better sound in visualization. Some call the use of sound to support visualization *sonification*. Scaletti and Craig[1] defined sonification as:

> a mapping of numerically represented relations in some domain under study to relations in an acoustic domain for purposes of interpreting, understanding, or communicating relations in the domain under study.

In fact some believe [2] that one should speak of *perceptualization* rather than visualization, since one should exploit all the senses when trying to help people make sense of complexity. Exactly what tasks and users would benefit from sonification, however, is unclear. In this section, we discuss some of the relevant results from cognitive psychology, as well as computer hardware and software concepts for use with sound.

5.1.1 Sound Basics

Sound consists of pressure waves traveling through air having a frequency of between 20 to 20,000 Hertz. Any sound can be characterized as the superposition of a series of sine waves having particular frequencies and amplitudes—according to Fourier analysis. However, as human beings, we do not consciously decompose sound into a Fourier series. Rather, like color, we perceive sound as having a certain characteristics. These characteristics include:

pitch Sounds can be as high as a piccolo or as low as a tuba. This corresponds to the frequency of the sound, and is typically measured in cycles per second or Hertz, though in music, the pitch is usually referred to as b-flat (b♭) or f-sharp (f♯).

loudness Sounds can be as soft as whisper or as loud as a supersonic airplane. This corresponds to the amplitude of the wave, and is typical measured as decibels.

timbre A saxophone playing an A above middle C (440 Hz), which is commonly used to tune an orchestra, does not sound the same as a violin playing the same note. The timbre is determined by the shape of the wave, i.e., the *waveform*. No simple scale exists to capture this quality of sound.

Data sonification in essence, encodes complex information into these perceptual features of sound, much in the same way that Bertin describes how data can be encoded into images. The next section discusses how sound can be generated on a computer.

5.1.2 Sound Hardware

Perhaps the most widely know standard for sound is the Musical Instrument Digital Interface or MIDI [3]. MIDI is essentially a networking protocol for linking electronic musical devices, computers and other devices. An electronic keyboard (or violin or drum, for example) can be used to control a variety of sound synthesizers and other devices using MIDI. MIDI is both a hardware and software specification. It is a hardware specification since it specifies the physical characteristics of the plugs and wires used to connect devices. It is a software specification since it also specifies the format of the digital messages transmitted through the plugs and over the wires.

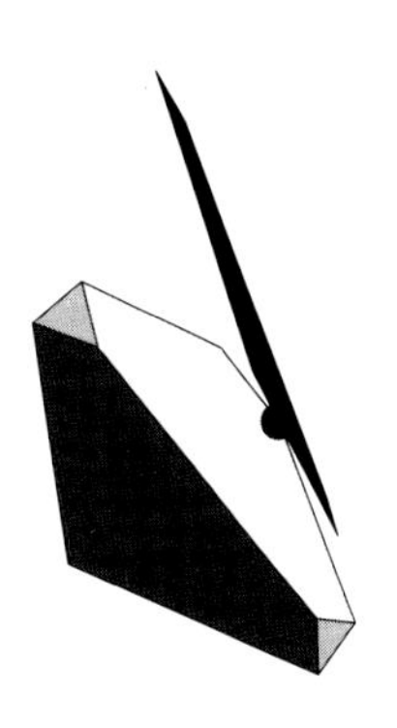

A digital signal processor (DSP) is a piece of hardware designed to quickly modify the characteristics of sound. They are also programmable, which allows them to be easily controlled by software. DSP's are now relatively inexpensively sold as add-in sound cards to personal computers, or bundled with them. They can be used to support both speech recognition and speech synthesis operations.

MIDI is intended for musicians making music; it was not originally designed to support sonification. DSP's seem better suited to support sonification applications. As Scaletti and Craig[1] wrote, "[MIDI synthesizers'] intended use as performance instruments in commercial music, however, can make them somewhat opaque and difficult to control; the very 'hardwiring' that makes them so fast also makes synthesizers less flexible than sound synthesis in software." The flexibility of DSP's and their increasing incorporation into computers makes them the long term choice for sonification.

5.1.3 Guidelines for Sound

Sound has been shown in some cases to be an effective aid to visualization. Bly [4] mapped data to various sonic parameters. She then tested 75 subjects on the effectiveness of the sonic map in conveying the information. She found that sound plus graphics was more effective than just graphics. Mezrich, Frysinger and Slivjanovski [5] found that sound was helpful in conveying a multivariate economic indicator.

Sound is perhaps best used for drawing the user's attention. As Wenzel [6], [7] says, "the ears tell the eyes where to look." The simple beep used to indicate an error is the quintessential example of this use of sound with computers. A police siren serves a similar function since the police car may be out of sight. Too many possible alerting sounds, particularly if they can occur simultaneously, can be counterproductive and even dangerous. During the Three Mile Island nuclear disaster [8] , over 50 audible alarms sounded simultaneously. By that point, the operators certainly knew that a problem was occurring, but the number of alarms hindered them from discovering the exact nature of the underlying problem.

Voice annotation is also now relatively easy, particularly since microphone input is becoming a common feature of personal computers. It is now possible to annotate any cell in a Microsoft Excel spreadsheet with voice annotation.

Others have proposed using scientific data as a the direct source of sonification, much as that same data is used to create visual representations. This type of sonification is much less well-explored, and hence few robust guidelines exist to guide the sound designer. In essence, one can encode data into any or all of the three basic sound characteristics—pitch, loudness and timbre. One could "play" any mathematical function of one variable. For example, Mathematica [9] provides a "play" command to play any mathematical function. The shape of the function can be interpreted as the shape of the sound to be played. The mathematical function could also be used to represent the distribution of frequencies in the sound to be played, rather than the pressure of the sound wave at any point in time.

Other than some very preliminary research, there is a general lack of theory of timbre perception to help guide sonification developers as to whether or not such sonifications would even be useful. Researchers (see [10] for a collection) have developed some intriguing examples, but how to use sound in general, if at all, to support perceptualization is not well understood.

5.1.4 Applications of Sonification

The use of sound for representing complex information has been explored since at least the early 1980's [11], [12]. In one experiment [11], spectra of known unknown chemicals were encoded as sounds. The sound of an unknown chemical was played, followed by the sounds of the known chemicals. In informal tests, subjects rarely misidentified the unknown chemical.

Bly [13] explored the use of sound for representing six-dimensional data. Subjects had to classify a particular data point into one of two categories based on perceived sound. She compared subjects performance when the data was presented as a scatter plot, using sound along, and using both a scatter plot and sound. Sound proved more effective than the graphics; sound and graphics were more effective than either sound or graphics alone.

Scaletti and Craig [1] describe several explorations on the use of sound for visualization. One example added sound to an animation of the evolution of Yellowstone National Park. The age of the forest in different areas was color-coded, and forest fires were colored red. Forest age was mapped to sound frequency; the amplitude of each frequency depended on the area of forest having the associated age. Forest fires were mapped to noise, the larger the fire, the greater the noise. No formal experimentation was conducted to determine the effectiveness of this use of sonification.

Another example described in [1] concerned the motion of two pendulums. Two pendulums start with nearly identical initial conditions, but their motion eventually becomes totally dissimilar. After several attempts at mapping different aspects of the simulated motion to different sonic parameters, they finally settled on mapping the displacement of each pendulum to frequency. A difference in displacement between the two pendulums was audibly detectable (as a difference in frequency) sooner than it was visible.

5.2 Touch

In addition to vision and sound, visualization has begun to explore the use of a third human sense, the sense of touch. Although the use of touch for visualization is far less well-developed than the use of graphics or even the use of sound, it is the subject of intense investigation.

5.2.1 Cognitive Psychology

The sense of "touch" actually has many different variations. Clearly, one is the fine *tactile* sense that allows one to distinguish between a silk scarf and coarse sandpaper. Another type is the *kinesthetic* sense which allows one to feel the difference in weight between a bowling ball and a ping pong ball. A third type, the *vestibular* sense, detects both linear and angular acceleration of one's body. The specific organs for detecting acceleration are located in the inner ear. A fourth, the *proprioceptive* sense, gives one the ability to determine the location of parts of our body, even when our eyes are closed. All of these different aspects relate to the way one perceives one's environment not from sound or light, but from physical contact with the environment [14].

5.2.2 Hardware Devices that Exploit the Motor System

In order to exploit the senses of touch, a variety of *haptic* ("of or relating to the sense of touch" [15]) input and output devices have been developed. Certainly the mice, trackballs, and joysticks that are now essential to user interfaces exploit the human motor system. To exploit the kinesthetic system even further, some of these devices have been enhanced to detect varying amounts of force input by the user [16].

Force *output* devices have also been developed, which convey a computed force back to the user. Such devices range from large robot arms first developed for manipulating nuclear materials [17] to more traditionally sized joy-sticks [16]. In some applications, force output allows tasks to be completed up to twice as quickly than with no force output provided [17].

Mice, joysticks and trackballs do not exploit the full expressiveness of the human hand. Entire languages, such as those for the deaf, rely solely on hand gestures, yet computers, until recently, have not been able to use them. Various types of gloves, rigged with sensors, have been developed to track the motion of the hand [18]. One type of glove is "wired" with optical fibers. As the fingers bend, the refractivity of the fibers change in predictable ways. Given the refractivity change, the angle of the joints can be calculated. The light-fiber technology has been extended to provide an input device for the entire body. Even more accurate measurement of hand movement is provided by gloves using mechanical hinges functioning as an exoskeleton. Sensors on the hinges very precisely measure the joint angles.

Both the light-fiber and the hinged systems only measure joint angles. One also needs to be able to detect the position of an object (the hands, for example), in space. The most widely used system for detecting the position of objects in three dimensions uses a set of three orthogonal electromagnets placed on the moving object [19], the *source*. The electromagnets are pulsed in sequence. A fixed *sensor* containing three orthogonal electromagnets, synchronized with the source, computes the location of the source based magnetic field detected. These devices have a range of about 50 inches, and are accurate to within 3 inches within that range [20].

Flight simulators and certain amusement park rides exploit the vestibular system, since the user's chair "rides" on computer-controlled hydraulics while viewing a display [21]. In a flight simulator, as the plane dives, the simulated cockpit is tilted using the hydraulic apparatus. This conveys a greater sense of realism to the pilot. Note that the hydraulics cannot simulate extreme motion such as a barrel roll. Furthermore, once the hydraulic jacks are extended their full amount, they must at some point be retracted. These limitations can lead to a disparity between the visual image seen and the motion produced by the hydraulics. This type of disparity can cause "simulator" sickness [22].

5.3 Summary

Sound and touch can augment human-computer visualization. Hardware and software are now emerging to allow the exploitation of these sensory channels. The introduction of graphical user interfaces has required rethinking how we deliver our tools as well as much rewriting of software. the introduction of sound and touch into computer displays may cause a similar transformation.

This and the previous two chapters surveyed the variety of representation formats that are available for purposes of visualization. Text and tables (Chapter 3), graphics and animation (Chapter 4), sound and touch (Chapter 5) are some of the basic formats used. Cognitive psychologists, graphic artists, computer scientists have developed theories, frameworks, guidelines, techniques and technologies to support these different representation formats. In the next chapter we discuss the different types of tasks involved in an optimization project.

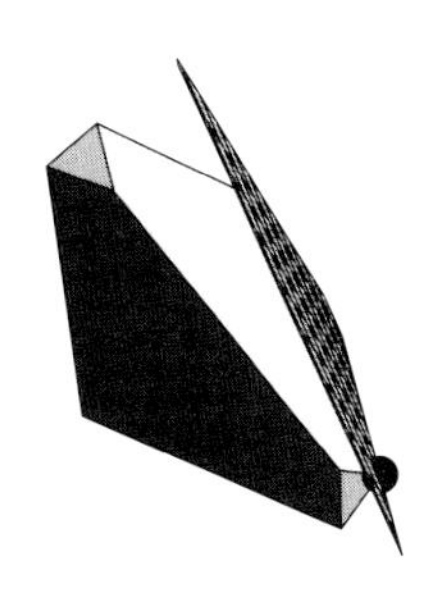

Bibliography

[1] Scaletti C, Craig AB. Using sound to extract meaning from complex data. Technical report:University of Illinois, CERL, 1993.

[2] Erickson T. Artificial realities as data visualization environments: Problems and prospects. In Wexelblat A, editor, *Virtual Reality: Applications and Explorations*. Boston:Academic Press, 1993.

[3] O'Donnell B. What is MIDI, anyway? *Electronic Musician*, January 1991.

[4] Bly SA. Presenting information in sound. In *Proceedings of CHI '85 Conference on Human Factors in Computing Systems*, pages 371–375, New York:1985. ACM Press.

[5] Mezrich J, Frysinger S, Slivjanovski R. Dynamic representation of multivariate time series data. *Journal of the American Statistical Association*, 1984;79.

[6] Wenzel EM, Fisher SS, Stone PK, Foster SH. A system for three-dimensional acoustic 'visualization' in a virtual environment workstation. In *Proceedings of Visualization '90*, pages 329–337, Los Alamitos (CA):1990. IEEE Computer Society Press.

[7] Wenzel EM, Foster SH. Realtime digital synthesis of virtual acoustic environments. *Computer Graphics*, 1990;24(2):139–140.

[8] Sanders MS, McCormick EJ. *Human Factors in Engineering and Design*. New York:McGraw-Hill, 6th edition, 1987.

[9] Wolfram S. *Mathematica: A System for Doing Mathematics by Computer*. Redwood City (CA):Addison-Wesley, 2nd edition, 1991.

[10] Kramer G, editor. *Auditory Display: Sonification, Audification, and Auditory Interfaces*. Reading (MA):Addison-Wesley, 1994.

[11] Lunney D, Morrison RC. High technology laboratory aids for visually handicapped chemistry students. *Journal of Chemical Education*, 1981;58(4).

[12] Yeung ES. Pattern recognition by audio representation of multivariate analytical data. *Analytical Chemistry*, 1980;52(7):1120–1123.

[13] Bly SA. Presenting information in sound. In *Proceedings of the CHI '82 Conference on Human Factors in Computer Systems*, pages 371–375, 1982.

[14] Boff K, Kaufman L, Thomas J, editors. *Handbook of Perception and Human Performance*, volume I–II. New York:John Wiley and Sons, 1986.

[15] Gove PB, editor. *Webster's Third New International Dictionary of the English Language*. Springfield (MA):Merriam-Webster, 1981.

[16] Ware C, Baxter C. Bat brushes: On the uses of six position and orientation parameters in a paint program. In *Proceedings CHI '89 Conference: Human Factors in Computing Systems*, pages 155–160, New York:1989. ACM.

[17] Brooks FP, Ouh-Young M, Batter JJ, Kilpatrick PJ. Project GROPE—haptic displays for scientific visualization. *Computer Graphics*, 1990;24(4):177–185.

[18] Zimmerman T, Lanier J, Blanchard C, Bryson S, Harvill Y. A hand gesture interface device. *SIGCHI Bulletin*, 1987;pages 189–192. special issue.

[19] Raab F, Blood E, Steioner T, Jones H. Magnetic position and orientation tracking system. *IEEE Transactions on Aerospace and Electronic Systems*, 1979.

[20] Bryson S, Levit C. Virtual wind tunnel: An environment for the exploration of three dimensional unsteady flows. *Computer Graphics and Applications*, 1992;12(4):25–34.

[21] Yan JK. Advances in computer-generated imagery for flight simulation. *IEEE Computer Graphics and Applications*, 1985;5(8):37–51.

[22] Casali J, Wierwille W. The effects of various design alternatives on moving-base driving simulator discomfort. *Human Factors*, 1980;22(6).

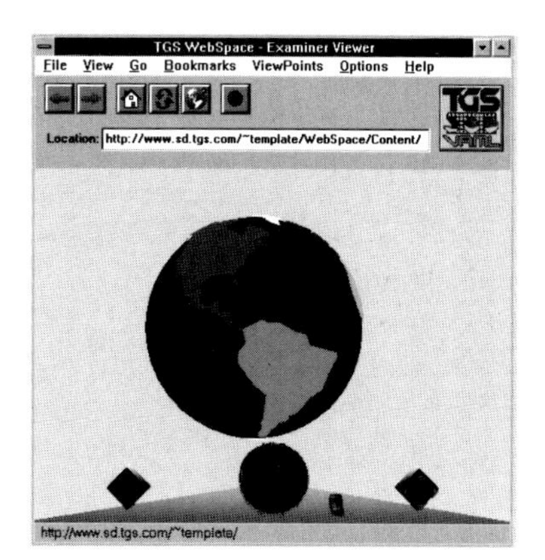

Chapter 6

Hypermedia and Virtual Reality

In the previous three chapters, we have surveyed different representation formats associated with three human senses. It is doubtful that the two other human senses—smell and touch—will soon be effectively used to "visualize" complex information. This chapter discusses two concepts—hypermedia and virtual reality—that integrate the different representation formats discussed in the previous three chapters.

6.1 Hypermedia

Printed matter presents information as a linear sequence of characters arranged hierarchically into words, sentences, paragraphs, sections and chapters. Footnotes, references, tables of contents, and indices allow the content of documents to be accessed in a non-linear fashion, but such devices are limited by the time needed to scan through a book. *Hypertext* allows quick transitions from one part of a document to any related, but "geographically" distant, part of a document. For example, while reading a document that cites another article, simply clicking the mouse on the citation, the cited article can be displayed. References (and other information) from that article could then be displayed, and so on. In essence, hypertext organizes information as a network connecting multiple, interrelated documents and their contents (which need not be only textual) and allows users to traverse the links among the documents quickly.

Hypertext was foreshadowed almost 50 years ago by Bush's [1] speculative article describing mechanisms to organize complicated information. Users need only point to hypertext *links* to jump to display additional information. Hypertext is now often used to provide on-line help, for example, in Microsoft Windows. Hypertext, of course, need not be limited to displaying only text. *Hypermedia* [2] provides nonlinear navigation capabilities for text, graphics, sound, and video. Although hypermedia may be more descriptive, "hypertext" now usually is meant to include graphics, sound and animation. We shall use both terms interchangeably.

6.1.1 World Wide Web

One notable example of hypertext is the World Wide Web [3]. Relying on the internet, the Web allows users to explore the information resources available on the internet using hypermedia. Text, pictures, animations, sound stored at thousands of computers around the world can be accessed.

Accessing the information requires a connection to the internet and an appropriate *browser*, a program running on the user's own computer that presents the hypermedia to the user and navigates the internet to find the requested information. Browsers include Mosaic, developed at the University of Illinois, and Netscape, sold by a commercial firm. "Surfing" the Web using such a browser has become the leading use of the internet surpassing even electronic mail.

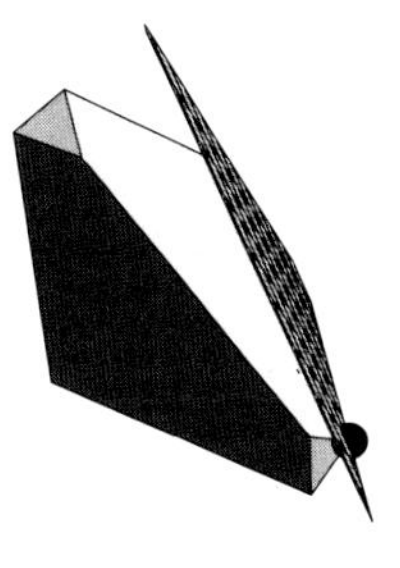

Below is a Web "page" containing answers to frequently asked questions about linear programming. By clicking on any of the underlined items, users can "jump" to see more information about that item. This particular page at the time of writing is located at `http://www.skypoint.com/subscribers/ashbury/linear-programming-faq.html`.

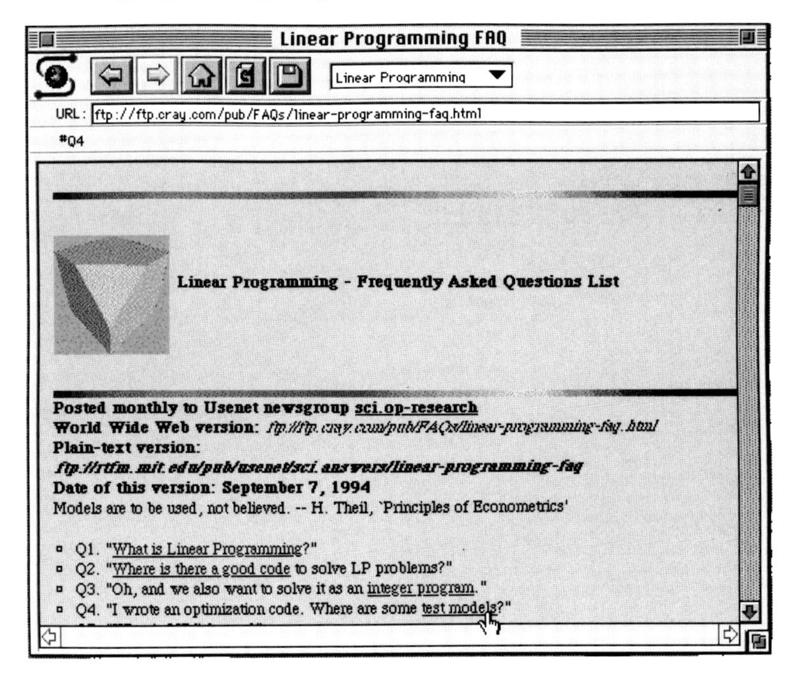

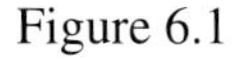

Figure 6.1

The Web also allows a surprisingly high degree of interactivity. Users can fill in forms, for example, to order flowers (the browser can automatically encrypt the buyer's credit card number), apply for admission to a university, or search catalogues of information available on the Web. Users can also point to items on a picture, a map for example, to trigger an appropriate action, to see more detail on the map for example.

The level of interactivity available on the Web is growing rapidly. For example, currently user's can download files containing digitized audio. Once the entire file is downloaded, the sound can then be played on the user's local computer. A recent development allows the audio to be played *while* it is being downloaded to the user's computer. Instead of waiting for the entire sound file to be copied, sound is heard almost as soon as the user selects they hypertext link [4].

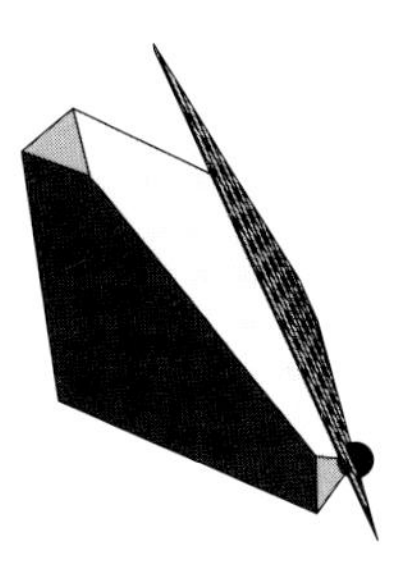

The Virtual Reality Modeling Language (VRML) defines a new standard for three-dimensional information. Combined with an appropriate browser, user's can download a VRML file and then navigate through a three-dimensional world on their local computer [5]. The example below shows a scene displayed in WebSpace [6], [7], a VRML browser. The scene [8] shows Earth; the icons at the bottom of the screen are actually controls for navigating through the the three-dimensional scene.

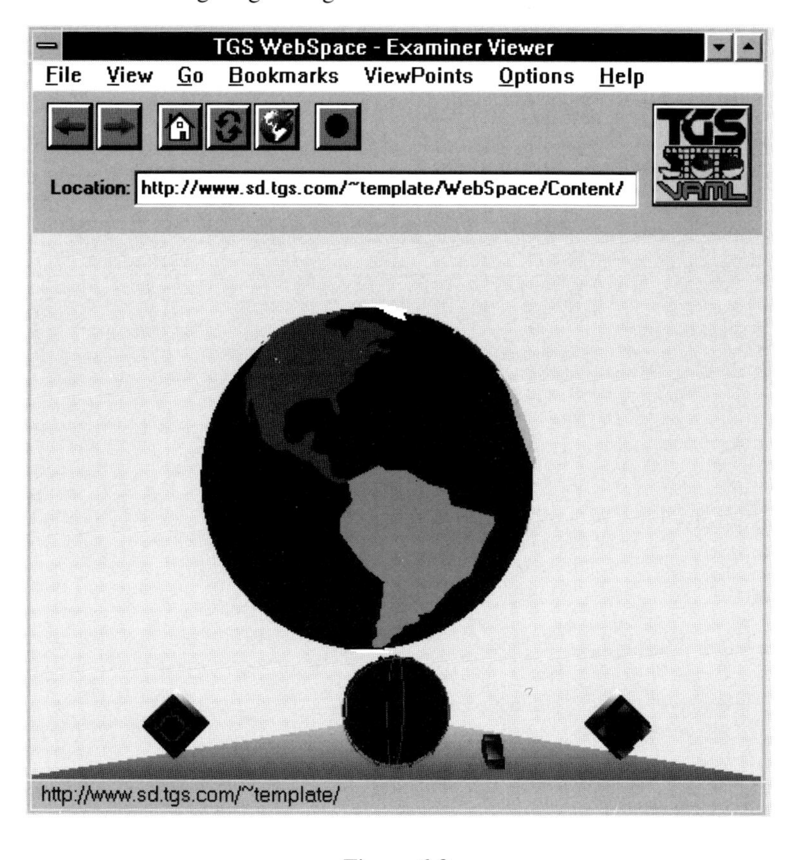

Figure 6.2

In order to play audio, see a digitized movie, or see a VRML file, users must have a browser that understands the particular information format, or a *helper* application that can display the information format. The helper application will be started by the browser, as needed. However, what if the user has not yet downloaded the appropriate browser? One attempt to provide a general solution to this problem, HotJava [9], defines a new programming language. When the user requests information, a HotJava program is downloaded to the user's computer and then executed.

In the example below, when users click on a picture showing a set of vertical lines, a sorting algorithm is illustrated through an animation. Different sorting algorithms can be animated simultaneously. The program for the sorting algorithm, as well as the program to animate it, is downloaded to the user's computer and then executed.

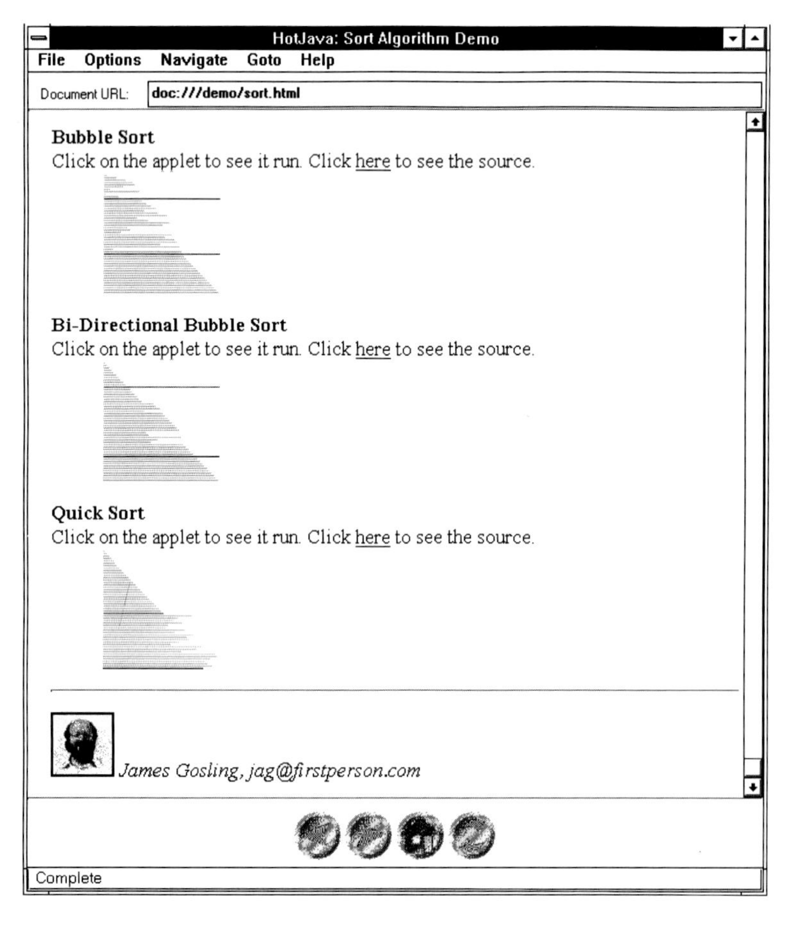

Figure 6.3

Of course, the security threat to the user's computer is potentially large. In order to prevent such security risks, HotJava limits the capabilities of its programming language.

At this point, it seems clear that the World Wide Web is here to stay. Its capabilities are being enhanced almost daily. Perhaps it will form one of the primary media through which optimization and other modeling technology will be delivered in the future.

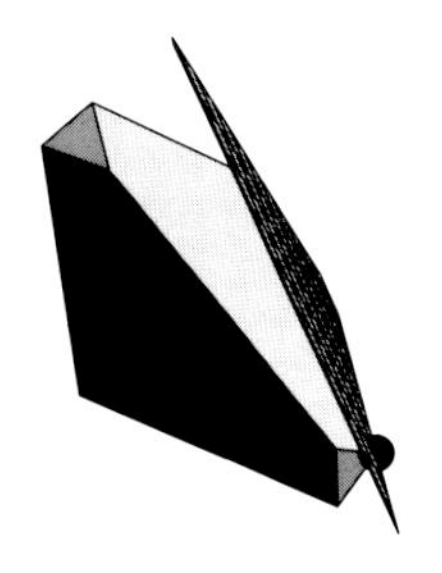

6.1.2 Guidelines for Hypertext

Not all information necessarily is appropriate to be represented as hypertext. Shneiderman [10] proposed three simple guidelines for determining if a particular set of information would be usefully represented as hypertext. They are:

1. There is a large body of information organized into numerous fragments.

2. The fragments relate to one another.

3. The user needs only a small fraction of the fragments at any one time.

From these recommendations, a traditional novel would probably *not* be a good candidate for hypertext, since a novel is usually read in a linear fashion (although one often jumps to the last page of a bad murder mystery to find out whodunnit). Mathematical models, in contrast, though, do seem to satisfy these basic recommendations. Mathematical models typically consist of a variety of equations (rule 1), that are linked together (rule 2). Although the fragments of mathematical models often cannot be separated for solution purposes, for example, simultaneous equation models, the modeler will certainly need to view small numbers of individual fragments at a time (rule 3). We discuss the use of hypertext for mathematical modeling more in Chapter 12.

6.2 Virtual Reality

Virtual reality has been the subject of a great deal of recent hype, though the concept of virtual reality has existed at least since 1965 [11]. Actually, the term "virtual reality" refers to a variety of different ideas and technologies that have been given labels including artificial worlds, artificial reality, augmented reality, and telepresence, among others [12], [13], [14]. Applications of Virtual Reality include medicine [15], fluid dynamics, space exploration [16], flight simulation [17], chemistry [18], [19] among others. One especially clever application attempted to desensitize people with acrophobia (fear of heights) [20]. Subjects were exposed to a virtual representation of a high place such as a bridge or building, although they were actually located in a standard room. Through progressive exposure to the simulated high place, their fear of high places gradually diminished.

Much of what is labeled virtual reality today consists merely of three-dimensional scenes displayed on a standard computer screen, with the user able to interactively explore the three-dimensional world. QuickTime VR [21] (the VR stands for virtual reality), for example, displays three-dimensional worlds that have been constructed from 360° photographic images. The Virtual Reality Modeling Language (VRML), discussed in Section 6.1.1 provides a similar capability for the World-Wide Web. However, virtual reality, at least as originally proposed, provided for a much richer sensory experience. We discuss some of these ideas in the next section.

6.2.1 Types of Virtual Reality

In essence, the goal of virtual reality is to synthesize stimuli that produce a realistic illusion of another world. Perhaps the most widely known form of virtual reality combines a head-mounted display (discussed in Section 4.3.4), stereo sound (discussed in Chapter 5), and a glove input device (discussed in Section 5.2.2). This type of virtual reality has often been called *immersive* since it completely replaces one's perception of the real world with artificial stimuli.

Augmented reality, in contrast, attempts to supplement one's perception of the real world through *non-immersive* means. One type of augmented reality, for example, uses a *see-through* head-mounted display [22]. The user can see the real world through the "lenses" of the glasses. The "lenses" are actually half-silvered mirrors that both transmit and reflect light. Light from the real world is transmitted through the lenses to the user; computer graphic images can be projected onto the lenses which are then reflected by the lenses into the user's eyes. This device allows artificial images to be superimposed onto the real world.

Another version of augmented reality projects synthetic images onto a screen [23], [24]. A television camera is focused on the user. Without having to wear any special devices such as gloves, the user merely gestures at the television camera. The gestures are interpreted by image processing software. This style of augmented reality was pioneered by Myron Krueger, who labeled the concept *artificial reality* [12], [13]. He [23] exploited this idea to create a "virtual desk." Documents are projected onto the surface of a physical desk. Using simple hand gestures, both paper-based and computer-based documents can be manipulated. For example, a user can use a hand gesture to indicate a portion of a paper document that should be digitized, for example, a table of numbers. Once digitized, another hand gesture requests that the numbers be totaled. The total is then added to the table. The total is then projected onto the existing paper document.

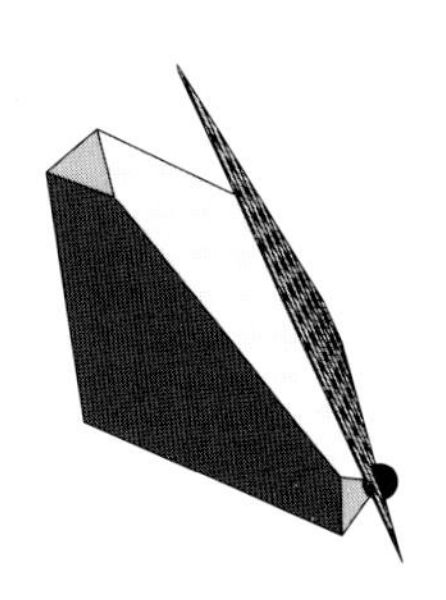

Augmenting the real world with computer sensors has also been explored by Moran and Anderson [25]. They propose to alter common tools such as blackboards and telephones so that they become more seamlessly integrated. Instead of requiring that users wear exotic clothing, the environment itself would be adaptable.

6.2.2 Limitations of Virtual Reality

Virtual reality, although the subject of active investigation, remains difficult to apply. The hardware and software remains expensive, difficult to use, and often has limited capabilities.

One of the key challenges to the development of virtual reality is the issue of *lag* [26]. Lag refers to the delay that occurs between the time an input (movement of the head, for example), and the time that the virtual reality system is able to respond (updating the image on the head-mounted display, for example). Lag is produced by delays in the input devices themselves (for example, the time for a magnetic tracker to detect the motion of the hand), as well as delays in the algorithms for generating images seen by the user.

For example, if one is viewing a complex scene containing thousands of polygons, the computer must quickly update the image in response to user movement. Optimization techniques are beginning to be used to reduce the lag. This work is discussed in Section 16.3.

6.2.3 Summary

Although virtual reality is an overused term, the ability to display realistic three-dimensional imagery, provide force-feedback, and allow users to express their wishes using all their skills is a lofty goal. From the initial concept in 1965, applications of virtual reality are becoming reality. Like the development interactive computer graphics, virtual reality has seen perhaps its greatest application in video games; it is now moving beyond that arena to be useful in areas including medicine, physics, chemistry, among others. We discuss optimization and virtual reality in chapter 16.

Bibliography

[1] Bush V. As we may think. *The Atlantic Monthly*, July 1945;pages 101–108.

[2] Newcomb SR, Kipp NA, Newcomb VT. The "HyTime" hypermedia/time-based document structuring language. *Communications of the ACM*, 1991;34(11):52–66.

[3] World Wide Web Consortium. The world wide web initiative: The project. Technical report, Office of Sponsored programs, E19-702 Massachusetts Institute of Technology 77 Massachusetts Avenue Cambridge (MA) 02139, USA:World Wide Web Consortium, 1995.
URL: *http://www.w3.org/hypertext/WWW/TheProject.html*

[4] Progressive Networks, Inc. RealAudio Home Page. Technical report:Progressive Networks, Inc., 1995.
URL: *http://www.realaudio.com/*

[5] Eubanks C, Moreland J, Nadeau D. VRML repository. Technical report, San Diego Supercomputer Center, PO Box 85608, San Diego (CA), 92186-9784:San Diego Supercomputer Center (SDSC), 1995.
URL: *http://sdsc.edu/vrml*

[6] Silicon Graphics, Inc. WebSpace home page. Technical report, Mountain View (CA):Silicon Graphics, Inc., 1995.
URL: *http://www.sgi.com/Products/WebFORCE/WebSpace/*

[7] Template Graphics Software, Inc. Webspace scenes and objects. Technical report, San Diego (CA):Template Graphics Software, Inc., 1995.
URL: *http://www.sd.tgs.com/~template/WebSpace/Content/stuff.wrl*

[8] Template Graphics Software, Inc. WebSpace information home. Technical report, San Diego (CA):Template Graphics Software, Inc., 1995.
URL: *http://www.cts.com/~template/WebSpace/*

[9] Sun Microsystems, Inc. HotJava Home Page. Technical report, 2550 Garcia Ave., Mountain View (CA) 94043-1100:Sun Microsystems, Inc., 1995.
URL: *http://java.sun.com/*

[10] Shneiderman B. *Designing the User Interface*. Reading (MA):Addison-Wesley, 2nd edition, 1992.

[11] Sutherland IE. The ultimate display. In *Proceedings of the IFIPS Congress*, volume 2, pages 506–508, 1965.

[12] Krueger M. *Artificial Reality*. Reading (MA):Addison-Wesley, 1983.

[13] Krueger M. *Artificial Reality II*. Reading (MA):Addison-Wesley, 1991.

[14] Rheingold H. *Virtual Reality*. New York:Summit Books, 1991.

[15] Sagar MA, Bullivant D, Mallinson GD, Hunter PJ. A virtual environment and model of the eye for surgical simulation. In *Computer Graphics, Proceedings of SIGGRAPH '94*, pages 205–212, Orlando (FL):July 1994. ACM SIGGRAPH, Association for Computing Machinery.

[16] McGreevy MW. *Virtual Reality and Planetary Exploration*, pages 163–197. Boston:Academic Press Professional, 1993.

[17] Yan JK. Advances in computer-generated imagery for flight simulation. *IEEE Computer Graphics and Applications*, 1985;5(8):37–51.

[18] Robinett W, Tayler R, Chi V, Wright WV. The nanomanipulator: An atomic-scale teleoperator. Technical report, Chapel Hill (NC):Computer Science Department, University of North Carolina, 1992.

[19] Brooks FP, Ouh-Young M, Batter JJ, Kilpatrick PJ. Project GROPE—haptic displays for scientific visualization. *Computer Graphics*, 1990;24(4):177–185.

[20] Kooper R. Virtually present: Treatment of acrophobia by using virtual reality graded exposer. Master's thesis, Delft, Netherlands:Technical University of Delft, 1994.
URL: *http://www.dcs.qmw.ac.uk/~kooper/Thesis/index.html*

[21] Apple Computer. QuickTime VR. Technical report, Cupertino (CA):Apple Computer, 1995. WWW page.
URL: *http://qtvr.quicktime.apple.com/*

[22] Feiner S, Shamash A. Hybrid user interfaces: Breeding virtually bigger interfaces for physically smaller computers. In *Proceedings of the ACM Symposium on User Interface Software and Technology*, pages 9–17, Hilton Head (NC):November 1992. Association for Computing Machinery.

[23] Krueger MW. An easy entry artifical reality. In Wexelblat A, editor, *Virtual Reality Applications and Explorations*, pages 147–161. Boston:Academic Press, 1993.

[24] Wyshynski S, Vincent VJ. Full-body unencumbered immersion in virtual worlds. In Wexelblat A, editor, *Virtual Reality Applications and Explorations*, pages 123–144. Boston:Academic Press, 1993.

[25] Moran TP, Anderson RJ. The workaday world as a paradigm for CSCW design. In *Proceedings of the Conference on Computer-Supported Cooperative Work*, New York:1990. ACM Press.

[26] Ware C, Balakrishnan R. Target acquisition in fish tank VR: The effects of lag and frame rate. In Davis WA, Joe B, editors, *Proceedings of Graphics Interface '94*, pages 1–7, Toronto:1994. Canadian Information Processing Society.

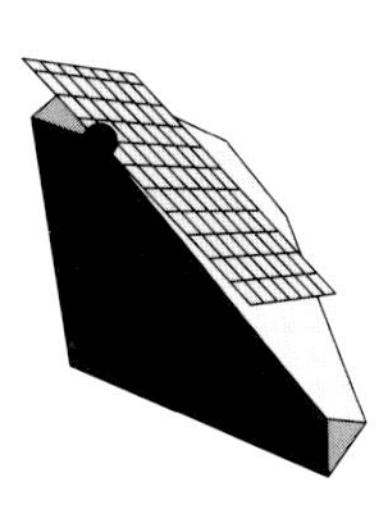

Part II

Visualization and the Modeling Life-cycle

Although we have stated that optimization research has concentrated on developing ever faster algorithms—with great success—a great deal of work *has* been accomplished introducing visualization to the optimization community. In this part of the book, we discuss how visualization has been used or might be used to support optimization. This part discusses how visualization has been used to support the tasks involved in an optimization modeling project. As discussed in chapter 1.6, we take a life-cycle approach to tasks involved in an optimization modeling project. That is, we model the modeling process as a series of stages from birth through life through death. At each of the stages, one must consider carefully the appropriate visualizations.

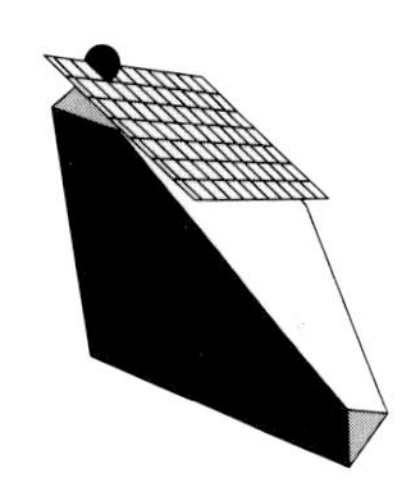

Part III then discusses a variety of different visualization formats that have been used in and outside of optimization. We concentrate especially on visualization formats that seem promising for application to optimization.

We do not claim to survey all optimization systems that are being sold or have been mentioned in the literature; the reader is referred to a recent survey that is much more comprehensive in this regard [1]. Rather, we concentrate on what we believe are major trends and highlights.

What could be usefully represented in support of optimization? The answer is simple: everything. One might need representations of problems, formal mathematical programming formulations, data, algorithms, ways of debugging algorithms, ways of debugging formulations, algorithm output, and summaries of algorithm output for presentation to decision makers.

We do not restrict our discussion to graphics. For example, a spreadsheet (except for its presentation graphics) consists of tables of numbers and text. We do not restrict ourselves to representations used exclusively for output. In the presentation graphics provided by a spreadsheet, for example, a bar chart encoding some data from the spreadsheet is only an output. The user cannot change the data by manipulating the bar chart with a mouse (At least a couple mathematical programming systems provide such a capability [2], [3]). In short, we believe that representation also involves input. How then will a user input the conceptual model, the formulation, and the data? How will a user interact with the algorithm and understand the solution? In discussing the different representations that seem appropriate to different phases of the modeling life cycle in this chapter, we also discuss how the user would input and interact with those representations.

The *key* challenge, then, is actually to provide an environment that can support the plethora of representation formats that are needed. We discuss possible solutions to this challenge in Chapter 18.

This part is organized according to the stages of the modeling life cycle discussed in the introduction, that is, conceptual models (Chapter 7), formulation (Chapter 8), execution (Chapter 9), solution analysis and final results presentation (Chapter 10). Many authors have presented different variations of the stages of the modeling life cycle [4]. The study of how to support all the stages in the modeling life cycle has been called *model management* [5], [6].

Bibliography

[1] Sharda R. Linear programming software for personal computers: 1992 survey. *OR/MS Today*, 1992;19(2):44–60.

[2] Chesapeake Decision Sciences. *MIMI/G User's Manual.* New Providence (NJ):Chesapeake Decision Sciences, 1993.

[3] Bisschop J, Entriken R. *AIMMS The Modeling System.* PO Box 3277, 2001 DG Haarlem The Netherlands:Paragon Decision Technology, 1993.

[4] Gass SI. Model world: Danger, beware the user as modeler. *Interfaces*, 1990;20(3):60–64.

[5] Dolk DR. A generalized model management system for mathematical programming. *ACM Transactions on Mathematical Software*, 1990;12:92–126.

[6] Shetty B, Bhargava HK, Krishnan R, editors. *Model Management in Operations Research*, volume 38. Basel, Switzerland:J. C. Baltzer, 1992.

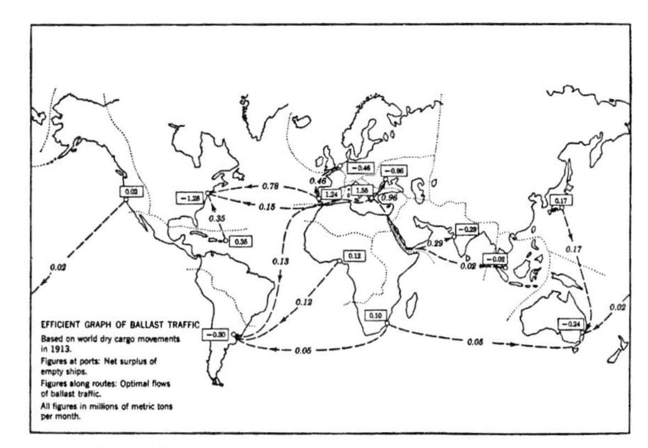

Chapter 7

Conceptual Models

The start of a problem-solving process involves translating ill-focused goals and ideas into more coherent, although still imprecise, descriptions of the problem to be attacked. The problem description must be understandable by the decision maker, who frequently has little training in optimization or modeling. The representations may be simple text describing the problem in natural language or they may involve formal or informal diagrams as well. An example of a conceptual description of a classic optimization problem, the Hitchcock-Koopmans transportation problem [1], [2], is given below.

> Each day a firm needs to move a particular product from several factories to its warehouses. Furthermore each factory produces a certain amount of the product each day, and each warehouse requires a certain amount of the product. The goal is to determine which factories should supply which warehouses so as to minimize cost. The cost to ship one unit of the product from each factory to each warehouse is known.

Figure 7.1

Note that Koopmans's original paper on the transportation problem included pictures illustrating the flow of product from factory to warehouse (below).

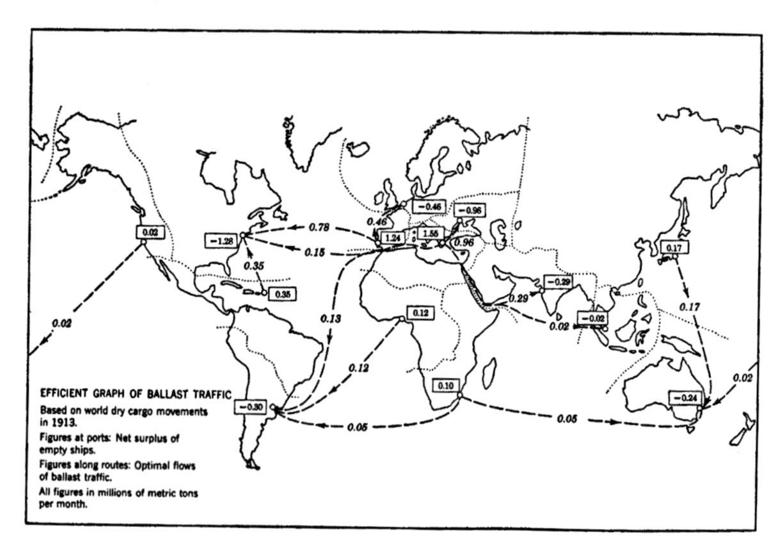

Figure 7.2

7.1 Visual Interactive Modeling

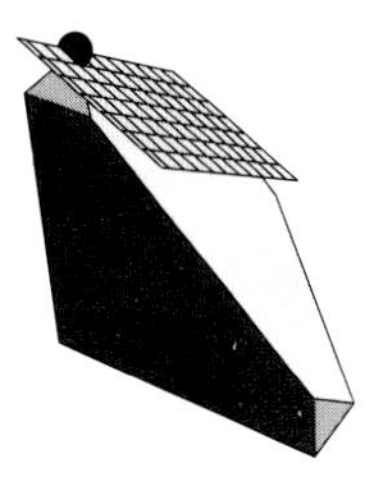

Several authors (Hurrion [3], Hurrion and Secker [4], and Bell [5], [6], [7] and Bell and O'Keefe [8]) have proposed a general methodology called *Visual Interactive Modeling* (VIM) for creating appropriate conceptual representations. VIM originated in the simulation community where it was originally called *Visual Interactive Simulation* (VIS). These authors believe that problem solving should concentrate heavily on developing an appropriate conceptual representation, with model formulation delayed until a precise conceptual representation (usually graphical) is developed. In particular, the conceptual representation should be the primary representation of the problem. Similar recommendations are made in [9] who calls the technique *symbiotic systems.*

Concentrating on the conceptual representation flies in the face of much practice, wherein the formal mathematical formulation is the primary representation of the problem. By representing the problem in its own terms, one helps ensure that the solution system presented to the decision maker corresponds to the actual problem. As Bell [10] notes, however, the challenge of the approach involves reconciling the conceptual representation with the mathematical representation. For VIM, commercial systems have been developed (for example, GENETIK [11]) that allow one to draw pictures for the conceptual representation, and then link algorithms to those pictures. These systems now provide links to mathematical programming systems as well.

For example, an extension to GENETIK called INCEPTA [12] allows problems to be analyzed using linear programming. An example of the application of INCEPTA to a multi-commodity, multi-stage transportation problem is shown below.

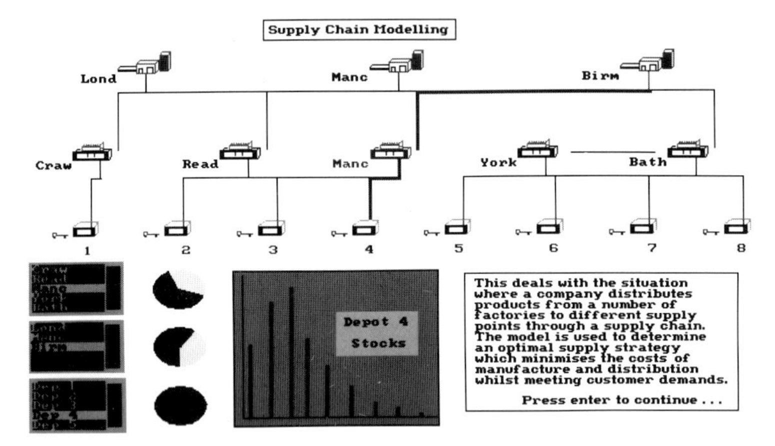

Figure 7.3

Several experimental studies of Visual Interactive Simulation (summarized in [8] have been conducted. Subjects are typically asked to use a VIS system to find the best setting for input parameters. As stated by Bell and O'Keefe [8] "performance of subjects has been mediocre." Fewer than 40% of the subjects produced the optimal answer in any of the experiments. From these results, one might then conclude that visual representations offer few if any advantages. At the same time, however, subjects who did find the optimal answer typically used animation extensively (as opposed to just static grapics) to explore the simulation model. Furthermore, how well would the subjects have performed if they used no visual interactive model at all?

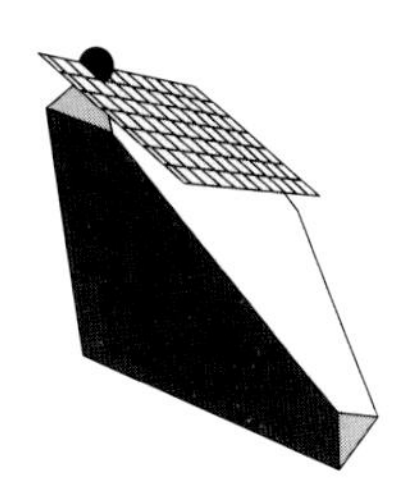

The VIM approach can be thought of in terms of a larger movement called *User Centered Design* [13], [14], which spans the fields of human factors, ergonomics, and industrial design. Advocates of user centered design believe that designers should begin, continue, and end the design process focused on users' needs. An oft-cited example concerns video cassette recorders. Although many such recorders provide highly developed features, many times those features go unused because users simply cannot figure out how they work. The engineers who developed the recorders became enamored with the bells and whistles, but actual users found them incomprehensible or superfluous. Mathematical programmers, including the author of this book, have often become so enamored with the powerful tools available that they lose sight of the underlying problem and the ultimate decision makers. Visual interactive modeling attempts to place the emphasis where it should be — on the people who actually have the problem.

The actual methodology for visual interactive modeling relies heavily on developing and evaluating prototype visual models. This is quite similar to what the user interface community has called *usability engineering* [15]. Usability engineering seeks to create easy-to-learn, easy-to-use, satisfying human-computer interfaces. It works to accomplish this through a series of formal or informal experiments with prototype interfaces. From the experiments, insights can be gleaned that can help to create a better human-computer interface. Tremendous improvements in human-computer interfaces can result with relatively low investment [16]. Similarly, Visual Interactive Modeling seeks to create complete, accurate and understandable (visual) models. This is accomplished by ongoing, close interaction with users who evaluate prototype visual models. Through this iterative process, one converges to an appropriate conceptual model—without constraining oneself *a priori* to a particular solution technique.

7.2 Conceptual Modeling Languages

Several researchers have proposed specific high-level languages for mathematical programming that represent problems at a conceptual level. The holy grail for this research thread has been the development of modeling tools that are easy to use. In Structured Modeling [17], [18], Geoffrion proposed a theory of modeling that provides several representations that attempt to represent models not just from optimization, but from the full spectrum of techniques used in OR. Relying on techniques from first-order predicate logic, Krishnan [19] proposed PM*, a high-level language for specifying linear programming problems. Originally developed for production, distribution, and inventory planning problems, models are specified not as a set of decision variables and constraints, but as a set of terms in the decision maker's own vocabulary. A graphical version of PM* has also been proposed [20].

7.3 Summary

Easy-to-use techniques for allowing modeling at a high level of abstraction are difficult to provide. One is caught between demands for generality and specificity — generality in order to model a variety of different problems, and specificity in that the resulting model should fit the problem like a glove. How can one provide powerful yet easy-to-understand modeling constructs? The approaches that have been described represent a start.

The philosophy espoused by the VIM community has merit, because VIM starts with the user and the user's problem, not with the model. Other alternatives, for example, Structured Modeling and PM*, attempt to impose a particular representation style on the problem holder. These two specific proposals lack the vagueness of the guidelines espoused by VIM. The constructs provided by these non-VIM techniques are meant to be sufficiently general and abstract, but one can use *just* those constructs to build models. It is naive to believe that one representation style will satisfy all users and all problems. VIM does not concentrate on a particular solution technique and is not attached to a particular set of modeling constructs, however general. Rather, it first attempts to represent the actual problem, usually visually. At that point, the model and solution technique can be developed. Unfortunately, the VIM technique makes it more difficult to assess the quality of the solution produced. In optimization, one at least can measure how fast an algorithm takes to find an optimal solution (though that is more difficult than might first appear [21]) or measure how far away an approximate solution is from the optimal solution. Taking the user-centered approach espoused by VIM, however, forces one to measure the quality of the solution produced by less objective measures such as user satisfaction.

Bibliography

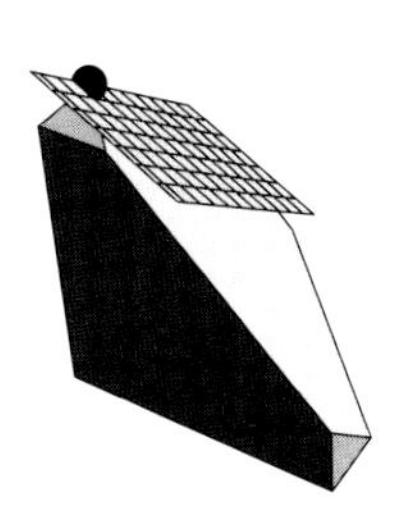

[1] Hitchcock FL. Distribution of a product from several sources to numerous localities. *Journal of Mathematical Physics*, 1941;20:224–230.

[2] Koopmans TC. Optimum utilization of the transportation system. *Econometrica*, 1949;17:3–4.

[3] Hurrion RD. Visual interactive modeling. *European Journal of Operational Research*, 1986;23:281–287.

[4] Hurrion RD, Secker RJR. Visual interactive simulation: An aid to decision making. *Omega*, 1978;6:419–426.

[5] Bell PC. Visual interactive modeling as an OR technique. *Interfaces*, 1985;15:26–33.

[6] Bell PC. Visual interactive modeling in operational research: Successes and opportunities. *Journal of the Operational Research Society*, 1985;36:975–982.

[7] Bell PC. Visual interactive modeling in 1986. In Belton V, O'Keefe R, editors, *Recent Developments in Operational Research*, pages 1–12. Oxford, United Kingdom:Pergamon Press, 1986.

[8] Bell PC, O'Keefe RM. Visual interactive simulation: A methodological perspective. *Annals of Operations Research*, 1994;53:321–342.

[9] Scriabin M, Farlette J, Kotak D, Matthews MA. Airport capacity planning symbiotic system (AIRSYM): Combining human and artificial intelligence. In *Canadian Operational Research Society Competition on the Practice of OR*, Quebec:May 1991.

[10] Bell PC. Visual aid: Interactive graphics come of age in the OR/MS community. *OR/MS Today*, 1992;19(4):24–27.

[11] Insight International Ltd. *Genetik Software Description*. Woodstock, Oxon, United Kingdom:Insight International Ltd, 1992.

[12] Insight International Ltd. *INCEPTA Tutorial*. Woodstock, Oxon, United Kingdom:Insight Logistics Ltd, 1993.

[13] Norman DA, Draper SW, editors. *User Centered System Design: New Perspectives on Human-Computer Interaction*. Hillsdale (NJ):Lawrence Erlbaum, 1986.

[14] Sedgwick J. The complexity problem. *The Atlantic*, 1993;271(3):96–104.

[15] Nielson GM, Shriver B, Rosenblum LJ. *Visualization in Scientific Computing*. Los Alamitos (CA):IEEE Computer Society Press, 1990.

[16] Nielsen J. *Usability Engineering*. Boston:Academic Press, 1993.

[17] Geoffrion AM. An introduction to structured modeling. *Management Science*, 1987;33:547–588.

[18] Geoffrion AM. The formal aspects of structured modeling. *Operations Research*, 1989;37(1):30–51.

[19] Krishnan R. A logic modeling language for model construction. *Decision Support Systems*, 1990;6.

[20] Jones CV, D'Souza K. Graph-grammars for minimum cost network flow modeling. Technical report, Burnaby (BC), V5A 1S6 CANADA:Faculty of Business Administration, Simon Fraser University, 1992.

[21] Barr RS, Hickman BL. Reporting computational experiments with parallel algorithms: Issues, measures, and experts' opinions. *ORSA Journal on Computing*, 1993;5(1):2–18.

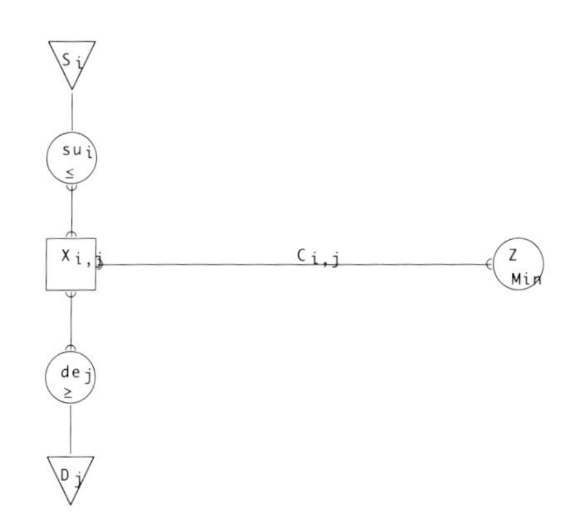

Chapter 8

Formulation

Once a conceptual model of the problem has been constructed (either formal or informal, visual or not), the next step in mathematical programming usually involves constructing a formal, mathematical representation of the problem, that is, a formulation.

One can distinguish between formulations for a large class of problems and formulations for a specific problem with all data values specified.

For example, one could specify a formulation for all Hitchcock-Koopmans transportation problems:

- Parameters
  - Let $\mathbf{N}^+$ represent the set of positive natural numbers.
  - Let $N \in \mathbf{N}^+$ indicate the number of *factories*.
  - Let $M \in \mathbf{N}^+$ indicate the number of *warehouses*.
  - Let $i \in \{1, \ldots, N\}$ represent a particular factory.
  - Let $j \in \{1, \ldots, M\}$ represent a particular warehouse.
  - Let $s_i \in \Re^+$ represent the *supply* at factory i.
  - Let $d_j \in \Re^+$ represent the *demand* at warehouse j.
  - Let $c_{ij} \in \Re^+$ represent the *cost* to ship one unit of product from factory i to warehouse j.
- Decision Variables
  - Let $x_{ij} \in \Re^+$ represent the quantity shipped from factory i to warehouse j.

$$\text{minimize total cost} = \sum_{i=1}^{N} \sum_{j=1}^{M} x_{ij}$$

subject to:

$$\begin{aligned}
\sum_{j=1}^{M} x_{ij} &= s_i \quad \forall i \qquad \text{Supply Limitation} \\
\sum_{i=1}^{N} x_{ij} &= d_j \quad \forall j \qquad \text{Demand Satisfaction} \\
x_{ij} &\geq 0 \quad \forall i, j
\end{aligned} \tag{8.1}$$

Figure 8.1

or for one specific Hitchcock-Koopmans transportation problem, as shown below. The formulation is written using LINDO [1]:

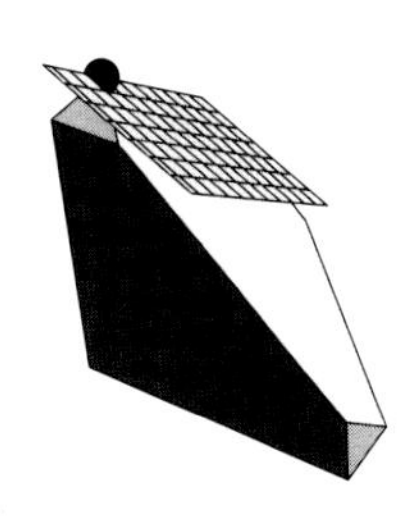

```
min 47 x11 + 52 x12 +34 x13 + 21 x14 +
  213 x21 + 38 x22 +41 x23 + 16 x24 +
  3 x31 + 67 x32 +23 x33 + 28 x34 +
  62 x41 + 84 x42 +80 x43 + 484 x44
subject to:
  x11 + x12 + x13 + x14 = 10
  x21 + x22 + x23 + x24 = 15
  x31 + x32 + x33 + x34 = 25
  x41 + x42 + x43 + x44 = 40
  x11 + x21 + x31 + x41 = 30
  x12 + x22 + x32 + x42 = 20
  x13 + x23 + x33 + x43 = 10
  x14 + x24 + x34 + x44 = 30
```

Figure 8.2

We shall call the former a *generic model*, and the latter a *model instance*. A generic model usually can be represented far more compactly than a model instance. Most authors recommend that a clean separation between the generic model and corresponding model instances be established. Such a separation facilitates changing either part without adversely affecting the other. The question is how to represent generic models and model instances, and moreover, how to establish a linkage between them.

Several different styles of formulation languages have been proposed. These include algebraic, block-structured, object-oriented, spreadsheet, and graphical formulations. Table 8.3 surveys some of the relevant references and systems. We discuss each in turn.

Style	*References*
Algebraic	AMPL [2], GAMS [3],MPL [4], LPL [5], SML [6], MODLER [7], [8]
Block-Structured	MIMI [9], PAM [10]
Object-Oriented	ASCEND [11], [12]
Spreadsheet	Microsoft Excel [13], Lotus 1-2-3, Lotus Improv [14], [15]
Visual	Activity-Constraint Graphs [1], [16]
	Netforms [17], [18]
	LPFORM [19]
	Entity-Relationship [20]
	Structured Modeling [21], [22], [23], [24]
Direct Manipulation	Modeling by Example

Figure 8.3

8.1 Algebraic Modeling Languages

Algebraic languages attempt to duplicate the algebraic notation already well known by many formulators of mathematical programming. Examples include GAMS [3] (Figure 8.4), AMPL [2] (Figure 8.5), and SML [6] (Figure 8.6). Note that these and other algebraic languages such as MODLER [7], [8], LPL [5] and MPL [4] all provide tools for manipulating sets and tables, for indexing over sets, and for expressing summations. Algebraic representations have often been called *row-wise*, since each constraint, or collection of constraints is specified directly.

At least one algebraic language, AMPL [25], [2], has been extended to allow model specifications in a column-wise fashion. AMPL has been further extended to include special constructs for problems such as minimum cost network flow models [25].

```
SETS
     I   plants    / SEATTLE,SAN-DIEGO /
     J   cust          / NEW-YORK,CHICAGO,TOPEKA / ;

PARAMETERS

     SUP(I)  capacity of plant i in units
       /    SEATTLE     350
            SAN-DIEGO   600  /
     DEM(J)  demand at customer j in cases
       /    NEW-YORK    325
            CHICAGO     300
            TOPEKA      275  / ;

TABLE COST(I,J)  cost in $/unit
                  NEW-YORK      CHICAGO      TOPEKA
    SEATTLE          2.5           1.7          1.8
    SAN-DIEGO        2.5           1.8          1.4   ;

VARIABLES
     FLOW(I,J)  shipment quantities in units
     Z  total transportation costs in thousands of dollars ;
POSITIVE VARIABLE FLOW ;

EQUATIONS
     COST         define objective function
     SUPPLY(I)    observe supply limit at plant i
     DEMAND(J)    satisfy demand at customer j ;
COST ..      Z  =E=  SUM((I,J), COST(I,J)*FLOW(I,J)) ;
SUPPLY(I) ..   SUM(J, FLOW(I,J))  =L=  SUP(I) ;
DEMAND(J) ..   SUM(I, FLOW(I,J))  =G=  DEM(J) ;
```

Figure 8.4: GAMS model [3] of the Hitchcock-Koopmans transportation problem. Note how data and model specification are interleaved. Recent extensions allow the data to be maintained in separate files.

```
set PLANT;    # plants
set CUST;    # customers

param sup {PLANT} >= 0;    # amounts available at plants
param dem {CUST} >= 0;    # amounts required at customers

   check: sum {i in PLANT} sup[i] = sum {j in CUST} dem[j];

param cost {PLANT,CUST} >= 0;    # shipment costs per unit
var Flow {PLANT,CUST} >= 0;    # units to be shipped

minimize total_cost:
   sum {i in PLANT, j in CUST} cost[i,j] * Flow[i,j];

subject to Supply {i in PLANT}:
   sum {j in CUST} Flow[i,j] = sup[i];

subject to Demand {j in CUST}:
   sum {i in PLANT} Flow[i,j] = dem[j];
```

Figure 8.5: AMPL model [2] of the Hitchcock-Koopmans transportation problem. The `check` statement performs a preliminary feasibility test for this model.

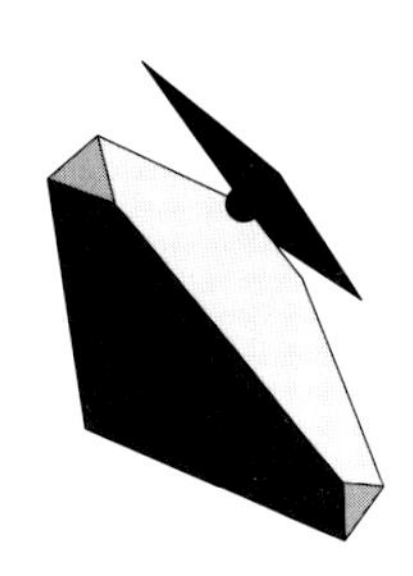

&SDATA *SOURCE DATA*

PLANTi /pe/ *There is a list of PLANTS.*

SUP(PLANTi) /a/ {PLANT} : $\Re^+$ *Every PLANT has a nonnegative SUPPLY CAPACITY measured in tons.*

&CDATA *CUSTOMER DATA*

CUSTj /pe/ *There is a list of CUSTOMERS.*

DEM(CUSTj) /a/ {CUST} : $\Re^+$ *Every CUSTOMER has a nonnegative DEMAND measured in tons.*

&TDATA TRANSPORTATION DATA

LINK(PLANTi,CUSTj) /ce/ {PLANT} × {CUST} *There are some transportation LINKS from PLANTS to CUSTOMERS. There must be at least one LINK incident to each PLANT, and at least one LINK incident to each CUSTOMER.*

FLOW(LINKij) /va/ {LINK} : $\Re^+$ *There can be a nonnegative transportation FLOW (in tons) over each LINK.*

COST(LINKij) /a/ {LINK} *Every LINK has* a TRANSPORTATION COST RATE *associated with all FLOWS.*

$(COST,FLOW) /f/ 1; @SUMij(COSTij * FLOWij) *There is a TOTAL COST associated with all FLOWS.*

T:SUP(FLOWi.,SUPi) /t/ {PLANT}; @SUMj(FLOWij) <= SUPi *Is the total FLOW leaving a PLANT less than or equal to its SUPPLY CAPACITY? This is called the SUPPLY TEST.*

T:DEM(FLOW.j,DEMj) /t/ {CUST}; @SUMi(FLOWij)=DEMj *Is the total FLOW arriving at a CUSTOMER exactly equal to its DEMAND? This is called the DEMAND TEST.*

Figure 8.6: Schema in SML [6] for the transportation problem. Note that SML models are not limited to optimization, but include database, probabilistic and other mathematical models.

8.2 Block-Structured Models

Another style of formulation involves a block-structured or process-oriented representation of the problem. Figure 8.7, adapted from MIMI [9], shows a block-structured formulation of the transportation problem. A block-structured formulation organizes the linear programming matrix into blocks, where each block represents the intersection of a collection of decision variables and constraints.

Figure 8.7 shows a model of a transportation problem that contains 1 column and 2 rows. The single column (labeled `T SOURCE SINK`) represents decision variables transporting material from each source to each sink. The first row (labeled `SUPPLY at SOURCE`) constrains the amount shipped from each source. It has a matrix coefficient of `+1` for all valid entries. The second row (labeled `DEMAND at SINK`) constrains the amount shipped to each sink to be larger than its demand. Other tables define the objective function coefficients and right-hand-side values. Greenberg and Murphy [26] described how the block-structured and algebraic representations can be integrated, since both attempt to represent the same fundamental object.

COEF ROW by COL	T SOURCE SINK
SUPPLY at SOURCE	+1
DEMAND at SINK	+1

Figure 8.7

Although block-structured languages represent primarily generic models, model instances can also be represented in tabular or matrix form. Moreover, the values in the table can be color coded. For example, negative values can be colored red and positive values can be colored green. Consider the example below from MPL [4]. Image courtesy of Maximal Software, Inc.

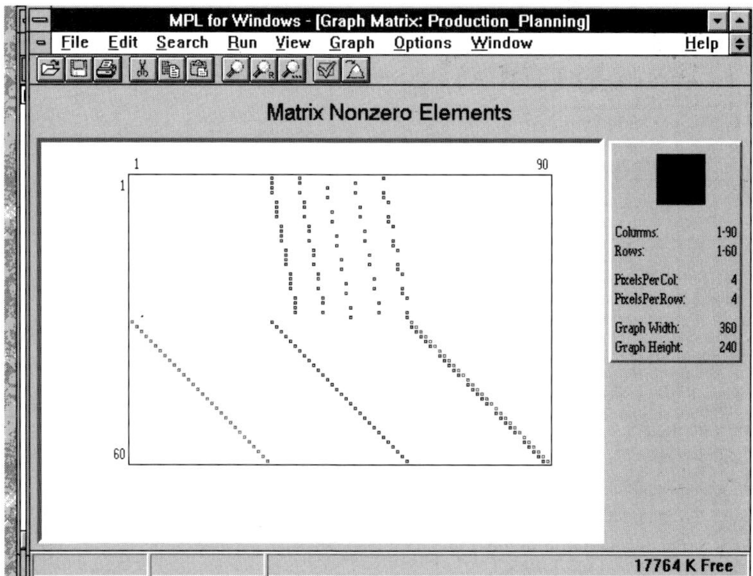

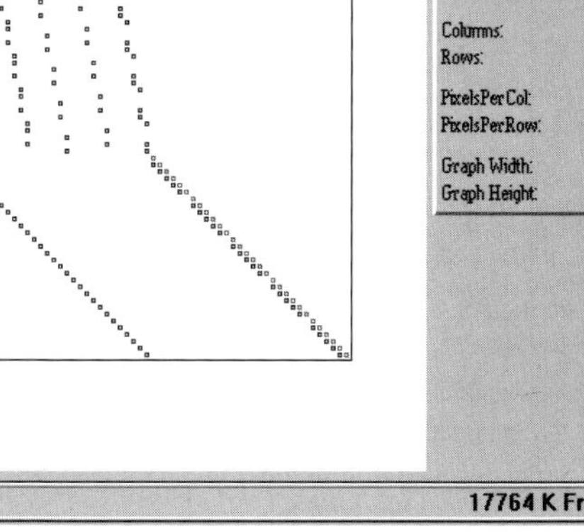

Figure 8.8

Such *matrix images* (as shown in the figure below courtesy of Chesapeake Decision Sciences) can represent large matrices. Drawing such images is actually more difficult than it might seem at first. The resolution of most computer screens is at best around 1000 by 1000 pixels. If the matrix to be drawn is larger than that, then what color should be drawn at each pixel? [27] computed a weighted average of the values in the matrix to determine the color of each pixel.

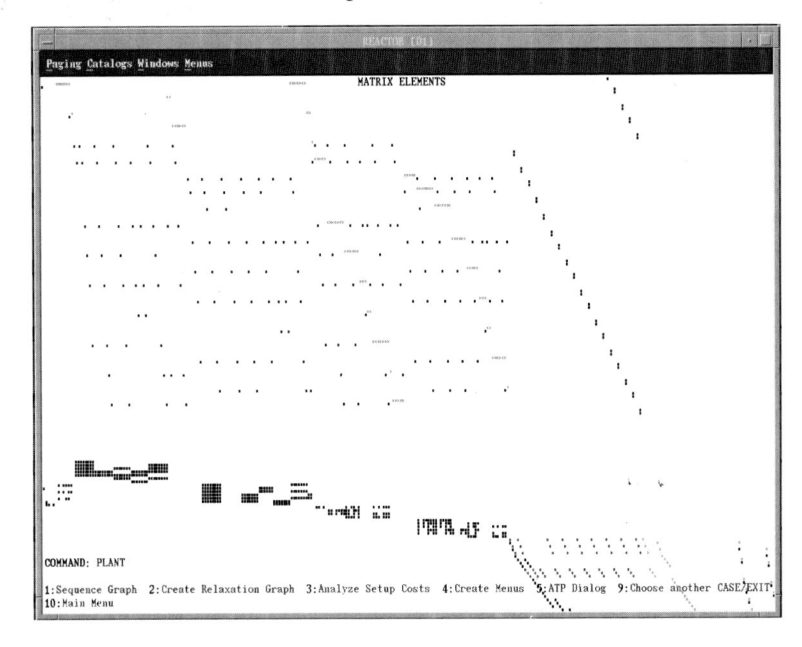

Figure 8.9

8.2.1 Object-Oriented Languages

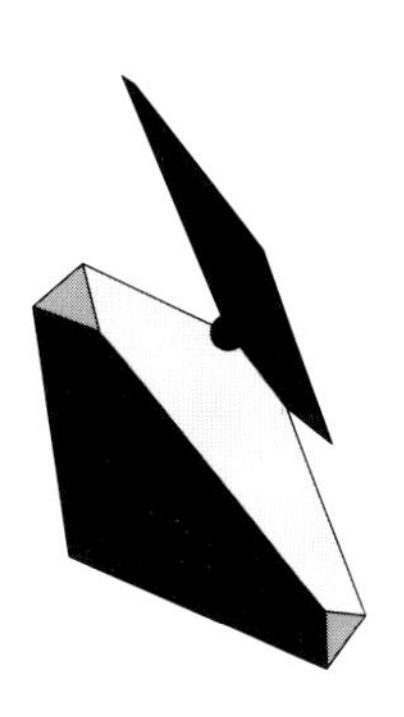

More recently, we have seen the development of object-oriented languages for optimization [28], [11]. Wegner provides a good survey of the basic concepts of object-oriented programming [29]. Object-orientation allows existing model components to be reused in a controlled fashion. An object contains not only data describing the object, but methods or procedures that can operate on the object's data. This *encapsulation* allows both the procedures and data associated with an object to be organized into one location. Furthermore, objects can *inherit* methods and data from other objects. For example, one could define an object storing the data and procedures associated with a a linear programming model. Another object, for the transportation problem could inherit some of the data and methods associated with the linear programming model, but could also *override* the method that actually optimizes the model, substituting a more efficient algorithm. Since the transportation model object is a type of linear programming model, however, anytime a linear programming model is solved, the appropriate solution method is automatically invoked. In other words, objects have a flexible yet controlled way of managing their types. This property is called *polymorphism.*

Object orientation seems natural for optimization since, for example, linear programming can be considered to be a special case of nonlinear programming, minimum cost network flow models are a special case of linear programming, and the transportation problem is a special case of a minimum cost network flow problem.

Below is a formulation of the transportation problem in ASCEND, an object oriented language [12].

```
IMPORT   transportation_atoms;
MODEL    plant;
         sup IS_A supply_capacity;
         customerId IS_A set OF integer
         maxCustomer IS_A integer

         CARD(customerId)<=maxCustomer;

         f[customerId],totalFlow IS_A flow;
         cost[customerId] IS_A unitCost;
         totalFlow = SUM(f[customerId])
         totalFlow <=sup;

         shipmentCost IS_A cost;
         shipmentCost = SUM(f[i]*cost[i]|i IN customerId);
END      plant;

MODEL    customer;
         dem IS_A demand;
         plantId IS_A set OF integer;
         f[plantId] IS_A flow;
         SUM(f[plantID])=dem;
END      customer;

MODEL    transportation;
         plantId, customerId IS_A set OF integer;
         p[plantId] IS_A plant;
         c[customerId] IS_A customer;
         FOR i IN customerId CREATE
           c[i].plantId:=[j IN plantId | i IN p[j].customerId];
         END;
         FOR i IN plantId CREATE
           FOR j IN p[i].customerId CREATE
             p[i].flow[j],customer[j].flow[i] ARE_THE_SAME;
           END;
         END;
         obj: MINIMIZE
           SUM(p[i].shipmentCost|i IN plantId);
END      transportation;
```

Figure 8.10

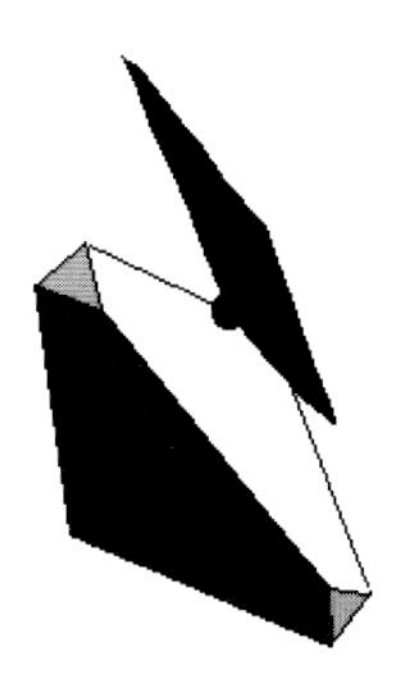

The first line of the ASCEND formulation imports some basic definitions useful to the transportation problem such as flow, supply_capacity and unitCost. The transportation model is then formulated as a collection of individual plant and customer models. The plant and customer models are essentially defined independently. For example, each plant model stores the plant's supply (sup), the list of customers it is allowed to supply (customerId), and the cost to ship material to each customer. Similarly, each customer model stores the customer's demand (dem), as well as the set of plants that can supply it (plantId), and the amount that each plant supplies to the customer (f[plantId]). The definition of the transportation model combines plant and customer models, (p[plantId] and c[customerId], respectively) and links their values together. For example, each plant stores the amount it ships to each customer. Similarly, each customer stores the amount shipped to it by each plant. In the transportation model, the ARE_THE_SAME construct ensures that these values are equal.

In [12], inheritance is used to create a refined model. Suppose, for example, that a forecasting model is used to determine the amount demanded by each customer. One would like to link the existing forecasting model to the existing transportation model elegantly. In Ascend, one can create a new model, trans_forecast, that refines the basic transportation model. The model simply refines the definition of demand for customers (REFINES is a keyword in ASCEND).

ASCEND represents perhaps the most well-developed object-oriented language for building models. It tailors the basic object-oriented concepts to the special needs of modeling, including optimization modeling. It explicitly promotes model reuse by allowing models to inherit and override parts of existing model definitions. Such modularity comes at some cost, though; the definition of the transportation model seems rather long compared to the definition available in algebraic languages.

8.3 Spreadsheets

Perhaps the most ubiquitous model formulation style can be found in spreadsheets. Recent spreadsheets incorporate linear and nonlinear programming algorithms. Once a spreadsheet is constructed, users need only indicate the cell whose value is to be optimized, the cells whose values can be changed (the decision variables), and the bounds to assign to other cells (the constraints), and the built-in algorithm will attempt to set the decision variables to the values that optimize the indicated cell.

A formulation in a spreadsheet need not be specified as a matrix of constraints, but can look exactly like a traditional spreadsheet. The figure below, for example, shows a spreadsheet model in Microsoft Excel for the transportation problem.

Solver2

	A	B	C	D	E	F	G
1	Example 2: Transportation Problem.						
2	Minimize the costs of shipping goods from production plants to warehouses near metropolitan demand						
3	centers, while not exceeding the supply available from each plant and meeting the demand from each						
4	metropolitan area.						
6			*Number to ship from plant x to warehouse y (at intersection):*				
7	*Plants:*	*Total*	*San Fran*	*Denver*	*Chicago*	*Dallas*	*New York*
8	S. Carolina	300	0	0	0	80	220
9	Tennessee	260	0	0	180	80	0
10	Arizona	280	180	80	20	0	0
12	Totals:		180	80	200	160	220
14	*Demands by Whse -->*		180	80	200	160	220
15	*Plants:*	*Supply*	*Shipping costs from plant x to warehouse y (at intersection):*				
16	S. Carolina	310	10	8	6	5	4
17	Tennessee	260	6	5	4	3	6
18	Arizona	280	3	4	5	5	9
20	*Shipping:*	**$3,200**	$540	$320	$820	$640	$880

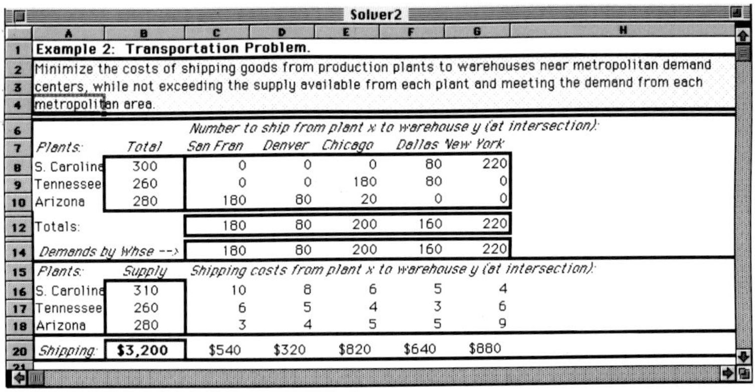

Figure 8.11

The optimization model, below, consists of the cell that computes the objective function ($B$20), the cells containing the values of the decision variables ($C$8:$G$10)) and constraints on the value of other cells. For example, the total amount shipped (computed in cells $B$8:$B$10) cannot be larger than their actual supply (stored in cells $B$16:$B$18). Note that the spreadsheet can function completely independently of the optimization algorithm. Users can change shipment amounts, supply capacity and demand amounts, and the total cost will be calculated automatically by the spreadsheet. Only when the user explicitly requests an optimal solution will the spreadsheet calculate it.

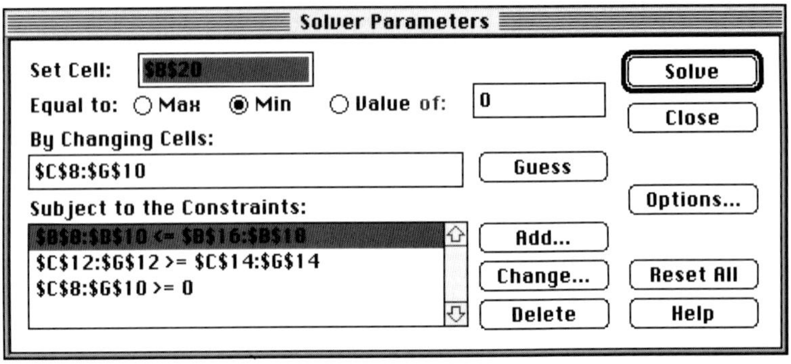

Solver Parameters

Set Cell: $B$20

Equal to: ○ Max ◉ Min ○ Value of: 0

By Changing Cells:

$C$8:$G$10

Subject to the Constraints:

$B$8:$B$10 <= $B$16:$B$18

$C$12:$G$12 >= $C$14:$G$14

$C$8:$G$10 >= 0

Solve | Close | Guess | Options... | Add... | Change... | Delete | Reset All | Help

Figure 8.12

Spreadsheets have recently begun to provide some of the higher-level indexing capabilities found in algebraic languages. For example, both Microsoft Excel [13] and Lotus Improv [14] allows multi-dimensional tables as discussed in Section 3.4.1. In addition, Lotus Improv [14], [15] supports algebraic-style formulas for defining the values of cells.

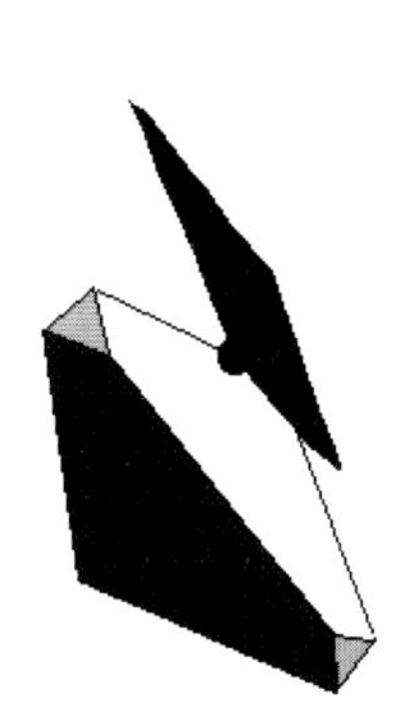

Improv has also been extended [30] to provide optimization capabilities (by translating the Improv spreadsheet into LINGO [31]. Below is a simple production planning model built using Improv. Note the algebraic style formulas at the bottom of the figure. The icons at the upper right, i.e., Min Max X(s) ≤ , allow users to specify the objective function, constraints, and to execute the optimization algorithm.

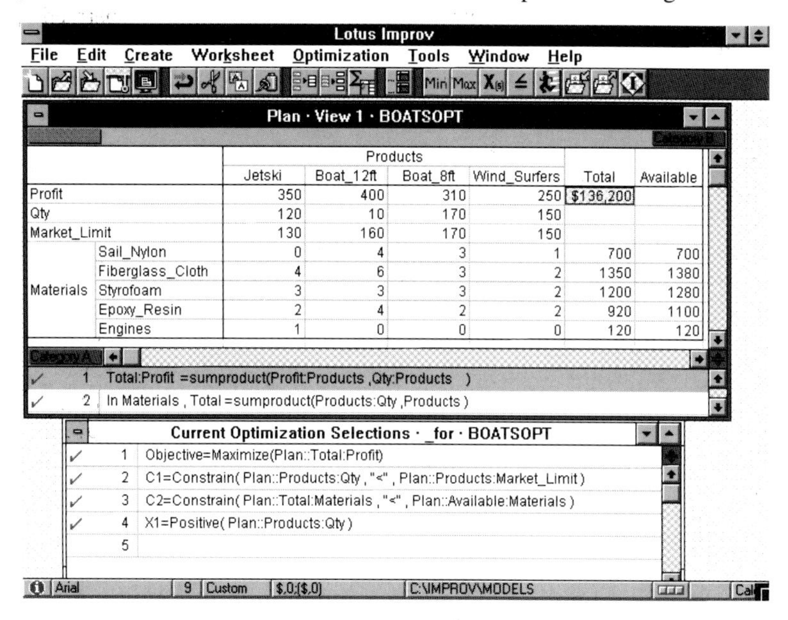

		Products					
		Jetski	Boat_12ft	Boat_8ft	Wind_Surfers	Total	Available
Profit		350	400	310	250	$136,200	
Qty		120	10	170	150		
Market_Limit		130	160	170	150		
Materials	Sail_Nylon	0	4	3	1	700	700
	Fiberglass_Cloth	4	6	3	2	1350	1380
	Styrofoam	3	3	3	2	1200	1280
	Epoxy_Resin	2	4	2	2	920	1100
	Engines	1	0	0	0	120	120

Figure 8.13

Quite sadly, Lotus has announced that it will no longer be enhancing Improv due to lack of market demand. Lotus stated that they intend to incorporate the unique features of Improv into a future release of their traditional spreadsheet, 1-2-3.

8.4 Visual Modeling Languages

The use of a small set of characters to represent individual phonemes was a significant development in human evolution. Textual languages, although incredibly powerful, are inherently one-dimensional. One character (word, sentence, paragraph) follows the other in simple linear sequence, although elaborate grammatic techniques are provided to allow one to link together ideas appearing in different places. With the development of hypermedia, such linkages are now even more fluid. A visual language, however, allows one to express ideas in two-dimensions. Links between two ideas can be made explicit: just draw a line or arrow connecting them.

Visual languages were perhaps the first recorded human languages; consider cave paintings, or even better, consider Egyptian hieroglyphics. With the development of textual languages, however, visual languages faded from common use. By the 1800's, however, visual languages began to reemerge for specialized scientific purposes. Electrical circuits are typically designed using a visual language with icons representing electrical components such as resistors and capacitors. In mathematics, the eminent philosopher and logician, Charles Sanders Peirce proposed a visual language for formal logic in 1896 that he called *existential graphs* [32], although it was never widely adopted. Computer software is often specified using visual language such as data flow diagrams and flow charts. The computer-science community has proposed a variety of visual programming languages [33], [34], [35]. Many computer simulation languages are visual as well (e.g., [36]).

Starting in the late 1970's, various authors proposed the use of graphical or visual languages to represent optimization models [37]. For optimization, most visual languages rely on graphs as the fundamental modeling construct. A variety of graphs and networks have been proposed:

Activity-Constraint Graphs	[1], [16]
Netforms	[17], [18]
LPFORM	[19]
Entity-Relationship	[20]
Structured Modeling	[21], [22], [23], [24]

Figure 8.14

8.4.1 Activity-Constraint Graphs

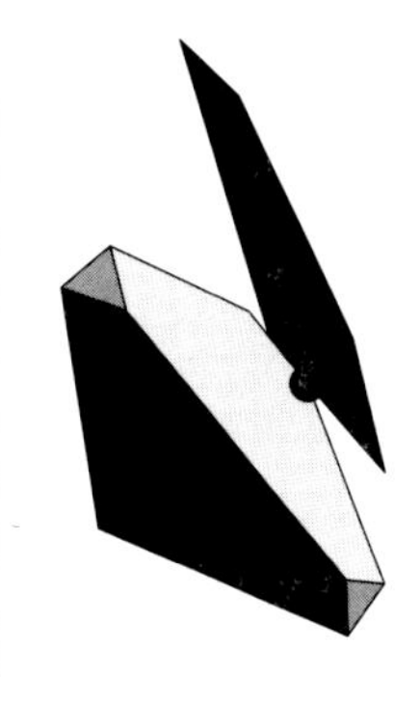

In many of these languages, each constraint and decision variable is represented either as a node or an arc. This style of representation originated with Schrage [1], who called it an *activity-constraint diagram*. In an activity-constraint diagram (below), square nodes represent decision variables, circular nodes represent constraints, and arcs represent the coefficients of decision variables in constraints. An activity-constraint diagram can model problems at both the generic (a) and instance (b) levels since a node can represent an individual decision variable (or constraint) or a collection of decision variables (or constraints). At the generic level, an activity-constraint diagram essentially translates a block-structured representation into a graph.

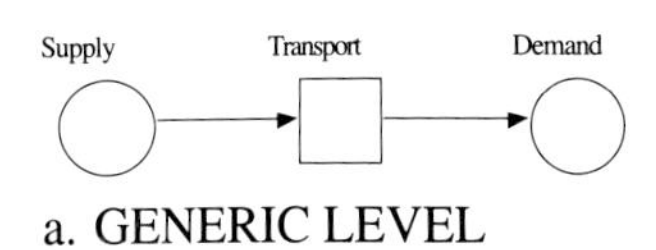

a. GENERIC LEVEL

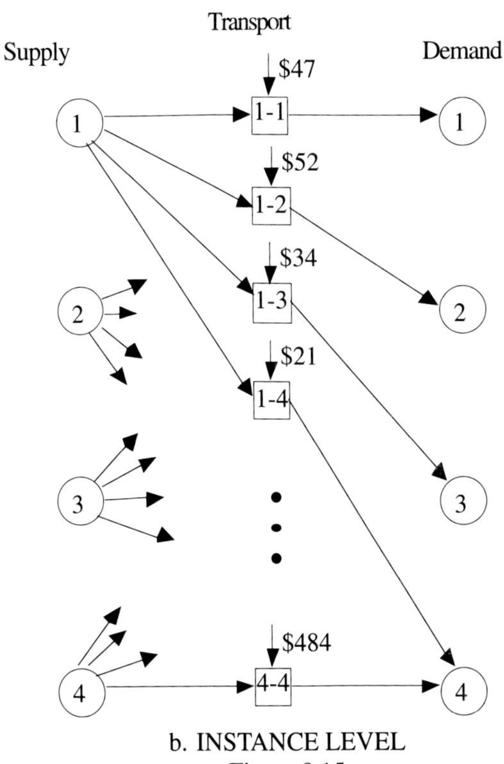

b. INSTANCE LEVEL
Figure 8.15

Collaud's system, gLPS [16], is based on activity-constraint diagrams. Below is the Hitchcock-Koopmans transportation problem formulated using gLPS [16]. Circles represent either constraints or the objective function; squares represent decision variables; triangles represent right-hand side values.

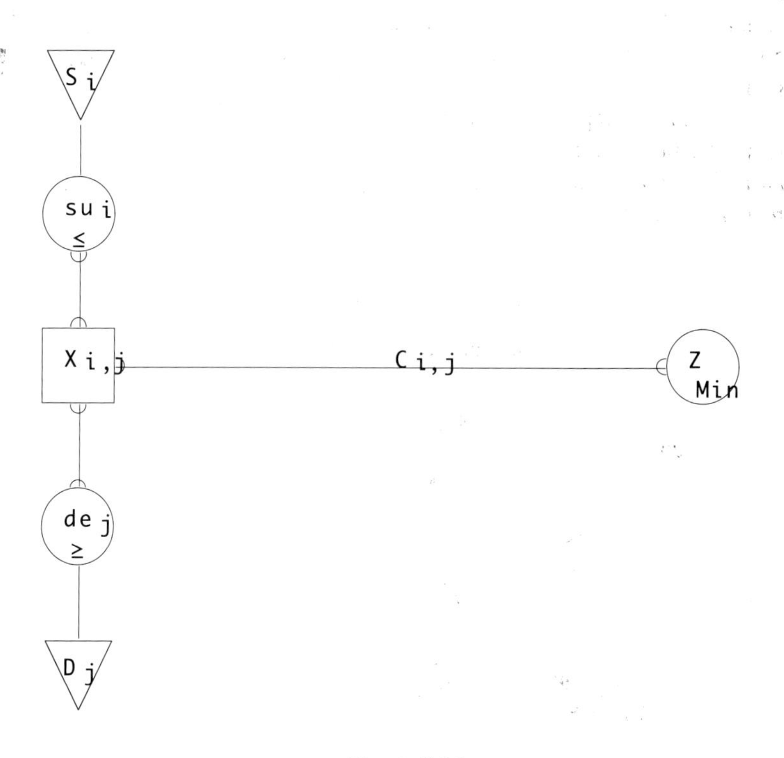

Figure 8.16

After users draw their models, they can then be automatically translated to algebraic form and then solved. An abridged algebraic formulation of Figures 8.16 and 8.18 is shown below. gLPS generates a subset of LPL, an algebraic language developed by Hürliman [5].

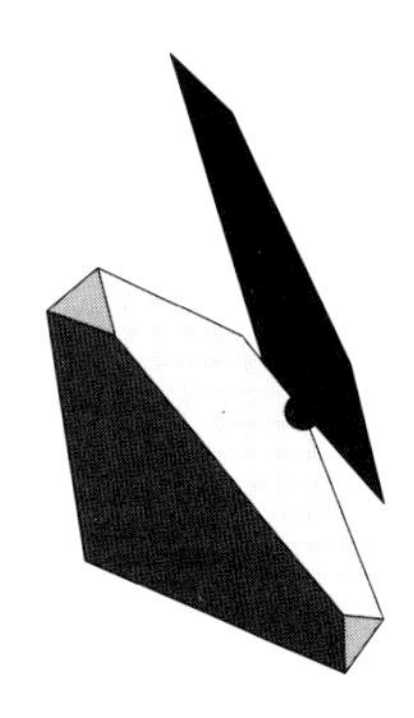

```
SET
   i = /1 2 3 4 5/;
   j = /1 2 3 4 5/;
COEF
     S{i}=
         /1 10
          ...  / ;
     D{j}=
         /1 15
          ...  / ;
     C{i, j}=
             /1 1 5
              1 2 34
              ...  / ;
VAR
   X{i,j} ;
MODEL
      su{i} :
                 SUM{j} X[i,j] <= S;
      de{j} :
                 SUM{i} X[i,j] <= D;
      Z:
                 SUM{i,j} C * X[i,j] ;
MAXIMIZE Z;
PRINT Z; X;
END
```

Figure 8.17

Note that nodes can represent a collection of other nodes. For example, one can expand the right-hand side nodes SU_i and DE_j, thereby seeing the actual supply and demand values directly on the graph.

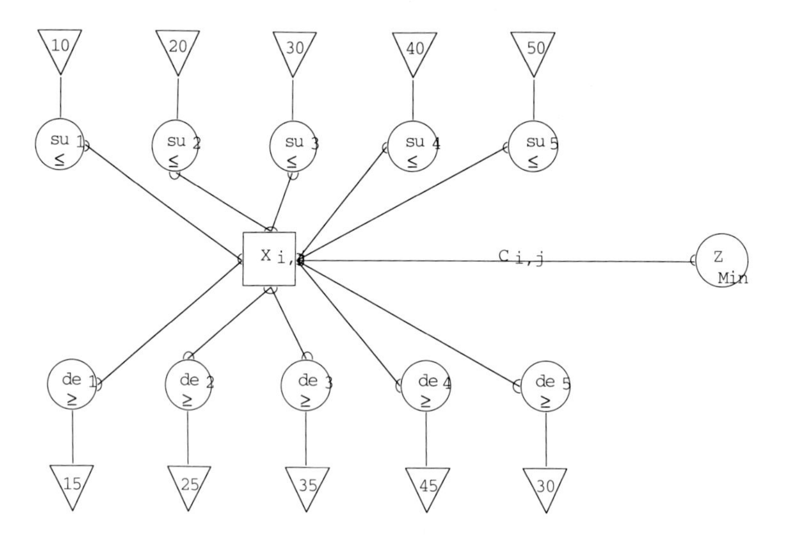

Figure 8.18

One can continue to expand other nodes. For example, if one expands the decision variable node X_{ij} over index i, one can begin to see which decision variables participate in each of the demand constraints.

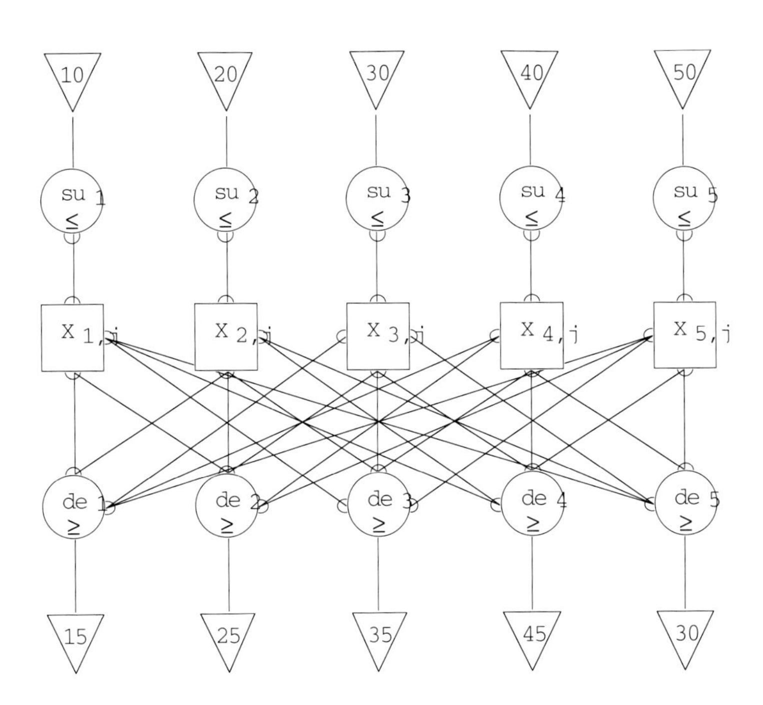

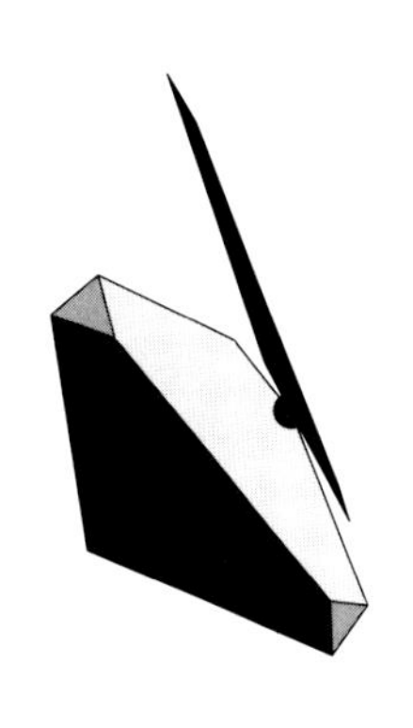

Figure 8.19

Note that the objective coefficient has been hidden in Figure 8.19. Otherwise, an edge connecting each decision variable node to the objective node would be included, which would hinder clarity. In general, this type expansion can produce a tangled pile of spaghetti. gLPS provides mechanisms for hiding and moving nodes, as well as expanding over only a few indices. Other techniques for displaying large graphs in a coherent fashion are discussed in Chapter 13.

8.4.2 Netform Representations

Glover, Klingman and McMillan [17] proposed *Netforms* as a representation technique for general modeling. Although often assumed to be targeted only to minimum cost network flow models, Netforms also allow full linear and integer programming models to be specified. For network flow problems (shown below), constraints are represented by nodes and decision variables are represented by arcs connecting the nodes. Perhaps the most ambitious attempt to apply Netform representations to minimum cost network flow models was made by Steiger, Sharda, and LeClaire [18].

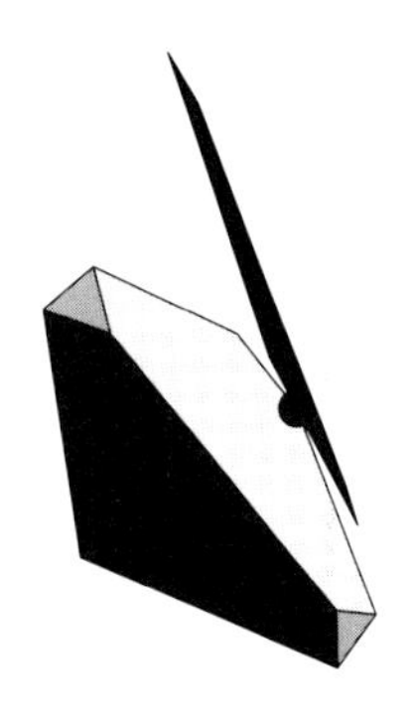

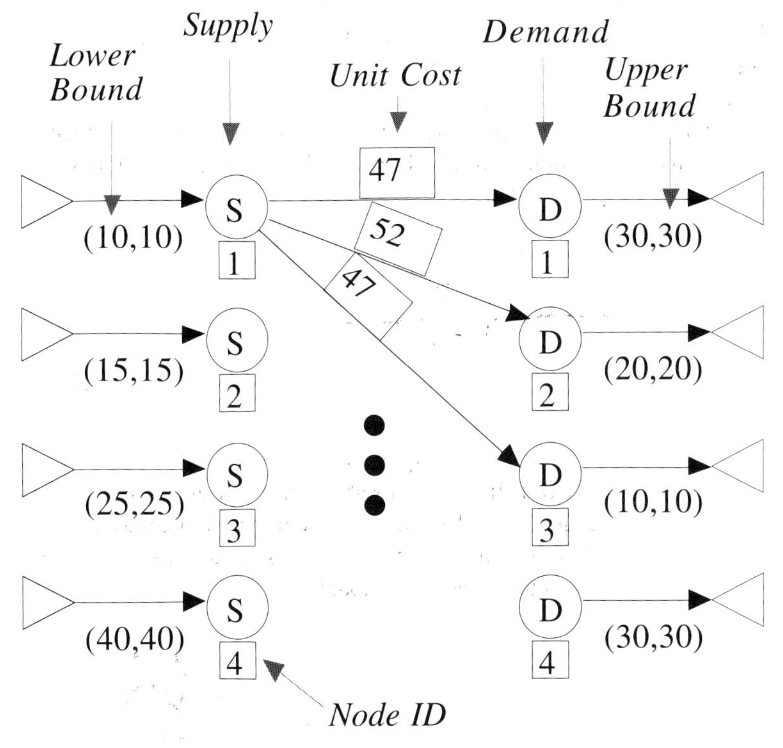

Figure 8.20

Linear and integer programming models are supported in a Netform representation through the introduction of additional descriptive information. For example, a particular model may insist that at most one out of several outgoing arcs can have positive flow. This constraint would be expressed as an additional restriction on the Netform representation.

8.4.3 Higher Conceptual Levels

Some graph-based languages attempt to represent mathematical programs at a higher conceptual level [38], [19]. Notable among those languages is LPFORM [19]. In LPFORM (see Figure 8.21), nodes represent decision variables and arcs represent either decision variables (such as transportation activities) or other logical constraints not directly represented in the linear program. Nodes can also represent existing sub-models that are being used in the larger model.

Below is an LPFORM graphical model of a simple Hitchcock-Koopmans transportation model. The optimal transportation pattern from oil wells to pumps is sought.

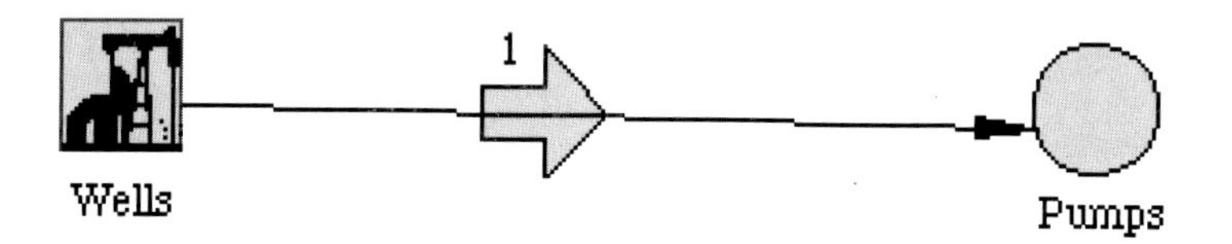

Figure 8.21

A notable feature of LPFORM is its ability to allow users to construct models in separate pieces and then combine them into a larger model. Below is another LPFORM graphical model; in this model, the transportation of oil from pumps to refineries is modeled.

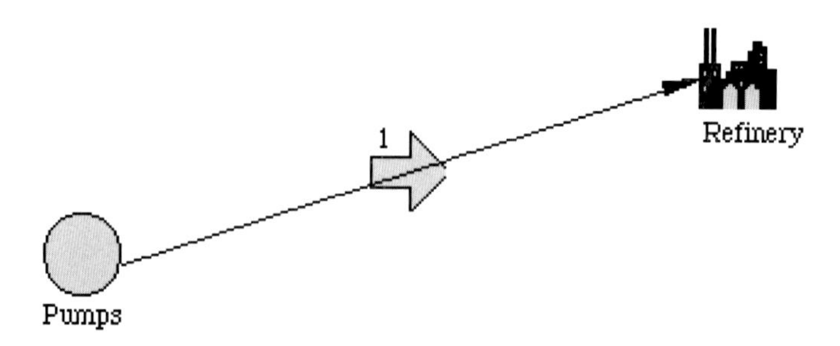

Figure 8.22

The models of Figures 8.21 and 8.22 can then be combined into the larger model shown below. LPFORM automatically identifies overlapping model components (in this case, the node labeled "Pumps") and combines them.

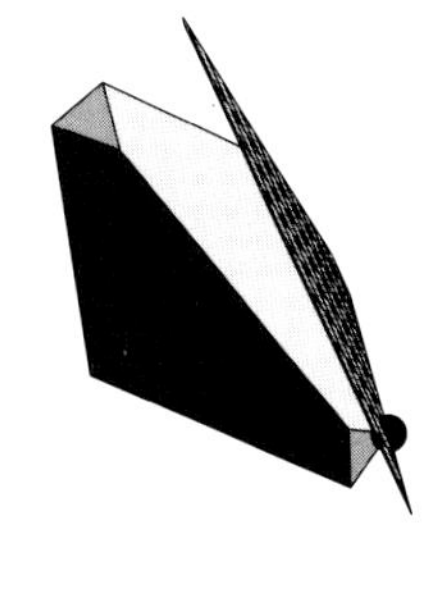

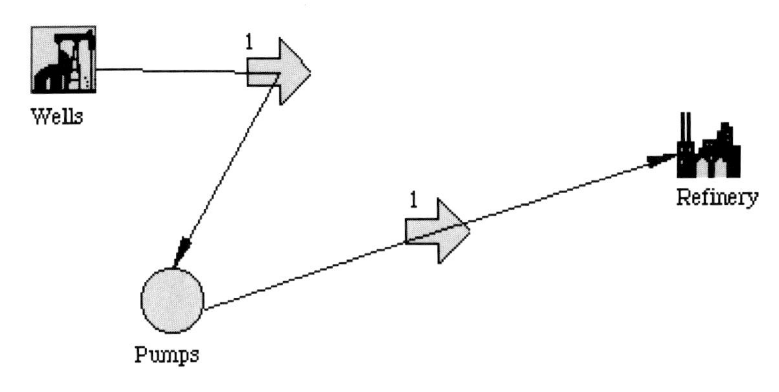

Figure 8.23

If one compares the graphical representations for the transportation problem in Figures 8.15 (activity-constraint), 8.20 (Netform), and 8.21-8.23 (LPFORM), they all involve a bipartite graph. This reflects the fact that half of the nodes are mainly associated with the sources of supply and the other half are associated with the sinks. At least for the simple transportation problem, these representations are sufficiently similar to suggest that a general form might exist. Murphy, Stohr, and Asthana [37], in fact, have shown that LPFORM generalizes both the Netform and activity-constraint representations.

8.4.4 Entity-Relationship Diagram Approaches

Choobineh [20] extended entity-relationship diagrams [39], a graph-based model for relational databases, to provide another graph-based language for mathematical programs (below). An entity-relationship diagram represents entities, such as sources and sinks, as squares. Simple relationships among the entities, such as the transport of material, are represented by diamonds connected to the entities. Entities and relationships can have attributes, represented as ellipses, connected to them. For example, each source has a capacity and each transport relationship has a unit cost.
Choobineh extended this basic model with additional node types to represent constraints, decision variables, and objective functions. Decision variables are represented as ellipses with light borders. For the transportation problem, the ellipse labeled "Flow" represents the decision variables. The objective is represented by a hexagon. Constraints are formed using nodes representing mathematical operators (triangles) and logical relationships (houses) such as greater than or equal to (ge). Since entity-relationship diagrams are commonly taught and widely used for database design, Choobineh's approach to providing a graphical language for linear programming may have a greater chance for success than other graphical approaches.

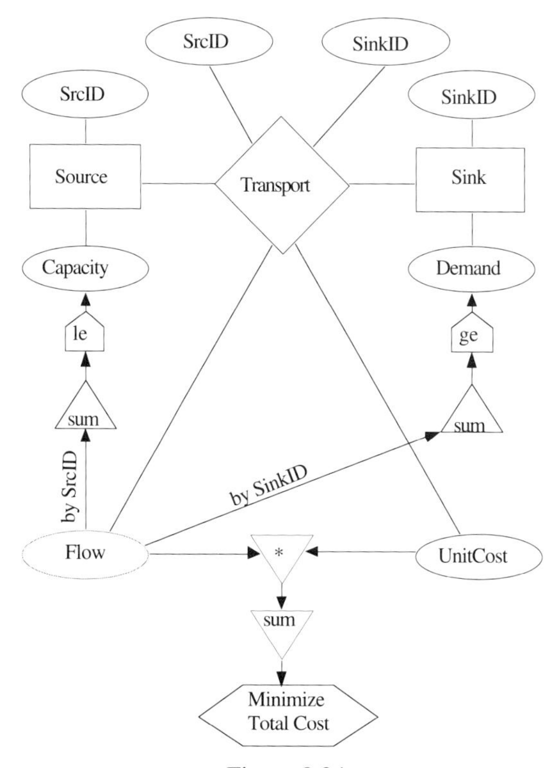

Figure 8.24

8.4.5 Structured Modeling

Structured Modeling [21], [22] represents an ambitious attempt to create a general theory for modeling, not just optimization modeling, but as many other types of mathematical and computer science models as possible. Note that the theory of Structured Modeling is independent of any particular modeling languages. One could create a variety of modeling languages that conform to the theory specified by structured modeling. For example, a very carefully designed textual language, SML, has been developed that provides an incredible rich variety of constructs for building models.

The theory of Structured Modeling also describes some graph-based representations. One type of graph, the *genus graph*, groups together individual model elements based on formal similarity properties. Compare this graph to the textual representation of SML in Figure 8.6. Each node represents a collection of elements in the model. Each arc $a \rightarrow b$ indicates that genus b is defined, at least in part, by genus a. In short, the genus graph illustrates the interrelationships among genera.

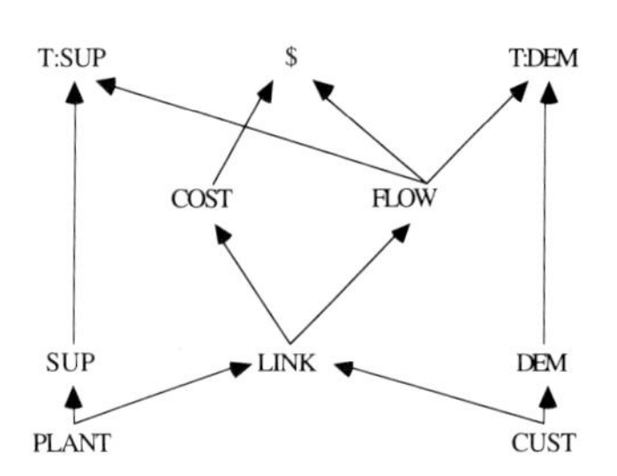

Figure 8.25

Several authors have developed prototype systems that allow modelers to draw genus and module graphs. Most of these systems add additional node and edge types to visually represent some of the relationships expressed in a Structured Model.

Sen and Chari [24], for example, extended structured modeling genus graphs by adding a few additional node types. They call this extended graph a *model graph.* The additional node types allow relational operators, logical primitives and propositions to be explicitly represented in the visual language. Below is an example model graph for the transportation problem. Compared to Figure 8.25, this graph contains three additional nodes, Supflow, CostXFlow and Demflow, which represent additional information about how the test and function attributes T:Sup, $, and T:Dem were calculated.

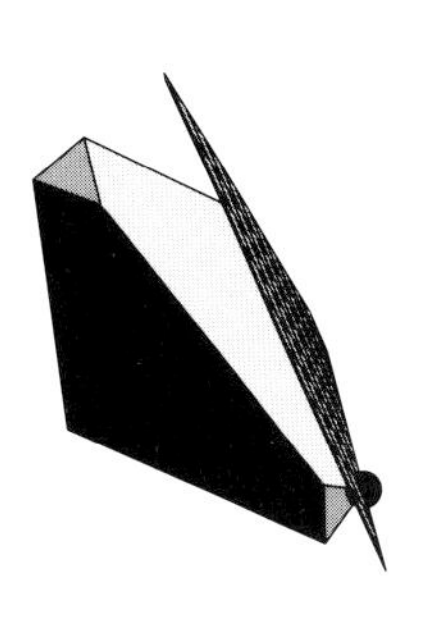

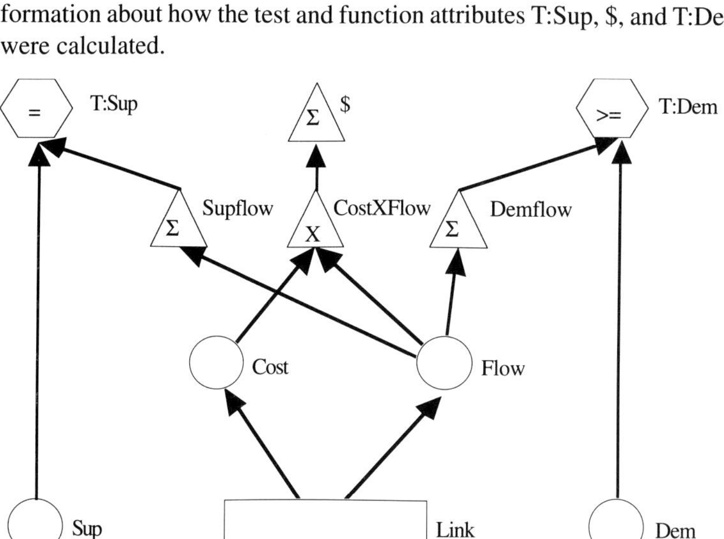

Figure 8.26

A legend identifying the node and edge types appears at right.

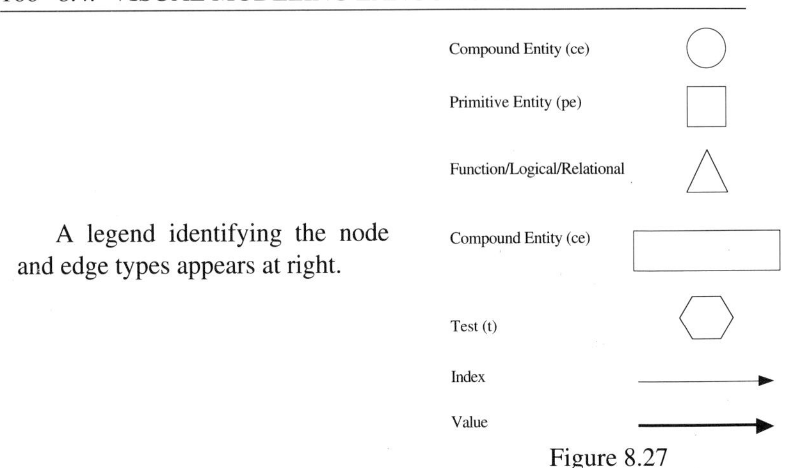

Figure 8.27

Hamacher [23] developed a system called IGOR (Integrated Graphical Modeling System for Operational Research). IGOR modifies structured modeling genus graphs based on entity relationship diagrams. The IGOR representation of the transportation problem appears below:

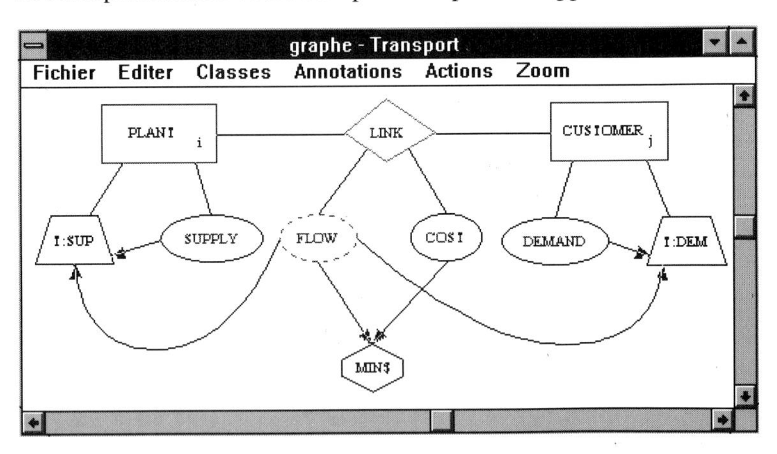

Figure 8.28

IGOR allows the user to filter the nodes displayed. Below, for example, only nodes of type Test and adjacent nodes are displayed:

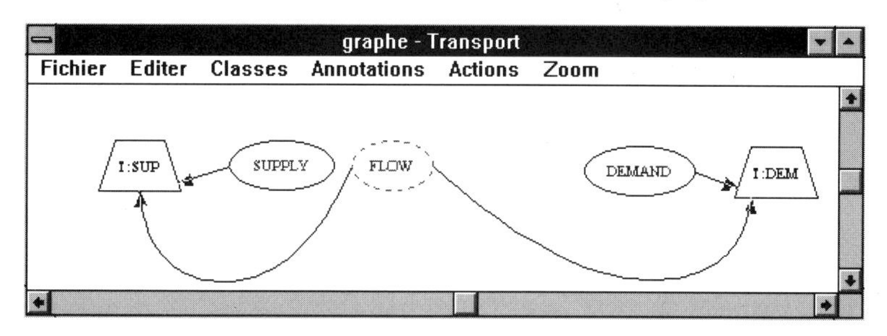

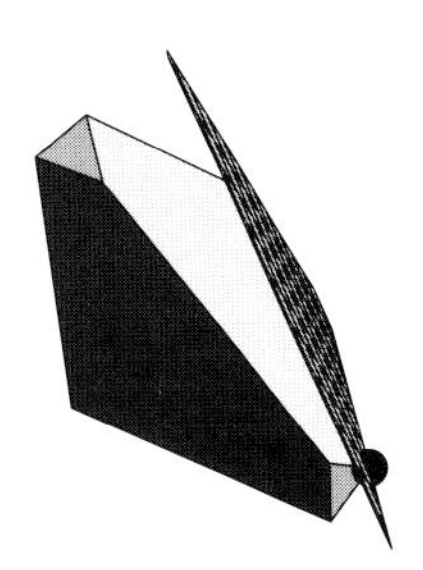

Figure 8.29

IGOR also allows nodes to be grouped into module nodes as shown below, where, the PLANT, I:SUP and PLANT genera are grouped:

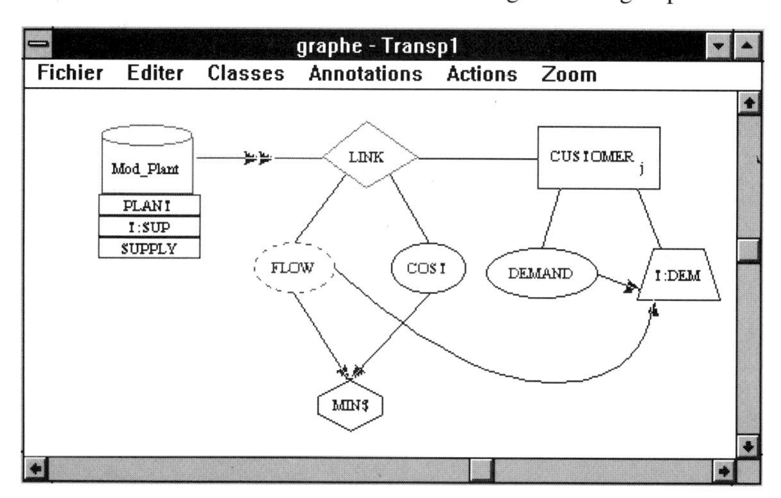

Figure 8.30

8.4.6 General Graph-Based Modeling

Given both the variety of proposals for graph-based languages for mathematical programming and the wide variety of other graph-based representations used for modeling such as project management and vehicle routing, instead of trying to invent the perfect graph-based language for all optimization problems, a more fruitful approach might seek to provide tools for working with a variety of different types of graphs. Many researchers have been pursuing this idea. We defer discussion of this work until Section 13.2.

8.4.7 Experimental Results

Although we are unaware of any researcher conducting detailed experimentation on the efficacy of any of the proposed visual languages for optimization modeling, experiments studying other visual languages have produced interesting result. Petre [40] conducted a series of experiments comparing visual languages such as electrical circuit diagrams and textual languages among both expert and novice users. Some of the results that she found include:

- "Graphics do not guarantee clarity." "Secondary notation" in visual languages such as the relative positioning of the nodes or the consistent drawing of similar parts play an important role in conveying meaning. The notation is called secondary because simply placing two nodes or icons near one another (though not having any explicit links) may represent some information about the relatedness of the nodes or icons.

  People often believe that visual languages provide greater freedom than textual languages, since in a visual language one can express ideas in two dimensions rather than the one linear dimension allowed by text. That freedom, it is believed, allows one to express ideas more easily. That freedom provided by visual languages can be an illusion, however, because it can lead to unstructured and confusing representations that are incomprehensible to others or even to the author over time.

- "Graphical readership is an acquired skill." Novices often performed better with textual representations than with visual representations, whereas experts often would perform better with visual as opposed to textual representations. Petre interpreted this as demonstration that graphics can be effective, but at least require substantial training. Moreover, experts are effective in their use of graphics because of their acquired skill at reading secondary notation; novices were less able to use the clues provided by the secondary notation. Nevertheless, graphical notations are often attractive to novices. To resolve this apparent contradiction, she recommends that graphical notations designed for novices rely on a limited set of constructs.

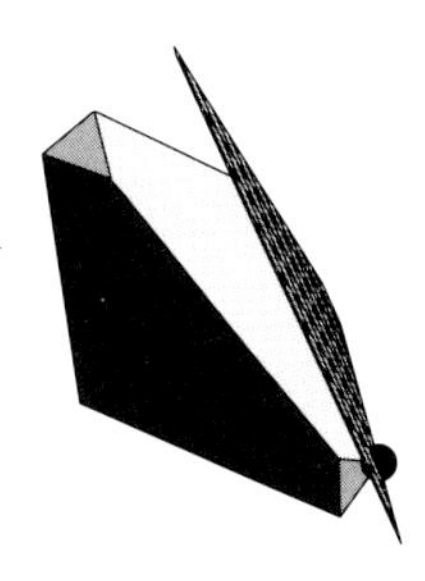

Among the visual languages for optimization modeling that have been proposed, many are targeted specifically to novice modelers, with the claim that traditional algebraic notation is often painfully opaque to them. Petre's results cast doubt on that assertion. Moreover, few of the proposed languages provide style guidelines for drawing models in their language. As Petre's results show, those guidelines can play a key, but unheralded role, in conveying meaning. Techniques to help draw graph-based languages are discussed in Chapter 13.

8.5 Direct Manipulation

To input graphical, such as the visual languages discussed in this chapter, one should use a *direct manipulation* interface. Direct manipulation interfaces have been defined [41], [42] as interfaces that increase the level of "semantic engagement" between the user and the computer. Rather than having to describe to the computer the desired action (for example, to delete a particular line of text), the user physically performs the operation directly. For graphics input, direct manipulation interfaces typically allow users to draw pictures of their problems using a mouse or stylus.

Angehrn [43] and Angehrn and Lüthi [44] proposed a different direct manipulation technique for formulating (and solving) models. Using their technique, known as *modeling by example*, modelers simply draw a picture of the current problem (actually, a graph).

Consider a telecommunications design problem, where the goal is to locate switching centers. a user would draw a picture of the existing network, assign attributes to the nodes and arcs in the network, and suggest locations for the switching centers.

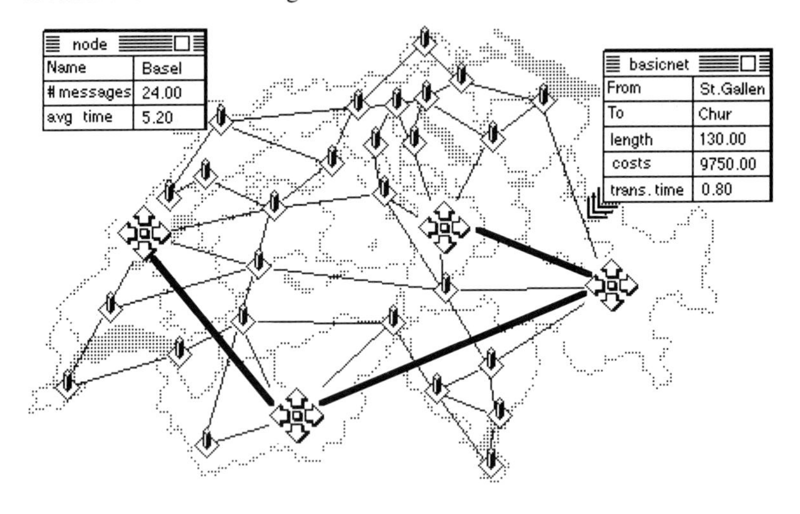

Figure 8.31

The system deduces the basic problem, selects and executes an algorithm, and then presents a solution, that is, a set of switching centers.

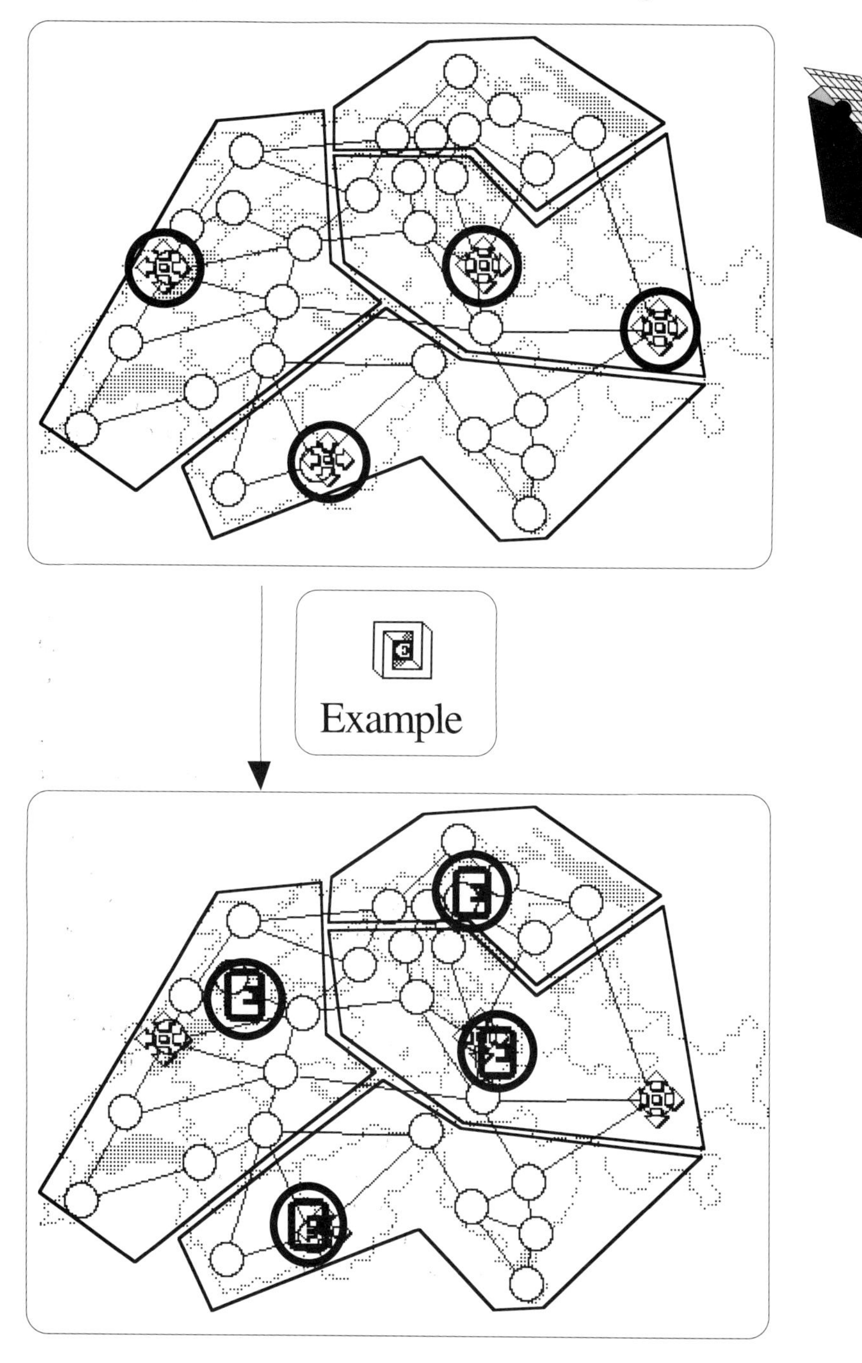

Figure 8.32

This technique seems similar to research in computer science on building user interfaces by example [45], [46], [47], [48]. In this line of research, a user interface builder draws the desired user interface, provides examples of how the user interface should respond to particular user actions such as moving the mouse, and the interface is then built automatically.

It is unclear how well modeling by example will work in actual practice. How can the modeler be certain that the inferred model accurately reflects the underlying problem? The modeler can of course simply look at the solution proposed by the current model. If the solution does not correspond to the actual problem, the modeler can indicate the source of the trouble, and the inference procedure can then be rerun to generate a new solution that *might* correspond better to the actual problem. There is no guarantee, however, that this iterative procedure will ever converge to an acceptable model yielding an acceptable solution. Modeling by example is an alluring idea that requires more research to establish its capabilities and its limits.

8.6 Summary

Given the variety of representation styles for formulations, the natural question arises: how do they compare? Great attention has been paid to this question by some authors [49], [50], [51], [37]. Although these comparisons have provided much food for thought, none of these representation formats are dominant. The authors of these representation formats almost certainly find their own format easiest to use. Many authors built the representation specifically for their own use.

Perhaps spreadsheets, given their widespread use, can be said to be the dominant representation format. Current spreadsheets, however, do not provide the sophisticated indexing abilities needed by mathematical programming models, nor do they provide a visual language for model building.

In short, if all of these representation formats have their admirers, it would seem prudent to allow users to model their problems in whatever form they find easiest. As stated by Greenberg and Murphy [51], in comparing three different modeling languages in detail:

> Many model builders from academic backgrounds think readily in algebra, while people with process industry backgrounds think in terms of processes [column-wise]...The next generation of modeling systems should be able to support all views. We found ourselves switching among the various lan-

guages with the same formulation when we wanted to learn different things about the model.

The challenge is to provide an ability to translate seamlessly from one form to another.

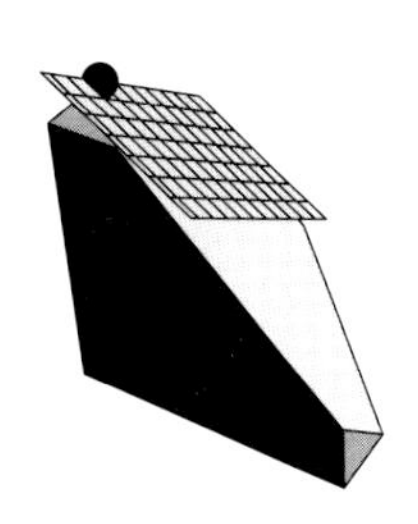

For visual languages in particular, although many have been proposed, none can be said to dominate. In fact, visual languages remain curiosities. Although this may be explained by their recent development, it may reflect fundamental problems with the idea of a visual language. Most optimization modelers were taught how to formulate problems using one particular representation style, often algebraic. After several courses and much practice, experienced optimization modelers find their favorite representation scheme quite natural. In short, optimization modelers are anchored to a few favorite types of representations for building their models.

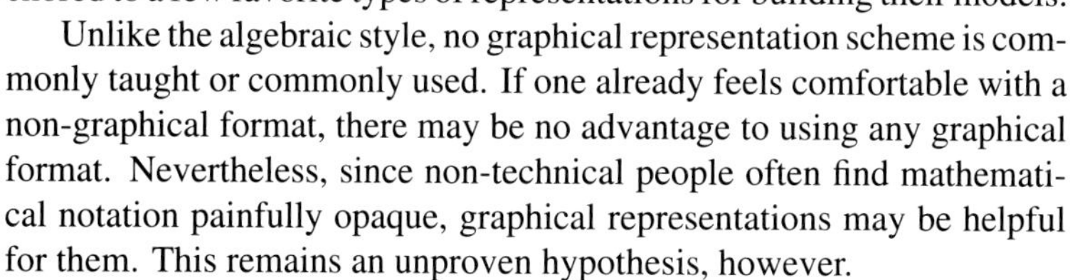

Unlike the algebraic style, no graphical representation scheme is commonly taught or commonly used. If one already feels comfortable with a non-graphical format, there may be no advantage to using any graphical format. Nevertheless, since non-technical people often find mathematical notation painfully opaque, graphical representations may be helpful for them. This remains an unproven hypothesis, however.

Bibliography

[1] Schrage L. *Linear, Integer and Quadratic Programming with LINDO*. Palo Alto (CA):Scientific Press, 1986.

[2] Fourer R, Gay DM, Kernighan BW. *AMPL: A Modeling Language for Mathematical Programming*. South San Francisco (CA):The Scientific Press, 1993.
URL: *http://www.iems.nwu.edu/ampl/ampl.html*

[3] Brooke A, Kendrick D, Meeraus A. *GAMS: A User's Guide, Release 2.25*. South San Francisco (CA):The Scientific Press, 1992.

[4] Maximal Software, Inc. *MPL Modeling System*. Arlington (VA):Maximal Software, Inc., 1994.
URL: *http://www.site.gmu.edu/~bjarnik*

[5] Hürliman T. LPL: A structured language for linear programming modeling. *OR Spectrum*, 1988;10:53–63.

[6] Geoffrion AM. The SML language for structured modeling. *Operations Research*, 1992;40(1):38–75.

[7] Greenberg HJ. MODLER: Modeling by object-driven linear elemental relations. *Annals of OR*, 1992;38:239–280.

[8] Greenberg HJ. *Modeling by Object-Driven Linear Elemental Relations: A User's Guide for MODLER*. Boston (MA):Kluwer, 1993.

[9] Chesapeake Decision Sciences. *MIMI/LP User's Manual*. New Providence (NJ):Chesapeake Decision Sciences, 1993.

[10] Welch JS. PAM—a practitioner's approach to modeling. *Management Science*, 1987;33(5):610–625.

[11] Piela P. *ASCEND: An Object-Oriented Computer Environment for Modeling and Analysis*. [dissertation], Pittsburgh (PA):Carnegie-Mellon University, 1989.

[12] Piela P, McKelvey R. An introduction to ASCEND: Its language and interactive environment. Technical report, Pittsburgh (PA):Engineering Design Research Center, Carnegie-Mellon University, 1992.

[13] Microsoft Corporation. *Microsoft Excel User's Guide Version 5.0*. Redmond (WA):Microsoft Corporation, 1994.

[14] Lotus Development Corporation. *Improv for Windows Release 2.0 Reference Manual*. Cambridge (MA):Lotus Development Corporation, 1993.

[15] Savage S. Lotus Improv as a modeling language. In *ORSA/TIMS Joint National Meeting*, San Francisco (CA):1992.

[16] Collaud G, Pasquier-Boltuck J. gLPS: A graphical tool for the definition and manipulation of linear problems. *European Journal of Operations Research*, 1994;72:277–286.

[17] Glover F, Klingman D, McMillan C. The Netform concept: A more effective model form and solution procedure for large scale nonlinear problems. Technical report, Springfield (VA):National Technical Information Service, US Department of Commerce, 1977.

[18] Steiger DM, Sharda R, LeClaire B. Graphical interfaces for network modeling: A model management system perspective. *ORSA Journal on Computing*, 1993;5(3):275–291.

[19] Ma PC, Murphy FH, Stohr EA. A graphics interface for linear programming. *Communications of the ACM*, 1989;32(8):996–1012.

[20] Choobineh J. A diagramming technique for representation of linear programming models. *Omega*, 1991;19(1):43–51.

[21] Geoffrion AM. Integrated modeling systems. *Computer Science in Economics and Management*, 1989;2:3–15.

[22] Geoffrion AM. The formal aspects of structured modeling. *Operations Research*, 1989;37(1):30–51.

[23] Hamacher S. A diagram representation for operations research problems. Technical Report 93-04A:Ecole Centrale Paris, 1993. *International Transactions on Operations Research*, forthcoming.

[24] Chari K, Sen T. A graphical approach to structured modeling: Model graphs and model instantiations. Technical report, Harrisonburg (VA):Department of Information and Decision Sciences, College of Business, James Madison University, 1993.

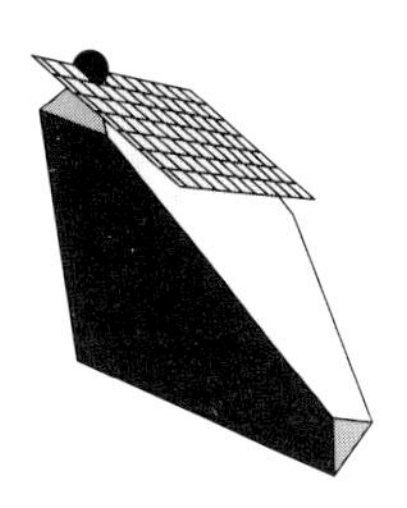

[25] Fourer R, Gay DM. Expressing special structures in an algebraic modeling language. Technical report, Evanston (IL):Department of Industrial Engineering and Management Sciences, Northwestern University, 1991.

[26] Greenberg HJ, Murphy FH. Views of mathematical programming models and their instances. *Decision Support Systems*, 1993;13(1):3–34.

[27] Bramley R, Loos T. EMILY: A visualization tool for large sparse matrices. Technical Report 412, Bloomington (IN) 47405-4101:Computer Science Department (IN) University, 1994.

[28] Hong SN, Mannino MV, Greenberg B. Measurement theoretic representation of large, diverse model bases: The unified modeling language: L_U. Technical report, Austin (TX):Department of Management Science and Information Systems, The University of Texas at Austin, 1991.

[29] Wegner P. Concepts and paradigms of object-oriented programming. *OOPS Messenger*, 1990;1(1):7–87.

[30] Primal Solutions, Inc. *Analytics User Manual.* Palo Alto (CA):Primal Solutions, Inc., 1994.

[31] Lindo Systems, Inc. *LINGO Optimization Modeling Language.* Chicago:Lindo Systems, Inc., 1992.

[32] Roberts DD. *The Existential Graphs of Charles S. Peirce.* The Hague:Mouton, 1973.

[33] Chang SK, editor. *Principles of Visual Language Systems.* Old Tappan (NJ):Prentice-Hall, 1990.

[34] Chang SK, Ichikawa T, Ligomenides PA. *Visual Languages.* New York:Plenum Press, 1986.

[35] Shu NC. *Visual Programming.* New York:Van Nostrand Reinhold, 1988.

[36] Conway RW, Maxwell WL, Worona SL. *User's Guide to XCELL Factory Modeling System.* Palo Alto (CA):The Scientific Press, 1986.

[37] Murphy FH, Stohr EA, Asthana A. Representation schemes for linear programming models. *Management Science*, 1992;38(7):964–991.

[38] Jones CV, Krishnan R. A visual, syntax-directed environment for automated model development. Technical report, Burnaby (BC), V5A 1S6 CANADA:Faculty of Business Administration, Simon Fraser University, 1992.

[39] Chen P. The entity-relationship model: Toward a unified view of data. *ACM Transactions on Database Systems*, 1976;1(1):9–36.

[40] Petre M. Why looking isn't always seeing: Readership skills and graphical programming. *Communications of the ACM*, 1995;38(6):33–44.

[41] Hutchins EL, Hollan JD, Norman DA. Direct manipulation interfaces. In Norman DA, Draper SW, editors, *User Centered System Design: New Perspectives on Human-Computer Interaction*, pages 87–124. Hillsdale (NJ):Lawrence Erlbaum, 1986.

[42] Shneiderman B. Direct manipulation: A step beyond programming languages. *IEEE Computer*, 1983;16(8):57–69.

[43] Angehrn AA. Modeling by example: A link between users, models and methods in dss. *European Journal of Operational Research*, 1992;55(3):296–308.

[44] Angehrn AA, Lüthi HJ. Intelligent decision support systems: A visual interactive approach. *Interfaces*, 1990;20(6):17–28.

[45] Maulsby DL, Witten IH, Kittlitz KA. Metamouse: Specifying graphical procedures by example. *Computer Graphics*, 1989;23(3):127–136.

[46] Myers BA. Visual programming, programming by example and program visualization; a taxonomy. *SIGCHI Bulletin*, 1986;17(4):59–66.

[47] Myers BA. *Creating User Interfaces by Demonstration*. Boston:Academic Press, 1988.

[48] Myers BA, Guise DA, Dannenberg RB, Vander Zanden B. Garnet: Comprehensive support for graphical, highly interactive user interfaces. *IEEE Computer*, 1990;23(11):71–85.

[49] Fourer R. Modeling languages versus matrix generators for linear programming. *ACM Transactions on Mathematical Software*, 1983;9:143–183.

[50] Geoffrion AM. Indexing in modeling languages for mathematical programming. *Management Science*, 1992;38(3):325–344.

[51] Greenberg HJ, Murphy FH. A comparison of mathematical programming modeling systems. *Annals of Operations Research*, 1992;38:177–238.

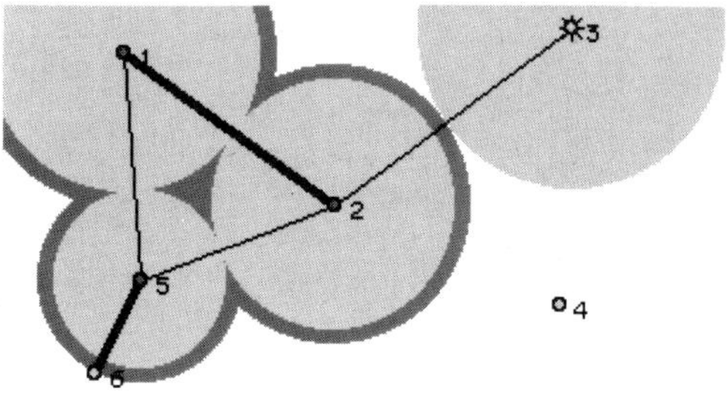

Chapter 9

Algorithm Execution

Once the conceptual model has been constructed, the model formulated, and the data collected, the algorithm must then be run. One might believe that the user's involvement only consists of typing the appropriate command to start the algorithm. However, when the algorithm does not converge or when the solution is claimed to be unbounded or infeasible, one often needs to explore the internal execution details of the algorithm. This again will require appropriate representations. These representations include human intervention during the solution process, visualizations to help understand the execution of the algorithm, as well as visualizations that provide new and novel insights.

9.1 Interactive Optimization

Fisher [1] proposed linking human perceptual abilities with a computer's computational speed to produce a hybrid form of optimization that he called *interactive optimization*. In this style of optimization, the user would intervene at particular points during the execution of the optimization algorithm and provide insights that the algorithm would find difficult to generate by itself. Brady, Rosenthal, and Young [2] provide an example of such a system applied to a facilities location problem. The algorithm drew various circular regions on the screen with the user having to identify a point in the intersection of the regions. The point chosen allows the system to further restrict the size of the region where the facility is to be located. The algorithm terminates when the region contains only a single point, that is, the optimal location of the facility. Another example comes from vehicle routing [3]. Users could specify new routes as starting points. The new routes then appear as new columns in the mathematical program.

9.2 Algorithm Animation

Another visualization technique has often been used to help algorithm designers understand the behavior of their algorithms. This visualization technique animates the internal execution of the algorithm, and has been dubbed *algorithm animation* [4], [5], [6], [7]. As an algorithm executes, a picture of the current state of the algorithm, perhaps showing the data structures, is updated when "interesting" events occur. Typical algorithm animations [8] for the simplex method, for example, show a polytope of either a three-variable linear programming problem, or a projection of a larger problem into three-dimensions (see Section 14.1.2). Each pivot of the algorithm is illustrated as a movement from one vertex to another along the polytope.

9.2.1 Algorithm Animation for Teaching

Algorithm animation has seen greatest success in teaching students about algorithms. For example, Edmond's "blossom" algorithm [9] for the minimum weighted matching problem is a classic, though complex, algorithm in combinatorial optimization. Jünger and Pulleyblank [10] developed an interesting graphical solution technique for which an algorithm animation system has been developed [11]. The minimum weighted matching problem can be formulated algebraically as:

- Given a set items of items N, $n = |N|$,with members i and j. Without loss of generality, assume that n is even. If n is odd, one can merely add an additional item such that $c_{n+1,i} = 0$ for all i.
- Let $x_{ij} = 1$, $i \neq j$, if i is matched to j and equals 0 otherwise.
- Let $c_{ij} = c_{ji}$ be the cost of matching i to j.

$$\begin{aligned} \min \quad & \sum_{ij} c_{ij} x_{ij} \\ \text{subject to:} \quad & \\ & \sum_{i \neq j} x_{ij} = 1 \quad \forall i \\ & x_{ij} \in \{0, 1\} \end{aligned} \tag{9.1}$$

One of course can view the set of items $N = \{1, \ldots, n\}$ as nodes in a complete graph, and a match between item i and j as selecting the edge connecting nodes i and j.

Edmond's algorithm is based on the observation that for any subset S of the n nodes, where $3 \leq |S| \leq n-3$ and $|S|$ is odd, then there must be one match between a member of S and $N \setminus S$. Each such S is called a *blossom*. A blossom constraint can be written as:

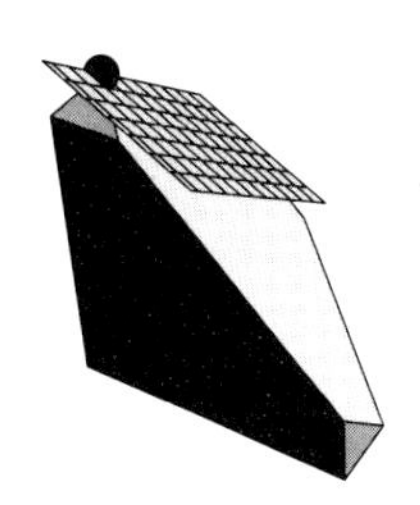

$$\sum_{i \in S;\, j \in N \setminus S} x_{ij} \geq 1 \quad \forall S \in \mathcal{S} \tag{9.2}$$

where $\mathcal{S}$ is defined as the set of all subsets $S \subseteq N$ where $3 \leq |S| \leq |N| - 3$ and $|S|$ is odd.

One can show that these constraints are in fact facets for the convex hull of integer solutions, meaning that a simple linear programming algorithm could be used to solve the problem. Edmond's actual algorithm is a complicated primal-dual approach that is usually taught in graduate level seminars (although complicated, Edmond's algorithm runs in polynomial time). The algorithm dynamically adds and removes constraints of the form (9.2).

Since Edmond's algorithm relies on a primal-dual approach, it is helpful to consider the dual to (9.1):

- Let π_i be the dual variable for node i.
- Let δ_S be the dual variable associated with the constraint on blossom S.

$$\begin{array}{llcll} \max & \displaystyle\sum_{i \in N} \pi_i + \sum_{S \in \mathcal{S}} \delta_S & & & \\ \text{subject to:} & & & & \\ & \pi_i + \pi_j + \displaystyle\sum_{S \in \mathcal{S};\, i \in S;\, j \in N \setminus S} \delta_S & \leq & c_{ij} & \forall i, \forall j \neq i \\ & \delta_S & \geq & 0 & \forall S \in \mathcal{S} \end{array} \tag{9.3}$$

Note that there is a constraint in the dual for each edge in the underlying graph.

Edmond's algorithm alternates between adjusting the primal and dual values of the problem. Instead of presenting the algorithm in detail, we present a series of snapshots from an algorithm animation developed by Goddyn [11] based on an idea first proposed by Jünger and Pulleyblank [12], [10].

The animation assumes that the nodes are positioned in Euclidean space. Consider the 6 node problem below.

Figure 9.1

The algorithm first chooses a node (in the example, node 1) and then increases the value of its dual variable, π_i. Graphically, the dual variable can be represented by the radius of a circle centered at the node. In the animation, the circle grows until it collides with another node, in this case, node 5. At this point, the dual constraint $\pi_1 + \pi_5 \leq c_{15}$ is satisfied, since $\pi_1 = c_{15}$.

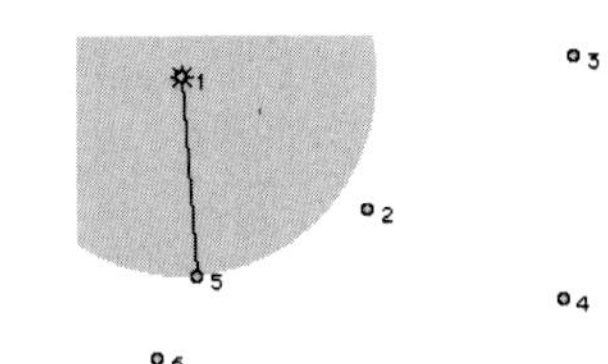

Figure 9.2

Node 1 is now matched to node 5, as indicated by the thicker black line at right.

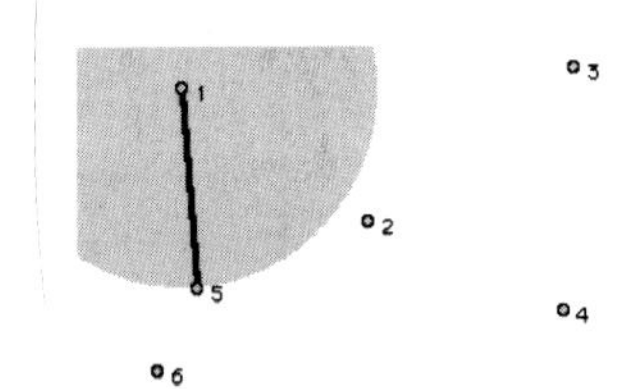

Figure 9.3

The algorithm then chooses a node with no match and increases its dual value. In the example, at right, the algorithm chooses node 2. The dual value, again represented by a circle, is increased until it collides with either another node or another circle. In the example, the dual value collides with the circle centered at node 1. This collision corresponds to the constraint $\pi_1 + \pi_2 \leq c_{12}$. The algorithm then pivots by increasing the value of π_2, decreasing the value of π_1, and increasing the value of π_5. Graphically, one sees the the circle 2 increasing, circle 1 decreasing and circle 5 increasing in size.

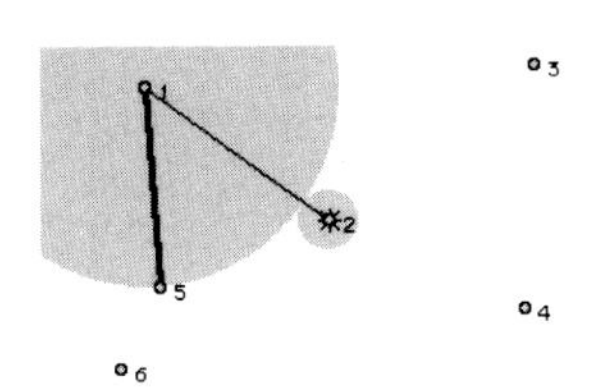

Figure 9.4

Eventually, circle 2 and circle 5 collide. At this point, three constraints in the dual are satisfied at equality, $\pi_1 + \pi_2 \leq c_{12}$, $\pi_1 + \pi_5 \leq c_{15}$, and $\pi_2 + \pi_5 \leq c_{25}$. These three constraints identify the blossom $S_1 = \{1, 2, 5\}$.

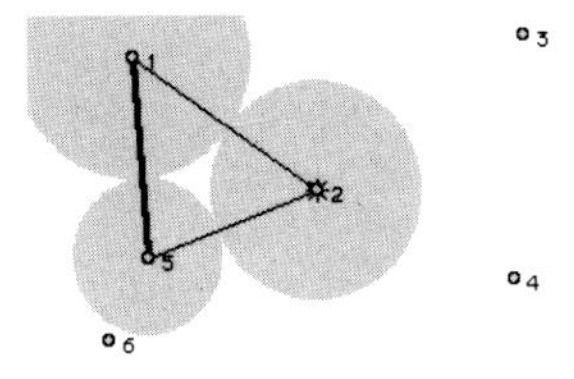

Figure 9.5

At this point, the dual variable, $\delta_{\{1,2,5\}}$ is increased until one of the constraints in which it participates becomes tight. Graphically, this is represented as a *moat* surrounding each of the circles centered at $\{1, 2, 5\}$. The moat grows until it collides with an unmatched node or with another circle. In this case, it collides with node 6. This means that $\pi_5 + \delta_{\{1,2,5\}} = c_{16}$. In other words, the dual constraint associated with edge $(5, 6)$ is satisfied at equality.

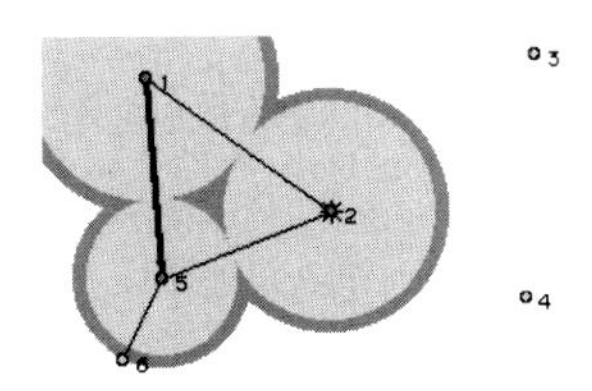

Figure 9.6

One can now increase the number of matched edges from 1 to 2, as shown in the figure at right.

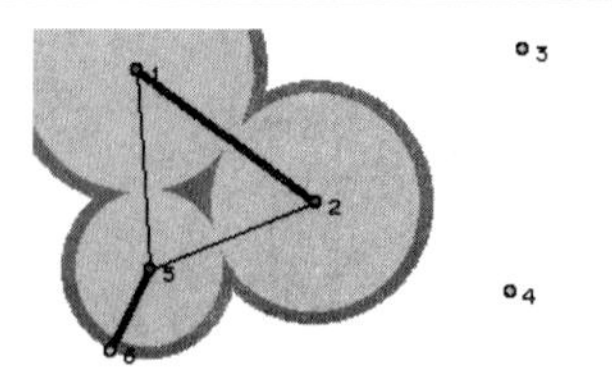

Figure 9.7

The algorithm then chooses another unmatched node (node 3), and increases its corresponding dual value, π_3. Graphically, circle 3 becomes larger. Circle 3 stops growing when it collides with the moat representing $\delta_{\{1,2,5\}}$. This corresponds to the dual constraint associated with edge $(3, 2)$ becoming tight, i.e., $\pi_2 + \pi_3 + \delta_{\{1,2,5\}} = c_{23}$.

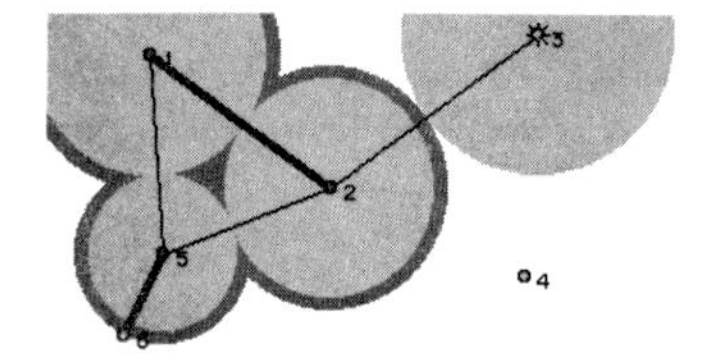

Figure 9.8

In the next step, blossom $\{1, 2, 5\}$ is shrunk, i.e., $\delta_{\{1,2,5\}}$ is set to 0. Dual values π_1, π_2, π_5 and most significantly, π_6 are adjusted in order to keep the relevant dual constraints satisfied (for edges $(1, 2)$, $(2, 3)$, $(2, 5)$, $(1, 5)$, and $(5, 6)$). At this point, however, the dual constraint associated with edge $(2, 5)$ is just about to become non-binding.

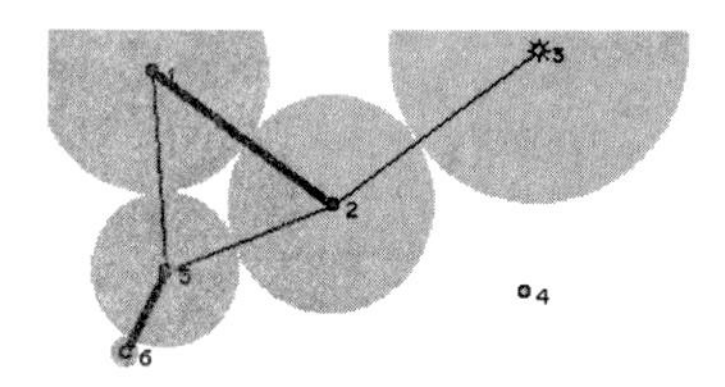

Figure 9.9

As π_3 is increased, π_2 is decreased, π_1 is increased, π_5 is decreased, and π_6 is increased, eventually, circle 3 collides with circle 1. This means that $\pi_1 + \pi_3 = c_{13}$, or equivalently, that the dual constraint associated with edge $(1, 3)$ is tight. This identifies the blossom $\{1, 2, 3\}$.

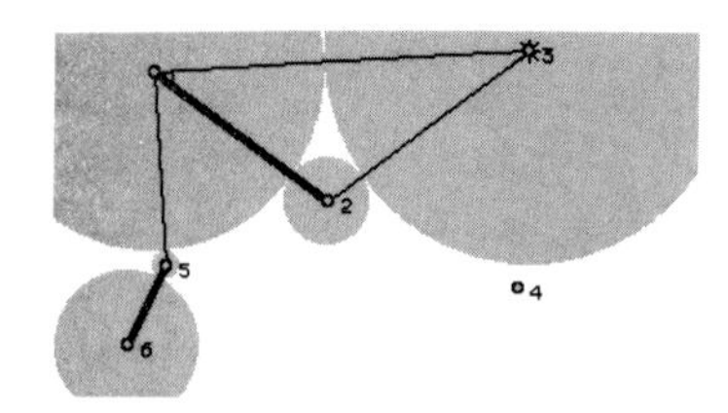

Figure 9.10

Since the primal constraint for blossom $\{1,2,3\}$ has become tight, the associated dual variable, $\delta_{\{1,2,3\}}$ becomes positive. This is represented by the moat surrounding the circles centered on nodes 1, 2 and 3. The moat is increased until it collides with the circle surrounding node 6. At this point $\pi_1 + \pi_6 + \delta_{\{1,2,3\}} = c_{16}$, or equivalently, the dual constraint associated with edge $(1,6)$ is tight. This then identifies the blossom $\{1,2,3,5,6\}$.

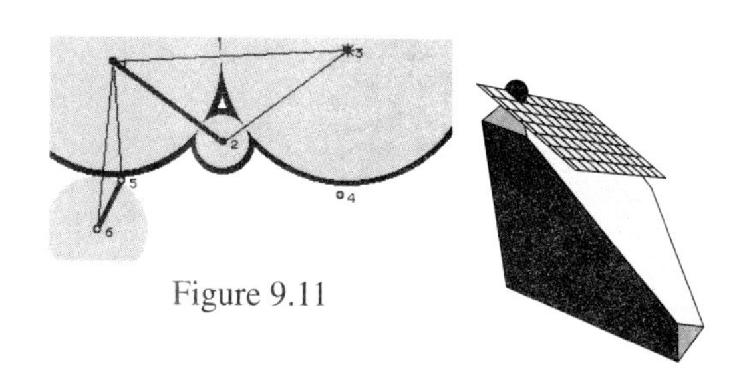

Figure 9.11

$\delta_{\{1,2,3,5,6\}}$ is now increased, which is represented at right by a moat. The moat increases in size until it collides with another circle or an unmatched node. In this case, the moat collides with node 4.

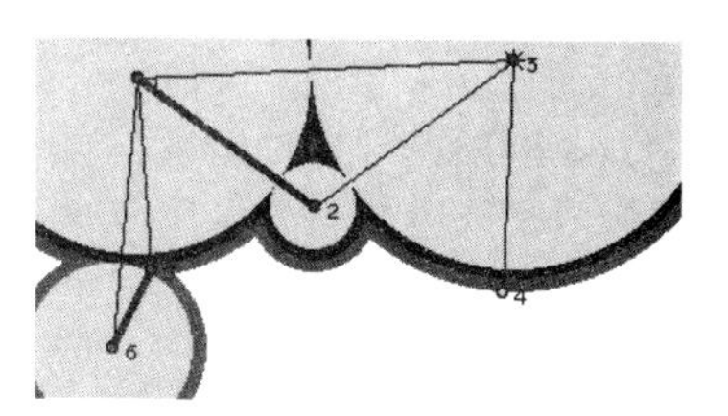

Figure 9.12

A simple augmentation occurs, since the blossom root (node 3) is unmatched. At this point, all nodes are matched. The final, minimum weighted matching appears below.

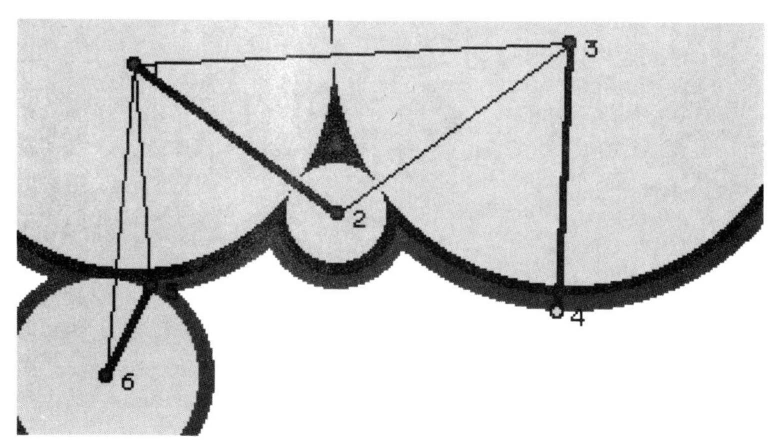

Figure 9.13

Why does this work? Note that the length of the matching is equal to the sum of the radii of the circles and the sum of the radii of the moats. In general, because of duality, the sum of the radii of the circles and the moats forms a lower bound on the minimum weight matching.

Although the exact details of the algorithm have not been presented (see [10], [12]), the algorithm animation gives one a sense of how the algorithm operates. This graphical representation has provided insights to help develop algorithms for spanning trees, Steiner trees and the Euclidean Traveling Salesman problem [12].

9.2.2 Algorithm Animation for Research

Algorithm animation is not limited to just teaching students about existing algorithms. It can help provide better insights to algorithm researchers. For example, Shannon, MacCuish and Johnson [13] used algorithm animation to study *push-relabel* algorithms [14], [15] for the maximum network flow problem.

Briefly, push-relabel algorithms relax the flow conservation constraints at each node, so that the flow into a node can temporarily be larger than the flow out of a node. Nodes with excess flow are called *active*. At any point in the algorithm, a subset of the edges will have excess capacity. The network consisting solely of edges with excess capacity is called the *residual network*. Associated with each node n is a label $d(n)$ which gives a lower bound on the distance from n to the sink in the residual network or to the source if no such path exists.

The algorithm works by carefully choosing an active node u and then *pushing* flow from the node over some attached edge (u, v) that has excess capacity where $d(u) = d(v) + 1$. Such edges are termed *admissible*. If node u is active, but has no admissible edges, it is relabeled. In particular, its label, $d(u)$, becomes $\min(d(v)) + 1$ over all nodes v such that edge (u, v) has excess capacity. This then produces at least one admissible edge. Pushing and relabeling a node until it is no longer active is called *discharging* a vertex.

Various schemes have been proposed for choosing the next node to discharge, as well as the admissible edge over which to push flow. Periodically generating more accurate labels for the nodes using a variety of schemes has also been proposed. Shannon, MacCuish and Johnson used their algorithm animation as a tool to study the performance of these different schemes.

In the algorithm animation (Figure 9.15), nodes and edges are colored as shown in the table below (the colors in the original image are listed, as well as the colors used in Figure 9.15)

	Original Color	*Color in Figure*	*Meaning*
Nodes	green	white	active with admissible edges
	blue	black	active with no admissible edges
	grey	grey	not active
Edges	green	white	admissible
	grey	grey	no flow
	red	black	saturated

Figure 9.14

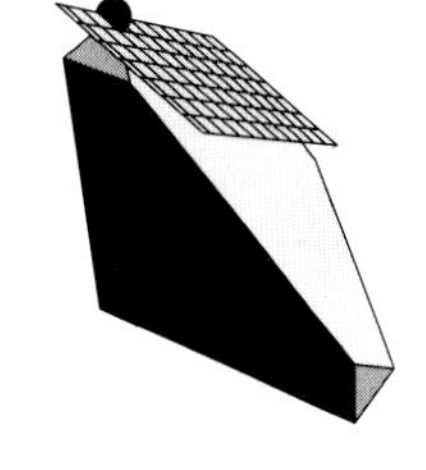

In the figure below, the height of each node u is proportional to $d(u)$, the lower bound on the distance from u to the sink in the residual network. This visually shows the flow moving downwards from source to sink. From the animation, the authors were able to identify a key reason why these algorithms sometimes perform poorly. Frequently, the algorithm spends a great deal of time merely pushing flow around a cycle in the network or dealing with a local cut. For example, once the global minimum cut has been established, the algorithms are often observed pushing flow around a cycle on the source side of the cut. Algorithms that used a relabeling scheme demonstrated this behavior less, but did not eliminate it entirely.

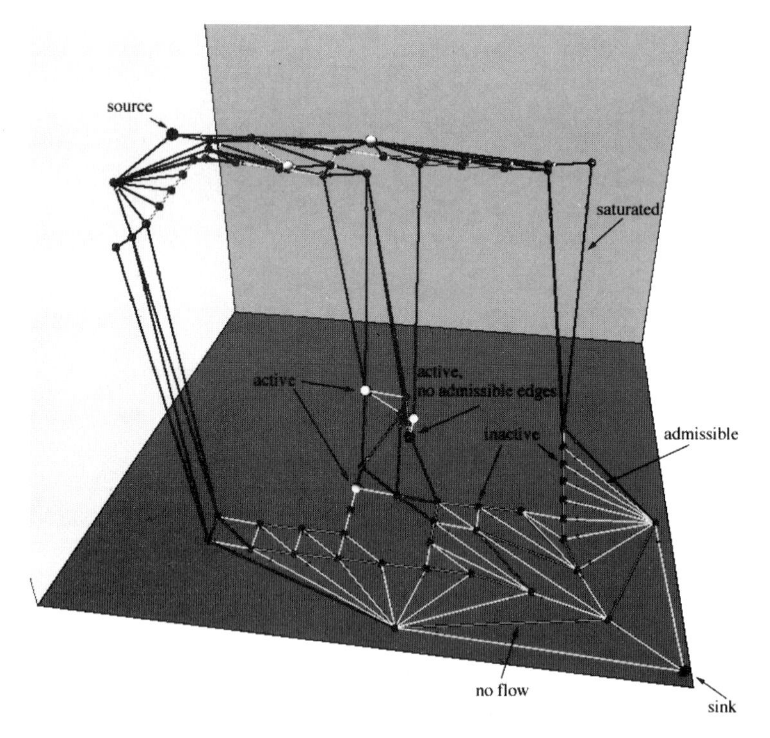

Figure 9.15

Other examples of algorithm animation for research studied how interior point method [16] for linear programming iteratively distorts the polytope (see Chapter 14). For an example from combinatorial optimization, Boyd, Pulleyblank, and Cornuejols [17] animated several different algorithms for the traveling salesman problem.

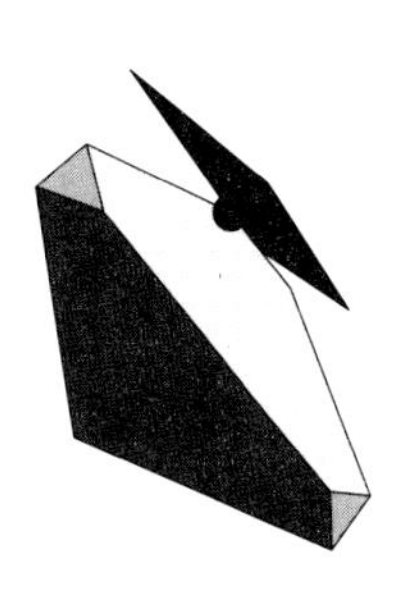

9.2.3 Toolkits for Algorithm Animation

If one wants to animate an algorithm, at least two algorithm animation toolkits are in the public domain, Zeus [6] and XTANGO [7], although they currently only run on Unix platforms. The toolkits assume that an algorithm designer has already implemented the algorithm in some traditional programming language. The toolkit then provides a set of subroutines that allows the author to annotate the algorithm at appropriate points, called *interesting events* in Zeus. At each one of the interesting events, the algorithm animation toolkit can update the display, to reflect the change in state of the algorithm.

Below, for example, is an algorithm animation for maximum flow in a network developed used Zeus [18].

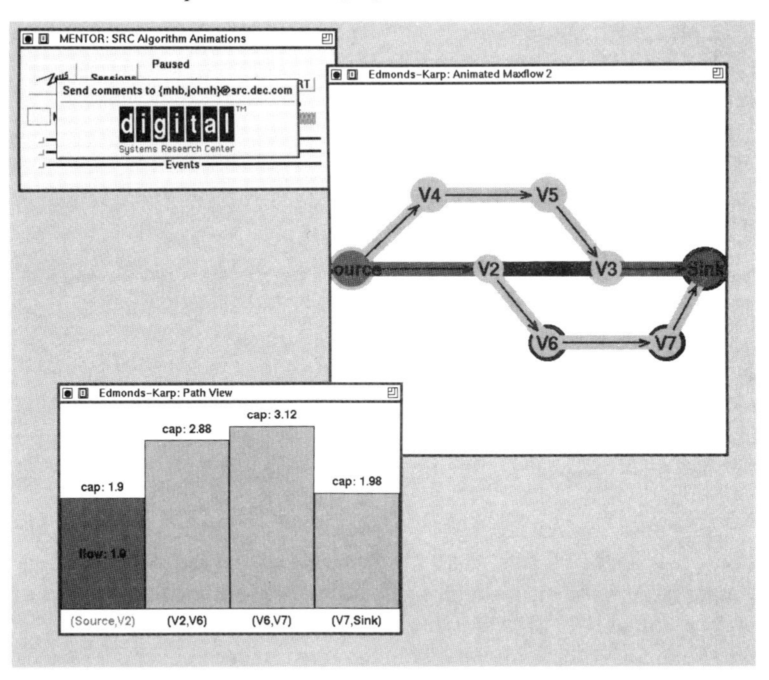

Figure 9.16

9.2.4 Algorithm Animation and Simulation

Algorithm animation is essentially the same as animation typically found in discrete event simulation [19], [20]. In such an animation, a picture of the underlying problem changes at interesting events in (simulated) time. Although most authors laud the ability of animation to debug and validate the simulation model, it does not remove the need for careful statistical analysis [19], [21], since a single animation may not include examples of important behavior.

Visual Interactive Modeling (VIM) discussed in Chapter 7 actually saw its principal first application as Visual Interactive Simulation (VIS). The VIS community makes pains to emphasize the difference between interactive animation and non-interactive animation. In an interactive animation, the model and data can be changed at any time, with the behavior of the model then observed. In non-interactive animation, the user cannot change the model or model data while the animation is proceeding — the animation merely represents the playback of a recorded "movie." For optimization algorithms, especially for large problems, playback of a movie may be all that is possible, since the time between significant events can be large. For example, if one is animating a linear programming problem for a large problem, each iteration may require several minutes (or longer), which is far too slow for real-time animation. One may have no alternative but to store a sequence of images, one for each iteration, for later playback after the algorithm is run.

9.3 Algorithm Theory

Visualization of algorithms can lead to new problems and new theory. Hubbard (as described in [22]) visualized the behavior of Newton's method for finding roots of nonlinear functions. He was interested in how the solution that resulted depended on the starting point. When he color coded each starting point based on the final solution that was produced, the image that resulted was a fractal. This example illustrates the power of visualization to provide new insights into the behavior of even well-established algorithms.

In another example related to optimization, conventional wisdom has stated that optimal solutions to combinatorial optimization problems are generally more sensitive to changes in the input data than are approximate solutions. As Bartholdi and Platzman [23] wrote about the traveling salesman problem, "optimal solutions are fragile in the sense that they can be exquisitely sensitive to changes in the data." For the planar case, Jones [24] considered how the tour changed when a single city is moved. If the city moves far enough in the correct direction, the tour will eventually change. So, for any tour, there is a region in which the tour will not change, as long as the moving city stays inside the region.

Different solution algorithms (optimal and approximate) can produce different sets of regions. By drawing the regions in which the tour remains unchanged, some insight can be gained on the behavior of different algorithms. In particular, as discussed in [24], most of the classic approximation algorithms, with some notable exceptions, produce less stable solutions than optimization algorithms.

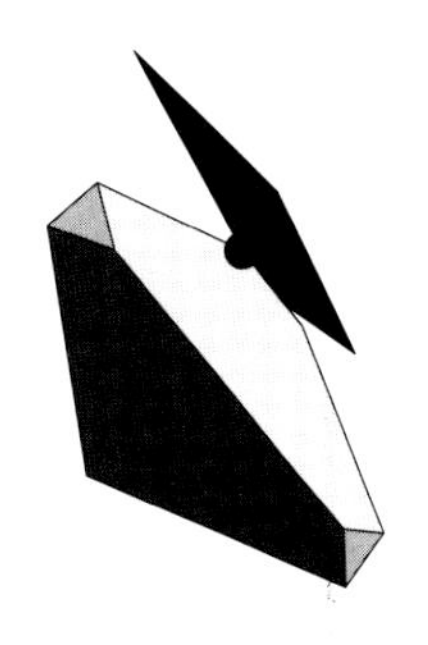

The figure below is associated with an optimization algorithm for the traveling salesman problem. Each region in the figure represents an area where the tour computed by the algorithm does not change, as long as city 1 stays inside the region.

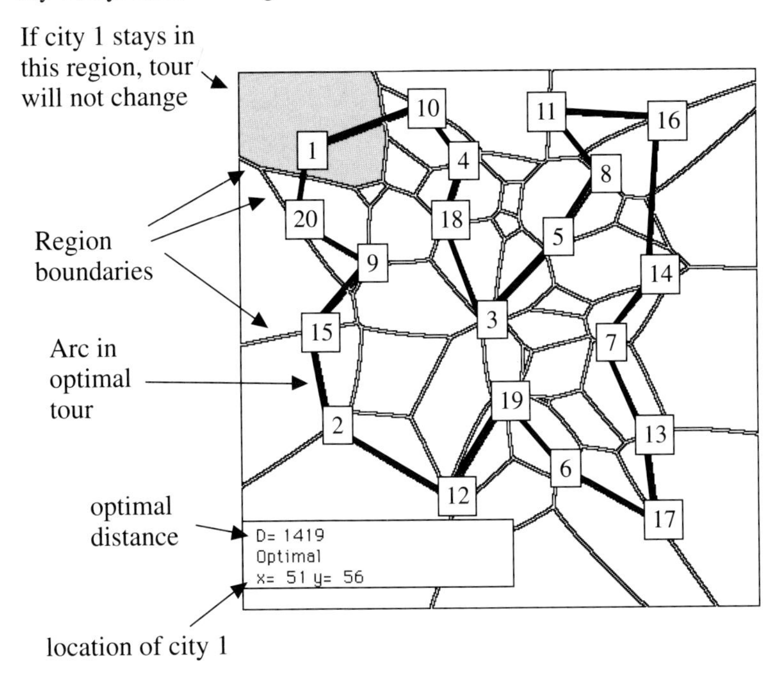

Figure 9.17

The figure below is associated with an approximation algorithm, FARTHEST INSERTION + THREEOPT, for the traveling salesman problem. Note that the regions for the approximation algorithm are far more complex than those for the optimization algorithm. It was only through the visualization that this result was uncovered.

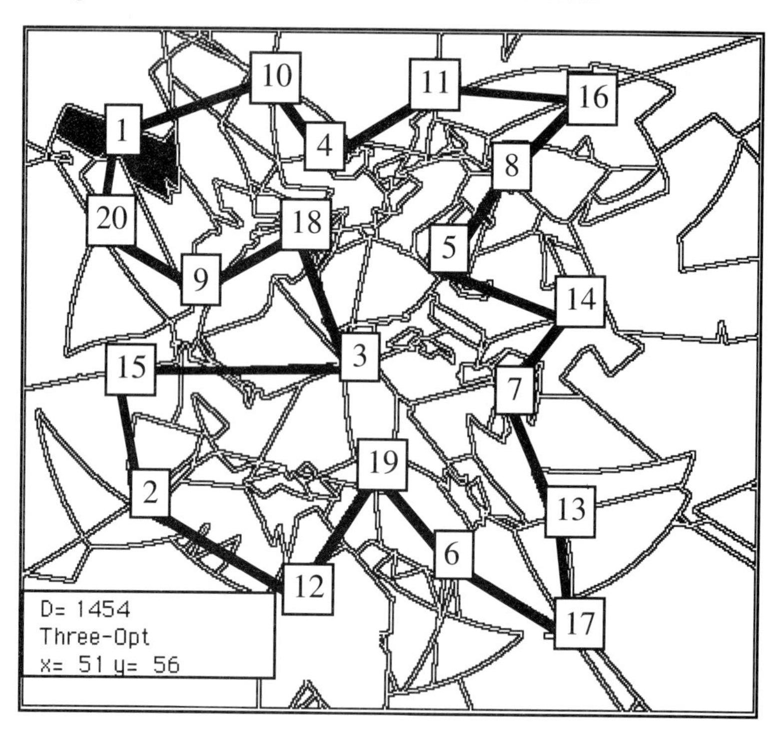

Figure 9.18

9.4 Summary

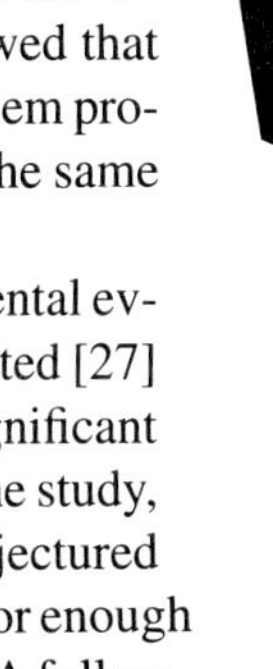

Although interactive optimization has often been used, is it useful? No definite answer can yet be given, although some results are beginning to appear. In an experimental comparison of interactive solutions produced by optimization algorithms for the planar traveling salesman problem, Mak, Srikanth, and Morton [25] showed that humans could produce better solutions than even the best heuristic algorithms. Taseen [26] showed that his particular example of the use of VIM for an optimization problem produced consistently better answers than a non-graphic version of the same problem.

As for algorithm animation, although widely touted, experimental evidence has been sorely lacking. In one study that has been conducted [27] actually found that algorithm animation does *not* provide any significant aid to understanding the behavior of algorithms. The authors of the study, who developed a major algorithm animation environment, conjectured that their study did not provide sufficient motivation to subjects, nor enough guidance as to what the subjects should look for in the animation. A follow-up study [28] showed that students learned more when they created their own algorithm animations rather than merely viewing already prepared animations.

The less than positive results may indicate that algorithm animation does not live up to the hype of its promoters, and thus should be abandoned. The wide use of the technique, however, suggests that it does have some useful applications. The challenge is to discover exactly how it is best delivered, to which audiences, and for what tasks.

Bibliography

[1] Fisher ML. Interactive optimization. *Annals of Operations Research*, 1986;5:541–556.

[2] Brady SD, Rosenthal RE, Young D. Interactive graphical minimax location of multiple facilities with general constraints. *AIIE Transactions*, 1984;15(3):242–254.

[3] Cullen FH, Jarvis JJ, Ratliff HD. Set partitioning based heuristics for interactive routing. *Networks*, 1981;pages 125–143.

[4] Bentley JL, Kernighan BW. A system for algorithm animation: Tutorial and user manual. Technical Report 123, Murray Hill (NJ):AT&T Bell Laboratories, Computer Science, 1987.

[5] Brown MH. *Algorithm Animation*. Cambridge (MA):MIT Press, 1988.

[6] Brown MH. Zeus: A system for algorithm animation and multi-view editing. Technical Report 75, Palo Alto (CA):Digital Systems Research Center, Digital Equipment Corporation, 1992.
URL: *http://www.research.digital.com/SRC/zeus/home.html*

[7] Stasko J. A practical animation language for software development. In *Proceedings of the IEEE International Conference on Computer Languages*, pages 1–10, Los Alamitos (CA):1990. IEEE Computer Society Press.
URL: *http://www.cc.gatech.edu/gvu/softviz/algoanim/algoanim.html*

[8] Lustig I. Applications of interactive computer graphics to linear programming. In Sharda R, Golden BL, Wasil E, Balci O, Stewart W, editors, *Proceedings of the Conference on Impact of Recent Computer Advances in Operations Research*, pages 183–189, New York:1989. North-Holland.

[9] Edmonds J. Paths, trees and flowers. *Canadian Journal of Mathematics*, 1965;17:449–457.

[10] Jünger M, Pulleyblank W. New primal and dual matching heuristics. *Algorithmica*, 1995;13:357–380.

[11] Goddyn LA. Visual algorithms, program for Macintosh and X-window, 1994.
URL: *mailto://goddyn@sfu.ca*

[12] Jünger M, Pulleyblank W. Geometric duality and combinatorial optimization. In Chatterji SD, Fuchssteiner B, Kulisch U, Liedl R, editors, *Jahrbuch Überblicke Mathematik 1993*. Braunschweig/Wiesbaden, Germany:Friedr. Vieweg & Sohn Verlagsgesellschaft mbH, 1993.

[13] Shannon GE, MacCuish J, Johnson E. *A Case Study in Algorithm Animation: Maximum Flow Algorithms*, volume 12, pages 97–118:American Mathematical Society, 1993.
URL: *http://cica.indiana.edu/projects/Graph.Maxflow/index.html*

[14] Cormen T, Leiserson C, Rivest R. *Introduction to Algorithms*:McGraw-Hill, 1990.

[15] Goldberg A, Tarjan R. A new approach to the maximum flow problem. *Journal of the ACM*, 1988;35(4):921–940.

[16] Gay DM. Pictures of Karmarkar's linear programming algorithm. Technical Report 136, Murray Hill (NJ):AT&T Bell Laboratories, Computer Science, 1987.

[17] Boyd SC, Pulleyblank WR, Cornuejols G. Travel–an interactive traveling salesman problem package for the IBM personal computer. *OR Letters*, 1987;6(3):141–143.

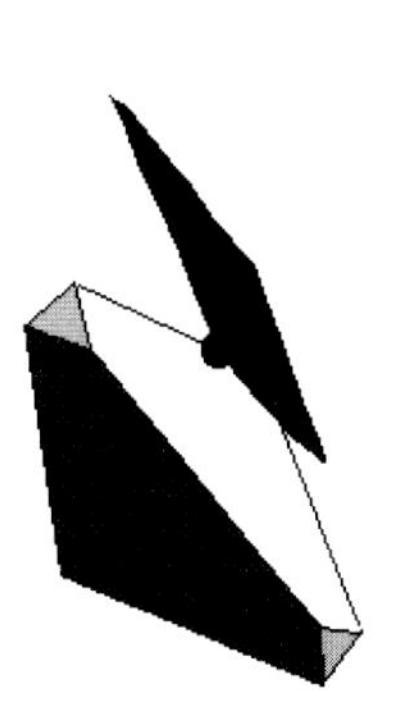

[18] Brown MH. The 1992 SRC algorithm animation festival. Technical Report 98, Palo Alto (CA):Digital Systems Research Center, Digital Equipment Corporation, 1993.
URL: *http://gatekeeper.dec.com/pub/DEC/SRC/research-reports/abstracts/src-rr-098.html*

[19] Bell PC, O'Keefe RM. Visual interactive simulation—history, recent developments, and major issues. *Simulation*, 1987;49(3):109–116.

[20] Grant JW, Weiner SA. Factors to consider in choosing a graphically animated simulation system. *Industrial Engineering*, 1986;18:37ff.

[21] Law AM, Kelton WD. *Simulation Modeling and Analysis.* New York:McGraw-Hill, 2nd edition, 1991.

[22] Gleick J. *Chaos: Making a New Science.* New York:Viking, 1988.

[23] Bartholdi JJ, Platzman LK. Heuristics based on spacefilling curves for combinatorial problems in Euclidean space. *Management Science*, 1988;34(3):291–305.

[24] Jones CV. The stability of solutions to the Euclidean traveling salesman problem. Technical report, Burnaby (BC), V5A 1S6 CANADA:Faculty of Business Administration, Simon Fraser University, 1992.

[25] Mak KT, Srikanth K, Morton A. Visualization of routing problems. Technical Report CRIM 90-03, Chicago:Department of Information and Decision Sciences, College of Business Administration, University of Illinois at Chicago, 1990.

[26] Taseen AA. *Visual Interactive Linear Programming: The Concept, an Example and an Empirical Assessment of its Value in Supporting Managerial Decision Making.* [dissertation], London, Ontario, Canada:University of Western Ontario, 1993.

[27] Stasko J, Badre A, Lewis C. Do algorithm animations assist learning? an empirical study and analysis. In *Human Factors in Computing Systems, INTERCHI'93 Conference Proceedings, Conference on Human Factors in Computing Systems, INTERACT '93 and CHI '93*, pages 61–66, Amsterdam, The Netherlands:April 1993.

[28] Lawrence A, Badre A, Stasko J. Empirically evaluating the use of animations to teach algorithms. In *Proceedings of the 1994 IEEE Symposium on Visual Languages*, pages 48–54, St. Louis, MO:October 1994.

Chapter 10

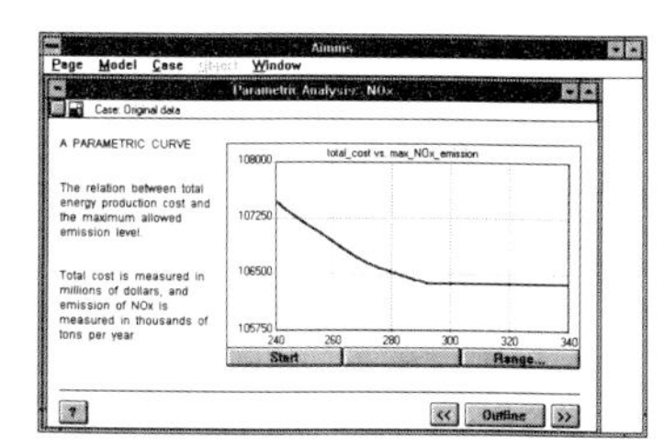

Solution Analysis

This section discusses representations useful for understanding the solutions produced by optimization algorithms. The first part of the chapter discusses sensitivity analysis, a well-established technique in optimization, to understand how solutions will change if the input data changes (Section 10.1). The second part of this chapter discusses representations suitable for particular application domains such as production planning and scheduling, as well as location and routing problems (Sections 10.3-10.5).

10.1 Sensitivity Analysis

Once an algorithm is run, a solution is produced. For linear programming problems, information useful in sensitivity analysis is also produced, including dual prices and ranges for the objective function coefficients and constraints.

In many cases, however, the solution is not correct. It may be that the problem is infeasible or unbounded. The algorithm may fail to converge. For nonlinear programming problems, although a local optimum may have been found, it may not be the global optimum.

The causes of these problems are legion: the input data can contain errors, constraints may be missing, or extraneous constraints may have been included. For nonlinear programs, the initial solution may not have been the best. For integer programs, adjustments to the branching method, the bounding method, or other parameters may need to be performed. For most problems, one is interested not just in an individual solution but in its behavior over a set of possible input parameters or over a range of possible cases.

In short, the completion of the algorithm represents the beginning of a long process of debugging and validation—understanding—the problem. Much research has been conducted to provide better analyses of solutions [1], [2], [3], [4], [5] and to diagnose infeasibilities [6], [7]. In this section, we discuss some of the visualization techniques that have been applied.

Several different types of plots are frequently used in sensitivity analysis. For example, plots of the value of the objective as a function of an input parameter (below) are commonly seen in textbooks, but are just emerging in commercial systems [8], [9]. The figure below was generated using AIMMS [9].

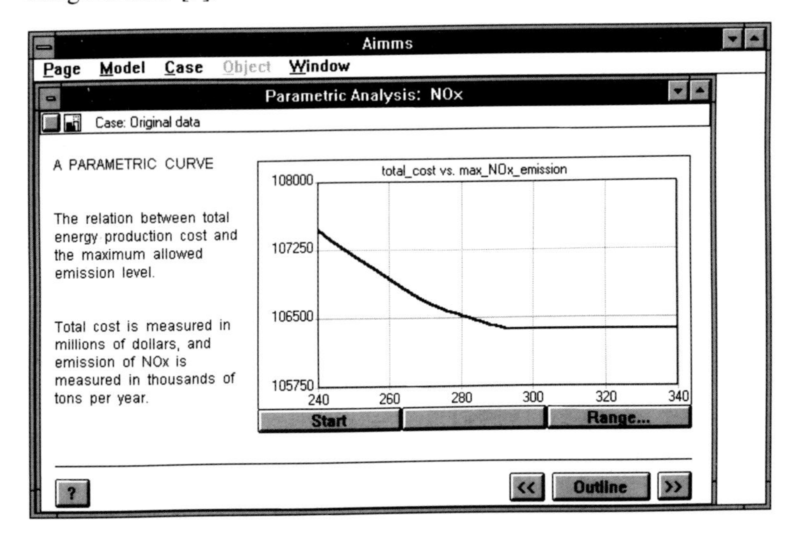

Figure 10.1

Perhaps the most well-developed tool for exploring the behavior of linear programming problems is Greenberg's ANALYZE [10], [11]. ANALYZE provides extensive facilities for displaying pieces of the linear programming problem. Since linear programming problems ultimately are transformed into some form of table or matrix, the pieces shown by ANALYZE are typically displayed as submatrices as shown at right. The '+'s represent non-negative matrix entries. The short row and column labels are a consequence of an eight-character limit on identifiers. The column label TNESW represents the transport (T) of material from the northeast (NE) to the southwest (SW) region. Similarly, the row label DNE ensures that the demand (D) by the northeast region (NE) is met.

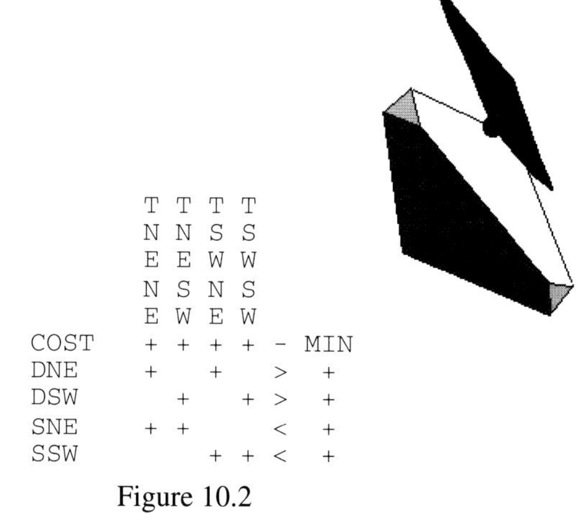

	TNENE	TNESW	TSWNE	TSWSW		
COST	+	+	+	+	-	MIN
DNE	+		+		>	+
DSW		+		+	>	+
SNE	+	+			<	+
SSW			+	+	<	+

Figure 10.2

ANALYZE also provides natural language explanations that automatically explain the meaning of variable names and constraints [12], among others (below). More sophisticated analysis of linear programming models is also provided. ANALYZE, for example, can produce a natural language explanation of the value of a shadow price. Currently, however, all ANALYZE output consists of text, including matrix imagessee Figure 10.2). At least some of ANALYZE's output could be usefully display graphically (for example, see Figure 8.9).

```
Row syntax has 2 classes
 A row that begins with S limits supply at some supply region.
 A row that begins with D requires demand at some demand region.

Column syntax has 1 class
 A column that begins with T transports from some supply region to
    some demand region.
```

Figure 10.3

10.2 Application Dependent Representations

We now turn from discussing general techniques for representing the sensitivity of a solution to representations designed for specific application areas. Several application areas have evolved representations useful in their particular domain. These include Gantt charts used for planning and scheduling (Section 10.3), geographic information systems (Section 10.4) of particular use for location and routing problems and Space-Time plots (Section 10.5).

10.3 Gantt Charts

We live in an era of management gurus promoting ideas such as business process reengineering, total quality management, lean production, among others. These ideas are widely promoted in best-selling books. This is not a new phenomenon. In fact, management gurus have been promoting new ideas in books since at least the late 1800's. One of these gurus was Henry Lawrence Gantt [13], [14].

Henry Lawrence Gantt was a disciple of another management guru, Frederick Taylor, whose "Scientific Management" advocated the use of the scientific method to improve manufacturing operations. In particular, Taylor emphasized the importance of accurate measurement of the current status of the manufacturing system. How can one hope to improve the operations of a manufacturing system without having accurate data describing its current performance? Taylor's (and Gantt's) ideas lead directly to the development of industrial engineering, and can be seen as providing many of the philosophical underpinnings of management science and operations research.

During World War I, the need for efficient manufacturing became critical, which lead to the generation of many new ideas. Gantt developed "simple chart system" that is now known as a *Gantt chart*.

A Gantt chart consists of a series of time lines, one for each product, employee, and machine—resource—in the factory. The time lines display the status of the resource (idle, busy, broken, on break, under/over quota) at any point in time. Clark [15] provides a variety of examples of and recommendations for Gantt charts.

Below, for example, is a Gantt chart [14] from 1909 showing productivity of workers in a weaving plant. The dark black squares indicate that workers have met their quota, whereas the grey squares indicate a quota was not met (cross-hatching indicates absences or assignment to other work). Gantt emphasized the importance of the visual tool stating, "Note the increase of black (meaning task achieved) and bonus earned) as time progresses" (at least for those workers who did not leave). In essence, by examining his chart, Gantt re-discovered the learning curve. Such *Gantt* charts have now become a standard visualization tool for classic optimization problems of production planning and scheduling.

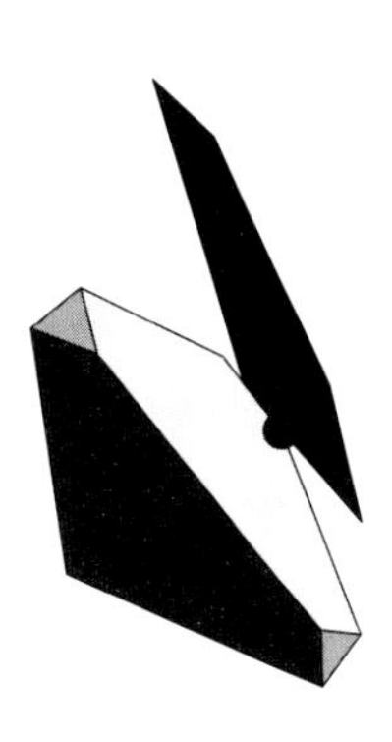

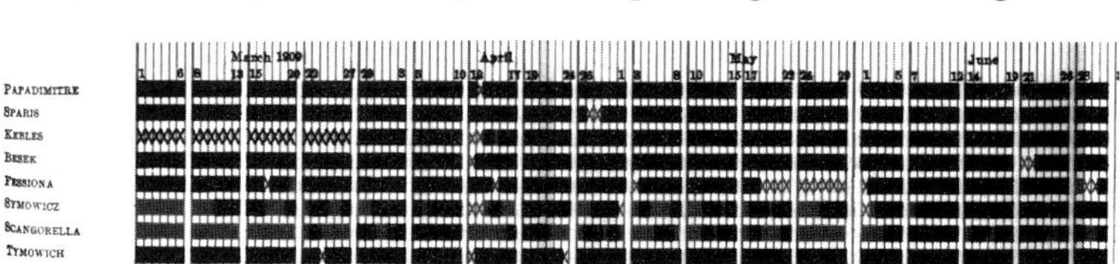

Figure 10.4

Clark [16] wrote the following testimonial to the value of this visualization tool not long after Gantt's death in 1919,

> Although they are only lines drawn on paper, where they are used production is increased, costs and inventories are reduced, special privilege is eliminated, initiative is stimulated, an organization is built up of men who "know," and workmen become interested in their work.

With the advent of interactive computer graphics in the late 1960's, interactive Gantt charting systems have seen ongoing development [17]. The figure below, for example, shows a a Gantt chart created using INCEPTA [18] from Insight Logistics:

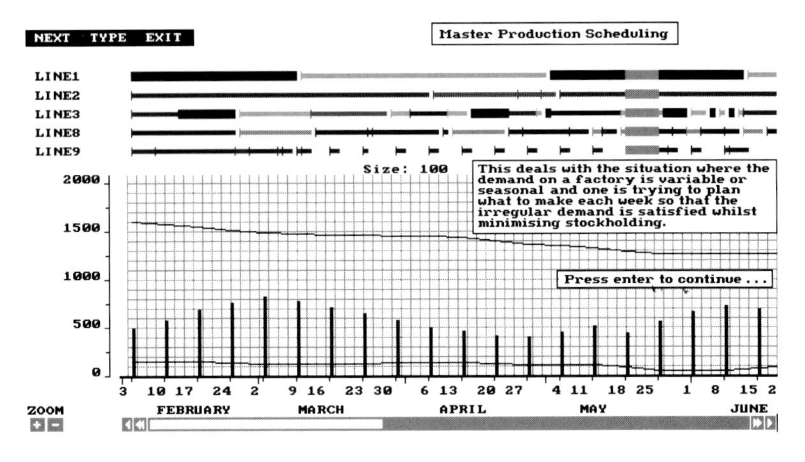

Figure 10.5

Below is another Gantt chart from a commercial system (MIMI)[19]. In this Gantt chart, the arrows represent the flow of product from several activities (operations) into a selected activity.

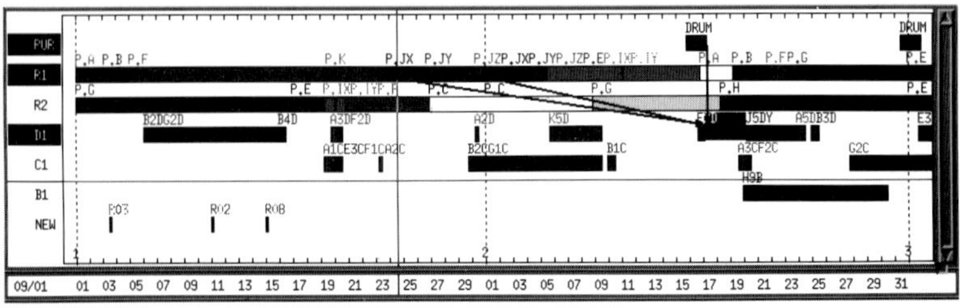

Figure 10.6

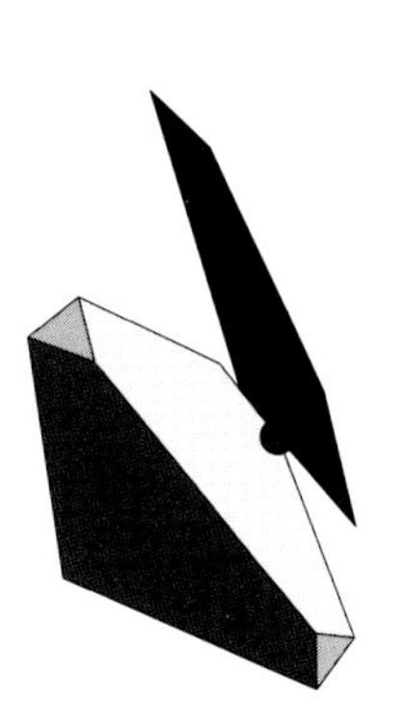

Although Gantt charts are an important visualization tool for scheduling applications, when many items are displayed, a Gantt chart can become extremely complex. Displaying status information for hundreds or thousands of activities (or more) soon overwhelms the cognitive capacity of most human beings. One can of course display subsets of the chart in multiple windows, zoomed in or out to various levels. One can also allow users query any point in the chart to see additional information (e.g., duration of an activity, amount of material produced).

Mackinlay, *et al.* [20] applied the fish-eye view idea to the Gantt chart to create a *Perspective Wall* (below). In a Perspective Wall, a user can zoom in on a portion of the Gantt chart in time. As with fish-eye views for networks and graphs, the rest of the Gantt chart does not disappear, but becomes smaller. In fact, the Perspective Wall shows a Gantt chart in three-dimensions. The area of interest is shown on a flat wall in front of the user, whereas parts of the Gantt chart to the left and right of the area of interest appear as if on walls that recede into the distance.

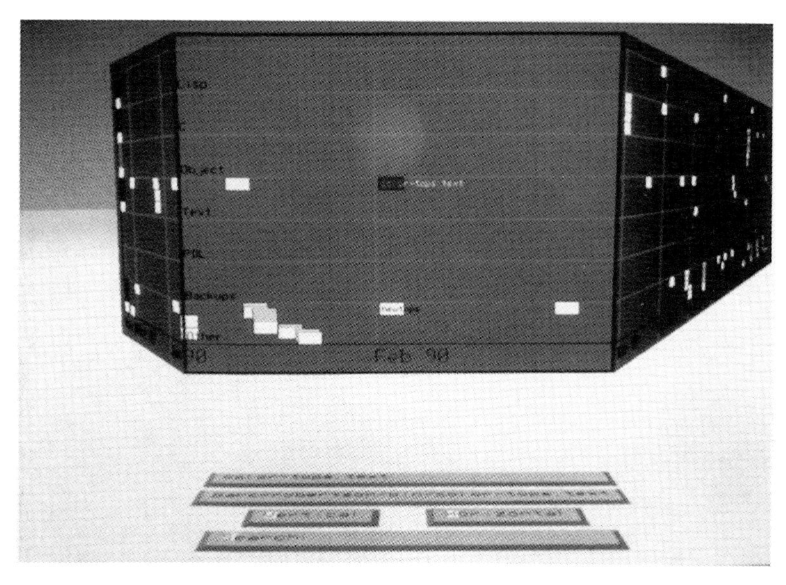

Figure 10.7

Jones [21] proposed a three-dimensional Gantt chart for a classic machine scheduling problem, the job-shop. The three axes corresponded to Time, Machines and Jobs. He showed that different two-dimensional projections of the three dimensional chart corresponded to traditional Gantt charts for the job-shop scheduling problem. His proposal has seen no practical application, however.

Gantt charts, although an old idea, provide a concise way to represent the status of time-constrained resources. As the number of resources increases, a Gantt chart can become overwhelming. A variety of techniques such as zooming, windowing, and filtering can help. The fish-eye technique seems quite promising, but we are unaware of any commercial system that has applied this idea.

10.4 Geographic Information Systems

Many optimization problems such as vehicle routing, facility location, among others can be readily visualized as maps. A general system for displaying two-dimensional spatial information is called a *Geographic Information System* (GIS) [22], [23], [24]. GIS's have been developed primarily by the geography community and have been used for purposes such as zoning, land-use planning, demographic analysis among others. Governments purchase them to manage their zoning, tax, and land-use databases; marketers use them to help identify new markets; miners use them to help identifying likely deposits of ore; individual consumers can even buy a single CD-ROM containing the entire road network of the United States.

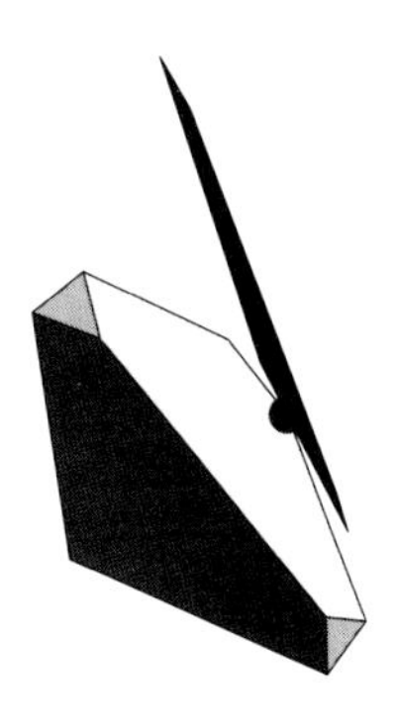

Most current GIS's can be viewed essentially as specialized databases. Spatial information is stored in a data structure and can be retrieved and displayed in a variety of ways. For example, below is a map from a system (OSARMS) [25] for responding to oil-spill emergencies. It shows a portion of the Gulf of Mexico including locations of oil-wells, skimmers, booms, as well as up-to-date satellite imagery showing the location of a spill.

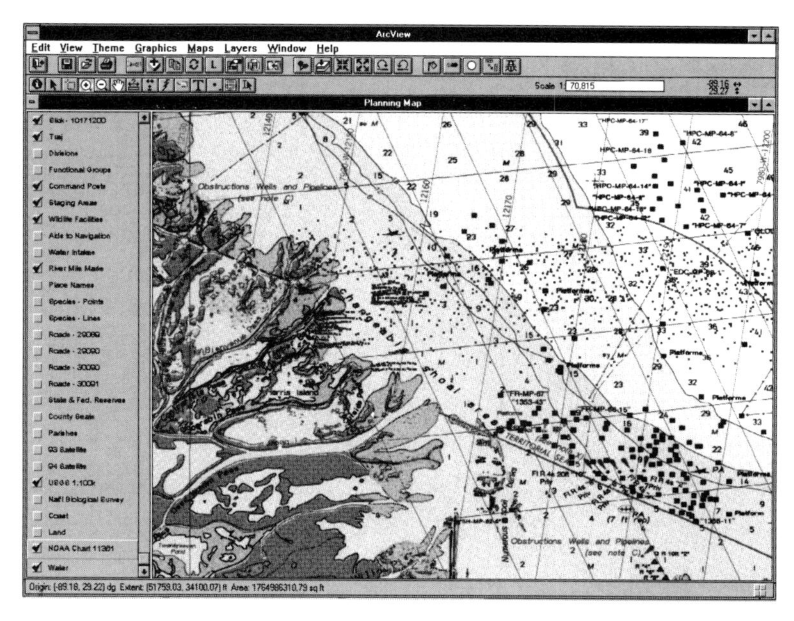

Figure 10.8

GIS's are now being expanded to include analytical models. OSARMS, for example, contains a model to predict the spread of a spill over time. Emergency response crews use the model's predictions to plan their cleanup efforts.

For certain problems often attacked by optimization, GIS's provide a variety of useful features. First, they provide access to a large database of spatial information. This is of particular use to optimization, since many optimization models such as vehicle routing, facilities location rely on spatial data. Before the availability of GIS's, a large amount of effort was required simply to create a road network database for a vehicle routing problem. Second, GIS's allow spatial data to be displayed in a large variety of ways. For vehicle routing, the road network can be displayed at any level of detail, e.g., one can display only secondary roads, or only divided highways. For facility location problems, one can color code population density, zoning classification, or any other area-based attribute.

The figure below shows a GIS specialized for routing hazardous materials [26]. Three routes between Wilbraham, Massachusetts and Mukilteo, Washington are displayed. The lightest gives the shortest time route, the darkest minimizes the size of the nearby population, and the other shows the shortest time route. Users can zoom in on any portion of the map to see the routes in more detail.

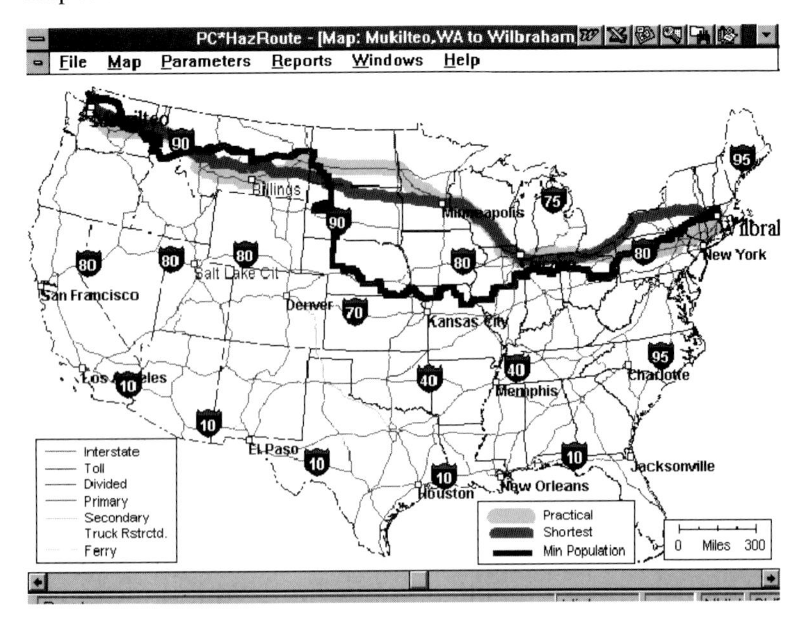

Figure 10.9

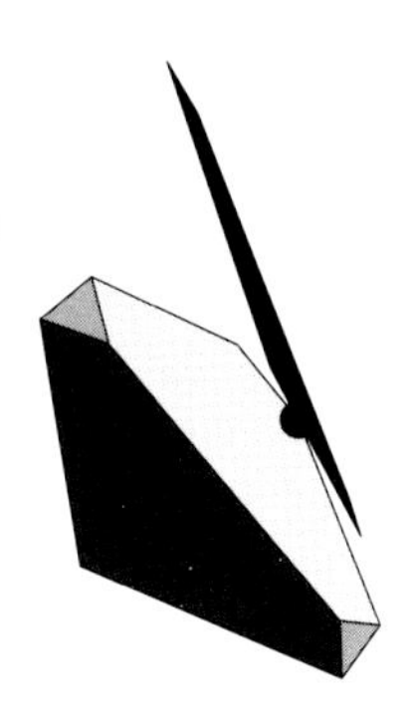

Bodin and Levy [27] discussed two visualization challenges for vehicle routing problems. One, they called the *clutter* problem. Large road networks, complicated routes are difficult to visualize. The clutter problem is found in other areas that want to display large networks. Approaches to the clutter problem are discussed in Chapter 13. For vehicle routing, however, many graph layout algorithms would be in appropriate since they move the nodes in order to reduce clutter. In vehicle routing problems, nodes correspond to locations fixed in space, and hence cannot be moved. Other solutions to the clutter problem such as fish-eye views are useful, however.

The second challenge concerns time, for example, vehicle routing with time windows, where a delivery must be made to a customer between specified times. One naive solution would merely animate the movement of vehicles, however, this does not represent the entire schedule with a single image, and requires time to display. Another solution would use a Gantt chart (with a time line for each vehicle), but the spatial relationships among different locations would be lost. Still another solution would use three-dimensional representations to display the third dimension of the problem (time). Each destination in a vehicle's route would be positioned along the third dimension. Note that three-dimensional representations of graphs and networks has been explored, most notably for visualizing expert system knowledge bases [28]. Given the difficulty that people have in perceiving three-dimensional representations, however, it is unclear how well this representation format would help a typical decision maker.

10.5 Space-Time Plots

Another solution for representing both spatial and temporal relationships is called a *space-time* plot. Patented by Ybry [29] in 1846, it consists of a simple two-dimensional plot. Along the horizontal axis, time is plotted; along the vertical axis, the location of the train. Diagonal lines represent the paths trains take in both space and time. Downward sloping lines represent trains moving in one direction; upward sloping lines represent trains moving in the opposite direction. The steepness of the slope gives an indication of the speed of the train. Horizontal segments represent points in a journey when a train is stopped at a station, or at a siding waiting for another train to pass. Note the similarity between Ybry's train schedule and Parallel coordinates discussed in Chapter 14.

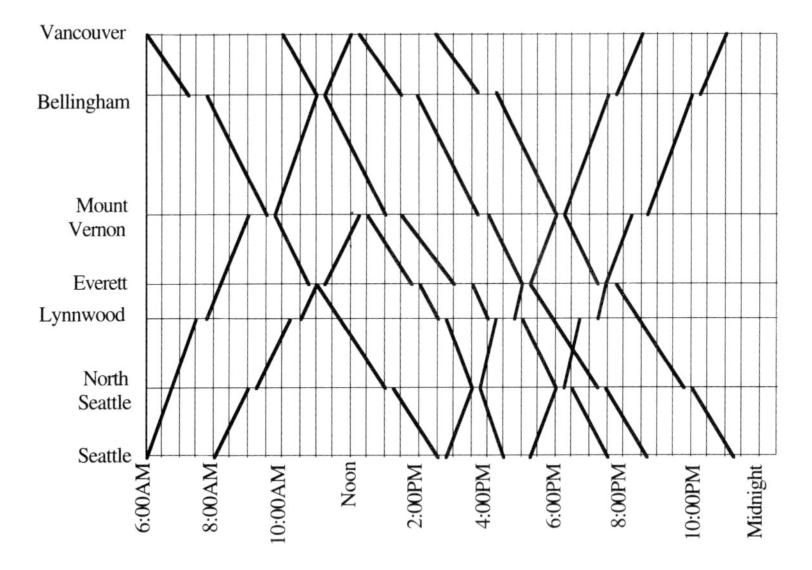

Figure 10.10

Of course, Ybry's representation may not be useful for many types of vehicle routing problems, since many vehicles do not travel along fixed tracks. Tufte [29] provides an extensive discussion and examples for how to visualize temporal information.

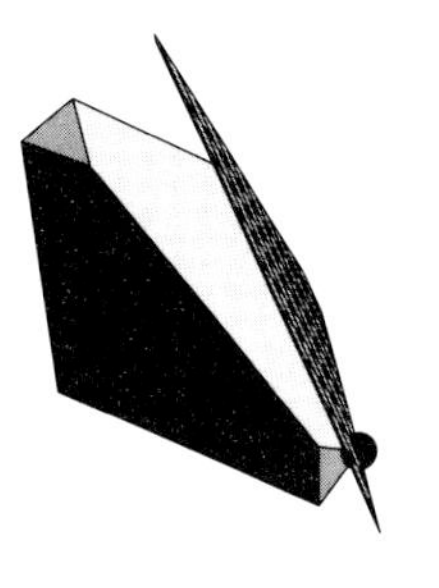

10.6 Summary

Models, like computer programs, are usually not written correctly the first time. Errors in the model itself, in the data, and in the basic concepts are usually not revealed until the model is built and solved. With current models often involving hundreds of thousands of decision variables and constraints, it seems imperative to provide "debuggers" to help uncover the problems in a model, similar to debuggers for computer programming.

For example, if an algebraic language is used to specify a problem, then the debugger should report its results in a form consistent with that algebraic language. If the model has been translated into a form required by a solver, much of the semantic information in the algebraic specification is lost. The debugger must somehow maintain linkages to the original formulation, much as a debugger for programming languages maintains linkage between the source code and machine code.

Debugging a model, although extremely important, does not guarantee the accuracy of the model. Models must be validated, that is, their behavior must be judged against a real-world standard. Does the output correspond to our expectations? If not, perhaps the model has uncovered an unexpected, but useful insight on the real problem. Or perhaps the model is simply inaccurate. In short, does the output make sense when applied to the actual problem? Debugging tools can certainly help probe the behavior of the model, but we seem less able to formalize validation. The model was built to represent a complex real-world problem. How can we be certain that the model represents the real-world with sufficient accuracy?

For specific application domains at least, useful representations are available. This is not to say that those representations cannot be improved. For example, a Gantt chart with hundreds or thousands of operations displayed will be difficult to understand. Similarly, Geographic Information Systems remain the subject of active investigation. For many other application domains, visual representations are not as common. Is their a useful visual representation of a balance sheet or marketing plan?

Note that the application specific representations discussed in this chapter display complex multi-dimensional data, albeit just a single solution point. It seems reasonable to explore displaying the entire multi-dimensional solution space. For example, for linear programming, it would seem desirable to be able to visualize the multi-dimensional polytope representing the feasible region. We consider that challenge in Chapter 14.

Bibliography

[1] Dantzig GB. *Linear Programming and Extensions.* Princeton (NJ):Princeton University Press, 1963.

[2] Fiacco AV. *Introduction to Sensitivity and Stability Analysis in Nonlinear Programming.* New York:Academic Press, 1983.

[3] Fiacco AV, editor. *Optimization with Data Perturbations*, volume 27 of *Annals of Operations Research.* Basel, Switzerland:J. C. Baltzer, 1990.

[4] Gal T. *Postoptimal Analyses, Parametric Programming, and Related Topics.* New York:McGraw-Hill, 1979.

[5] Hoesel CPMV, Kolen AWJ, Kan AHGR, Wagelmans APM. Sensitivity analysis in combinatorial optimization: A bibliography. Technical report, PO Box 1738, 3000 DR, Rotterdam, The Netherlands:Econometric Institute, Erasmus University Rotterdam, 1989.

[6] Chinneck JW. Localizing and diagnosing infeasibilities in networks. Technical Report SCE-90-14, Ottawa, Canada:Department of Systems and Computer Engineering, Carleton University, 1990.

[7] Chinneck JW, Dravnieks EW. Locating minimal infeasible constraint sets in linear programs. *ORSA Journal on Computing*, 1989;3(2):157–168.

[8] Maximal Software, Inc. *MPL Modeling System.* Arlington (VA):Maximal Software, Inc., 1994.

[9] Bisschop J, Entriken R. *AIMMS The Modeling System.* PO Box 3277, 2001 DG Haarlem The Netherlands:Paragon Decision Technology, 1993.

[10] Greenberg HJ. ANALYZE: A computer-assisted analysis system for linear programming models. *OR Letters*, 1987;6(5):249–259.

[11] Greenberg HJ. *A Computer-Assisted Analysis System for Mathematical Programming Models and Solutions: A User's Guide for ANALYZE.* Boston (MA):Kluwer, 1993.

[12] Greenberg HJ. A natural language discourse model to explain linear programs. *Decision Support Systems*, 1987;33:333–342.

[13] Gantt HL. *Organizing for Work.* New York:Harcourt, Brace and Howe, 1919.

[14] Gantt HL. *Work, Wages, and Profits.* New York:The Engineering Magazine, 1910.

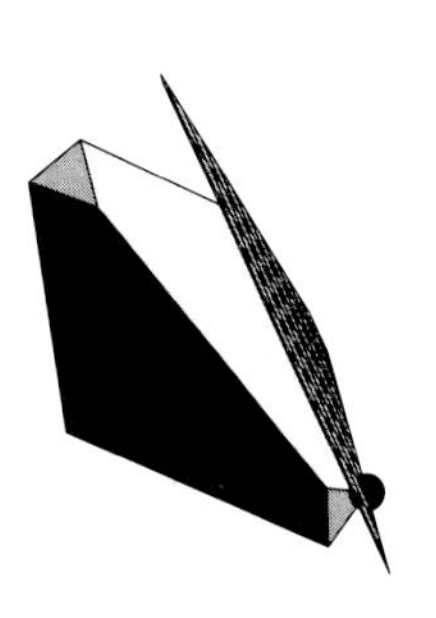

[15] Clark W. *The Gantt Chart.* London:Sir Isaac Pitman and Sons, 3rd edition, 1957.

[16] Clark W. *The Gantt Chart.* New York:The Ronald Press Company, 1st edition, 1923.

[17] Garman M. *Solving Combinatorial Decision Problems via Interactive Computer Graphics with Applications to Job-Shop Scheduling.* [dissertation], Pittsburgh (PA):Carnegie-Mellon University, 1970.

[18] Insight International Ltd. *INCEPTA Tutorial.* Woodstock, Oxon, United Kingdom:Insight Logistics Ltd, 1993.

[19] Chesapeake Decision Sciences. *MIMI: Manager for Interactive Modeling Interfaces: User's Manual.* New Providence (NJ):Chesapeake Decision Sciences, 1993.

[20] Mackinlay JD, Robertson GG, Card SK. The perspective wall: Detail and context smoothly integrated. In Robertson SP, Olson GM, Olson JS, editors, *Proc. ACM Computer-Human Interaction '91 Conference on Human Factors in Computing Systems*, pages 173–179, New York:1991. ACM Press.

[21] Jones CV. The three-dimensional Gantt chart. *Operations Research*, 1988;36(6):891–903.

[22] Bonham-Carter GF. *Geographic Information Systems for Geoscientists: Modelling with GIS.* Tarrytown (NY):Pergamon (Elsevier), 1994.

[23] Laurini R, Thompson D. *Fundamentals of Spatial Information Systems.* London:Academic Press, 1992.

[24] Worrall L, editor. *Geographic information systems : developments and applications.* London:Belhaven Press, 1990.

[25] Environmental Systems Research Institute, Inc. Oil spill response comes of age. *ArcNews*, April 1995.
URL: *http://www.esri.com/headlines/arcnews/win9495articles/oil.html*

[26] ALK Associates, Inc. *ALK's PC*HazRoute Version 2.0 User's Guide.* 1000 Herrontown Road, Princeton (NJ) 08540:ALK Associates, Inc., 1995.

[27] Bodin L, Levy L. Visualization in vehicle routing and scheduling problems. *ORSA Journal on Computing*, 1994;6(3):261–269.

[28] Fairchild KM, Poltrock SE, Furnas GW. Semnet: Three-dimensional graphic representations of large knowledge bases. In Guindon R, editor, *Cognitive Science and its Applications for Human-Computer Interaction*, pages 201–233. Hillsdale (NJ):Lawrence Erlbaum Associates, 1988.

[29] Tufte ER. *Envisioning Information.* Cheshire, Connecticut:Graphics Press, 1990.

Part III

Visualization for Optimization

In the previous part, we explored visualization from the perspective of the tasks involved in a mathematical programming project. This part of the book discusses some of the different formats that are available for representing optimization models and their solutions. Future optimization systems will allow a variety of representations to be viewed and manipulated simultaneously. Chapter 11 discusses the use of text for optimization. Chapter 12 discusses the use of hypertext and hypermedia for optimization. Chapter 13 discusses the use of networks and graphs to represent optimization models and solutions. Of particular importance are algorithms for drawing graphs. Optimization problems typically model problems consisting of thousands of dimensions. Chapter 14 discusses a variety of techniques that have been proposed for representing multi-dimensional information. Chapter 15 discusses the use of animation to support optimization modeling and solution. Moreover, it discusses how optimization has been used by the visualization community to help construct animations. Chapter 16 discusses the use of sound, touch and virtual reality for optimization. Although there is much research in the visualization community on these topics, little has been applied to optimization, although optimization has been applied to virtual reality.

Generic tools for constructing graphical visualizations are then discussed in Chapter 17. This chapter foreshadows a more general discussion of how to allow for multiple simultaneous representations of models and solutions that is found in Chapter 18.

Chapter 19 provides guidelines to the academic who wishes to pursue research in this area; developing a successful research program in this area is difficult but not impossible; the chapter concludes the book with some fearless predictions concerning the outlook for visualization and optimization in the next few years.

Chapter 11

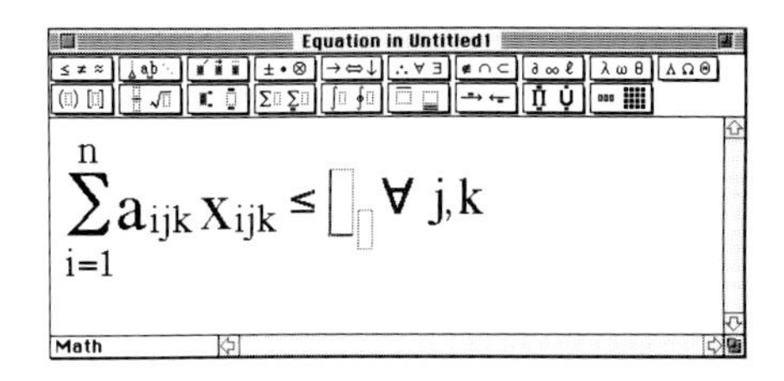

Text

The principal type of visualization used for optimization is text. Models are formulated using mathematical symbols such as $\sum$, $\leq$ and $\geq$. Algorithms are specified as computer programming languages. The output from algorithms often consists of tables of numbers. Although the use of graphical representations for optimization continues to expand, text continues to play a significant role.

11.1 History

We take for granted the mathematical notation now in common use. That notation did not arise overnight. There were many proposed notations, that are now abandoned, as mathematics developed. For example, Arabic numerals displaced Roman numerals at least for mathematics though not for identifying Super Bowl games.

Many of the mathematical symbols commonly used in optimization were invented in a four-hundred year period, from 1450 to 1750 (Figures 11.1 and 11.2) [1]. As Boyer and Mezbach write (p. 264), "we now take for granted our symbolic notations...with little thought for the slowness with which these developed in the history of mathematics." Before 1600, symbols for equality ($=$), addition ($+$), subtraction ($-$), and decimal numbers were introduced. By 1750 symbols for inequality relations ($<$, $>$), naming of variables ($\ldots, x, y, z$) and parameters ($a, b, c, \ldots$), summation $\sum$, integration $\int$, functions $f(x)$ and infinity ∞ were invented, some of them by such giants as Descartes, Liebnitz and Euler. Other symbols were invented more recently. In the 1800's, matrix notation was developed by Cayley and symbols for set union $\cup$ and intersection $\cap$ were

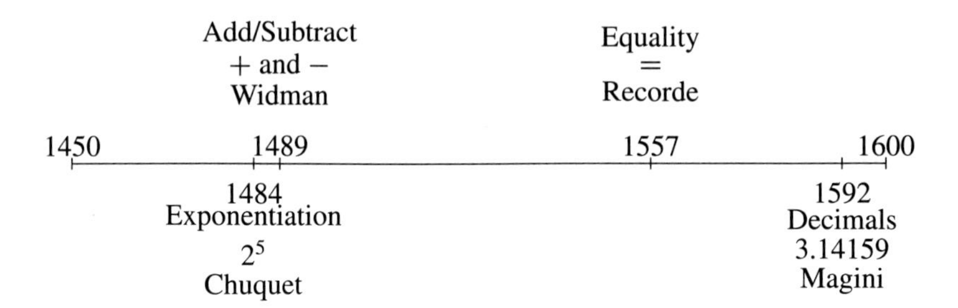

Figure 11.1: Time line illustrating the development of common mathematical notation between 1450 and 1600 [1] .

introduced by Grassman. In the early 1900's, symbols for the logical quantifiers $\forall$ and $\exists$ were introduced by Russell and Whitehead in their monumental *Principia Mathematics* [2].

Many optimization modeling systems simply use appropriate character strings in place of these symbols. In GAMS [3], for example, $\sum$ is replaced by the string `SUM` and a variable x_{ijk} is written as `X(I,J,K)`.

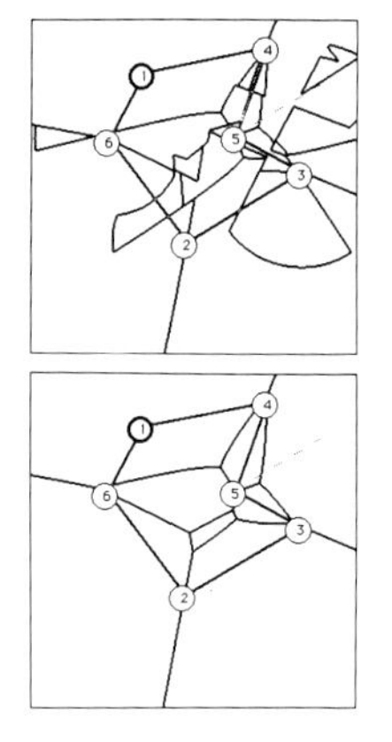

Naming
$a, \ldots$ knowns
unknowns, $\ldots, z$
Descartes
1628

Integral
$\int$
Liebniz
1676

π
Jones
1706

Function
$f(x)$
Euler
1734

1600 — 1750

1631
Inequality
$>, <$
Harriott

1655
Infinity
∞
Wallis

1748
Sum
$\sum$
Euler

Figure 11.2: Time line illustrating the development of common mathematical notation between 1600 and 1750 [1].

Whereas textual languages can simulate algebraic syntax, they do not duplicate it. It is now possible, though, for users to enter algebraic symbols directly at least for text-editing purposes. Below, for example, is the standard mathematical text editor provided with Microsoft Word. The user is just about to enter the subscripted value for the right-hand side of the inequality. The two thin rectangles indicate positions for the name of the value and its subscripts, respectively. The menu above the window allows users to enter square roots, summation, integral, and other mathematical syntax.

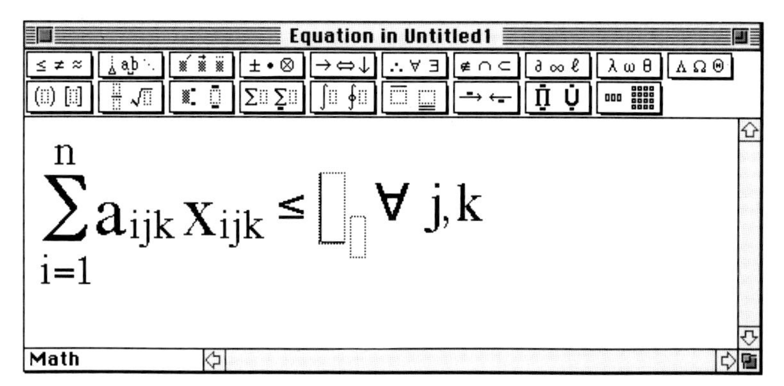

Figure 11.3

Although we are unaware of any optimization system that accepts such formatted text directly, those capabilities should break down the barrier between the traditional algebraic formulation and the actual modeling language. Geoffrion [4] in Structured Modeling Language (SML) showed how different text styles could be used to highlight different aspects of information. An example SML model for the transportation problem is shown in Figure 8.6.

11.2 Natural Language

Some authors have developed software that can generate automatically natural language expressions that explain the behavior of mathematical programs. We have already discussed ANALYZE [5] (see Figures 10.2-10.3). The system of Kimbrough *et al.* [6], [7] also generates text automatically (see Figure 11.4).

The figure below shows the results of running the **asset** model with two scenarios, **swath** and **hydrofoil**. The user has also asked for more information about the variable **f_c**, which is displayed in the lower window. The text describing **f_c** was generated automatically.

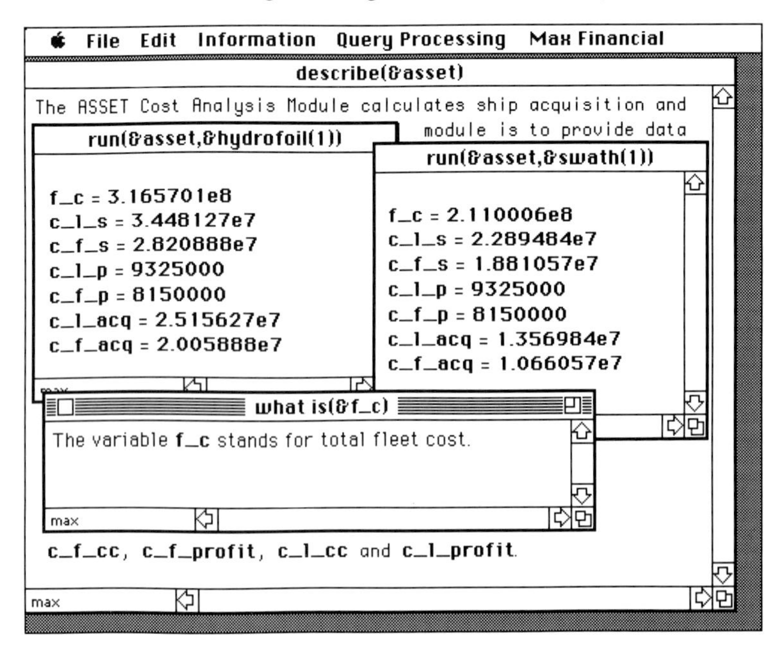

Figure 11.4

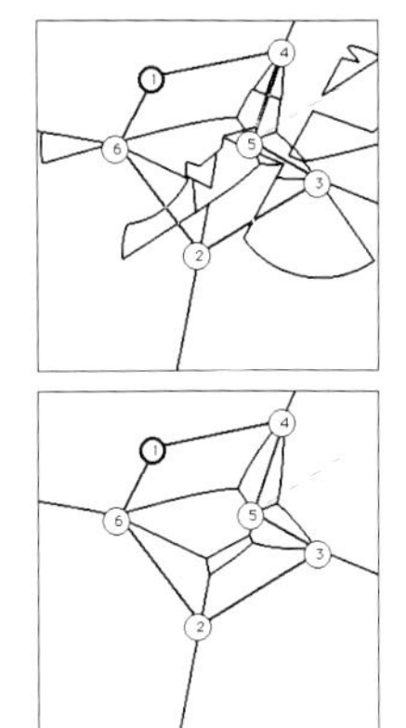

11.3 Symbolic Mathematics Systems

Once mathematical expressions are entered into a computer system, not only can numerical values be calculated, but the expressions themselves can be manipulated. Symbolic integration and differentiation, for example, are not calculations per se, but manipulations of the mathematical expressions themselves. Languages for performing symbolic manipulation, for example, MACSYMA [8], Mathematica [9], and Maple [10], provide extensive facilities for performing symbolic differentiation, symbolic integration, and equation solving. Mathematica includes linear and nonlinear programming algorithms for example (Figures 11.5-11.6). For nonlinear programming, this type of manipulation seems especially useful. Symbolic mathematics systems also generally provide extensive capabilities for producing graphical (and even audio) representations.

Below is an example of the use of Mathematica for optimization. The example shows how Mathematica can minimize a non-linear function, in this case, $(x-2)^4 + (x-2y)^2$. Mathematica also provides extensive graphical output capabilities including contour plots as shown below (from [11]).

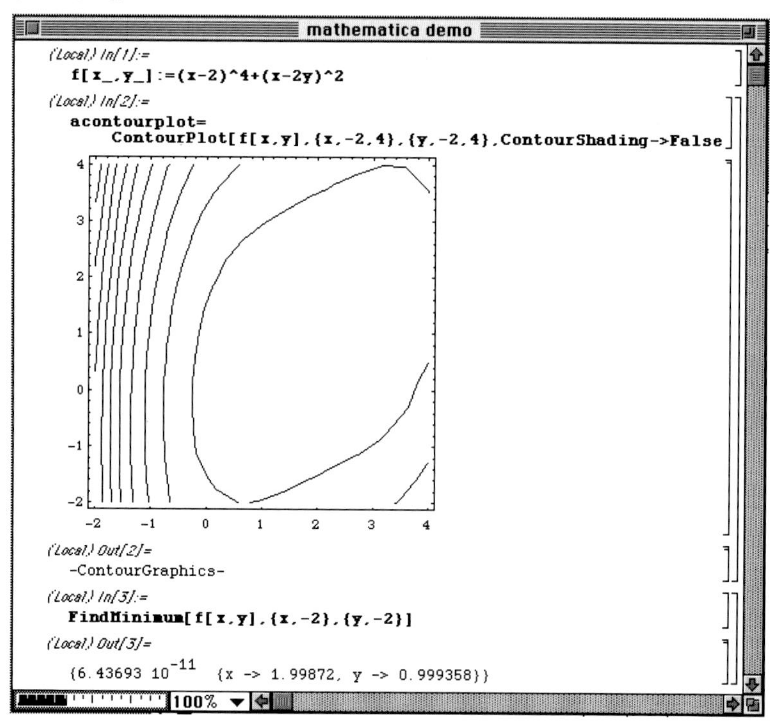

Figure 11.5

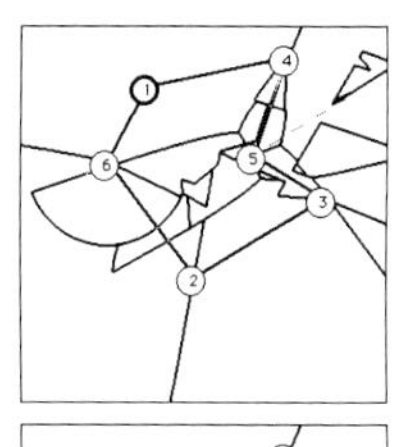

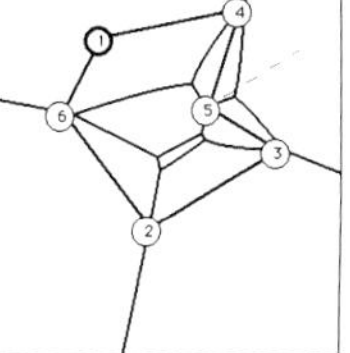

Below is a more advanced example of the use of Mathematica for optimization. By using additional Mathematica capabilities, variable `l` stores a list of the points (x, y) visited by the non-linear optimization algorithm. The subsequent plot overlays the trajectory of the algorithm on the original contour plot of figure 11.5.

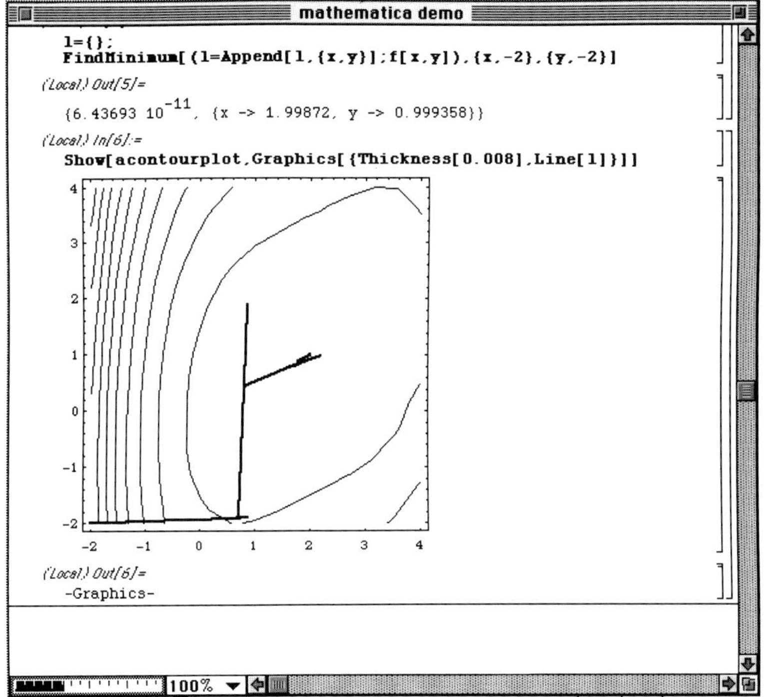

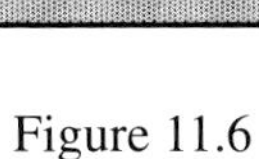

Figure 11.6

Below is an example of Maple, another symbolic mathematics system, applied to optimization. Maple, like Mathematica, can provide a rational (i.e., exact) solution to linear programs with rational coefficients.

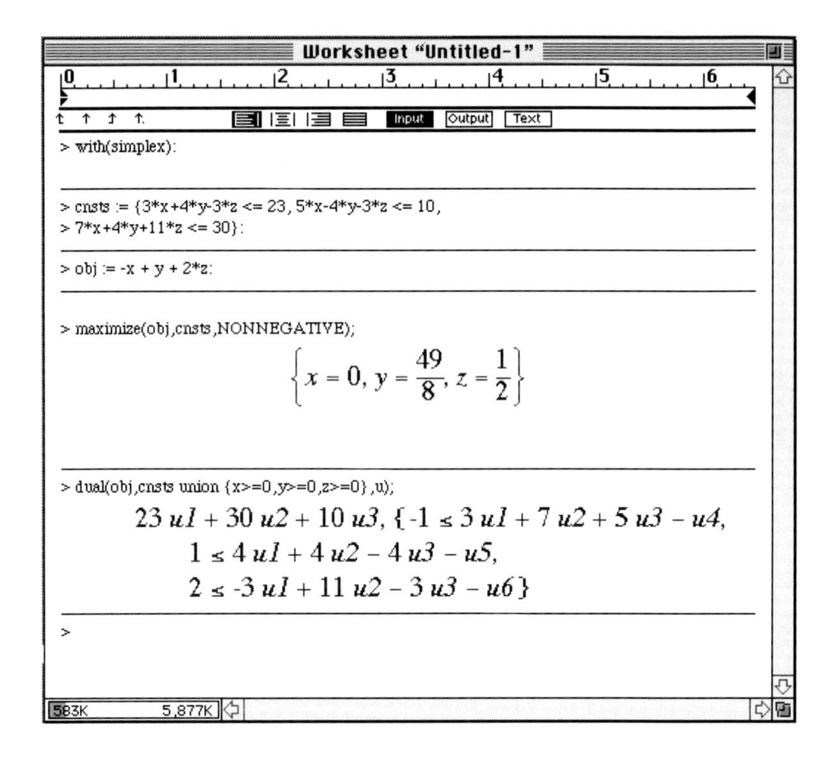

Figure 11.7

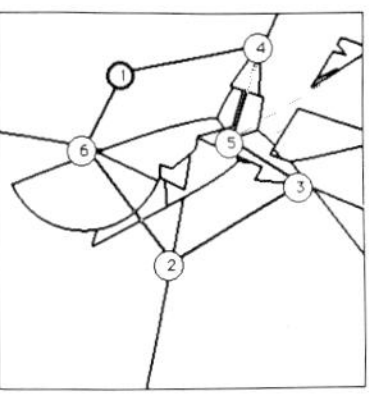

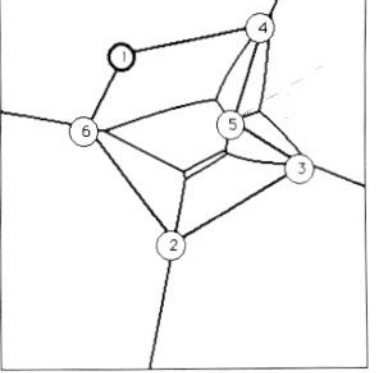

11.4 Syntax-Directed Editors

For the input of text, *syntax-directed editors* have been proposed. They provide editing operations tailored specifically to a particular language. For example, the system in [12] provides simple editing operations to add and remove if-then-else and do-while constructs from Pascal, while ensuring that variables are declared before they are used. Such syntax-directed editing techniques have been applied to mathematical programming and other modeling languages [13], [14]. .

Many current modeling languages rely on plain ASCII text created using a text editor. Although this choice greatly enhances portability, it denies users access to more sophisticated text formats and text input techniques. The ability to use multiple fonts in multiple sizes is slowly becoming easier. On Unix systems, X Window [15] provides standard mechanisms for working with complex fonts. Microsoft Windows and the Apple Macintosh also provide standard font manipulation capabilities.

Syntax-directed editing should help to eliminate syntactic errors in models. Most algebraic modeling languages, for example, assume a standard text editor for building the model. The model is then "compiled," with syntax errors uncovered. The modeler must then iterate through a revise-compile loop before the solution algorithm is ever run. A syntax-directed editor would help the modeler uncover syntax errors more quickly, thereby providing valuable assistance in the model-building process.

11.5 Optimization for Text

Optimization has been usefully employed in typesetting to help determine where to break a sequence of characters into lines and a set of lines into pages. In fact, the typesetting program used for this book, TEX by D. Knuth [16] (actually a set of TEX macros called LATEX by L. Lamport [17]), uses a dynamic programming algorithm to determine where to break strings of characters into lines. The exact details of the algorithm are too complex to describe here (see [16]), but a sketch can be provided.

Consider a sequence of characters that will form a paragraph. TEX identifies a series of *feasible breakpoints*, $b_0, \ldots, b_n$, where a linebreak can be inserted. Break b_0 represents the beginning of the string of characters. TEX also defines *demerits* for a given breakpoint. Demerits are assessed based on how loose or tight a particular line is typeset, whether or not a hyphen has to be inserted, among other criteria. The demerits for a particular breakpoint, though, depend on the previous breakpoint chosen. Define $d(b_i, b_j)$ to be the demerits assigned to breakpoint $i \geq j$ given that b_j was the previous breakpoint chosen. Now define $g(b_i)$, the optimal value function, to be the minimum number of demerits for all breakpoints up to and including $g(b_i)$:

$$g(b_i) = \begin{cases} 0 & i = 0 \\ \min_{j<i} d(b_i, b_j) + g(b_j) & i = 1, \ldots, n \end{cases} \tag{11.1}$$

Given the optimal value function, the dynamic programming algorithm is now defined. TEX's optimization algorithm for linebreaking provides a great deal of flexibility.

For example, below is a paragraph typeset in TEX (originated by W. J. Youden, cited in Tufte [18], p. 143) whose shape was specified to be a normal density function with $\sigma = 0.8$in. Below that is the same paragraph with $\sigma = 0.6$in.

The
Normal Law
of error stands
out in the experience
of mankind as one of the
broadest generalizations of
natural philosophy. It serves as
the guiding instrument in researches
in the physical and social sciences and in
medicine, agriculture and engineering. It is an
indispensable tool for the analysis and the interpretation
of the basic data obtained by observation and experiment.
$\sigma = 0.8$

The
Normal
Law of er-
ror stands out
in the experience
of mankind as one
of the broadest gen-
eralizations of natural
philosophy. It serves as
the guiding instrument in
researches in the physical and
social sciences and in medicine,
agriculture and engineering. It is
an indispensable tool for the analysis
and the interpretation of the basic data
obtained by observation and experiment.
$\sigma = 0.6$
Figure 11.8

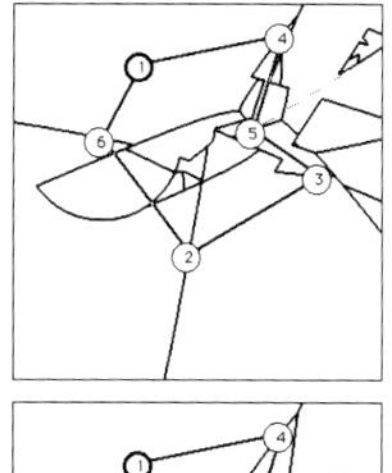

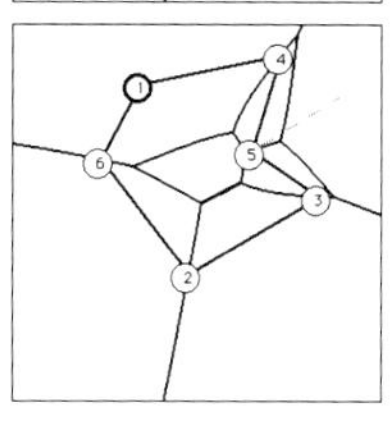

11.6 Summary

The use of multiple fonts and text styles can play a useful role in modeling systems for optimization. The emergence of direct manipulation interfaces for inputting mathematical text offers the possibility for allowing mathematical modelers to work directly with the notation that is most familiar to them. Symbolic mathematics systems provide useful tools for performing symbolic manipulations of mathematical formulae. Optimization has also been usefully employed in typesetting systems to choose line breaks for text layout. Text will remain perhaps the most important "visualization" format for quite some time; a picture is not always worth a thousand words.

Bibliography

[1] Boyer CB, Merzbach UC. *A History of Mathematics*. New York:John Wiley and Sons, 2nd edition, 1991.

[2] Whitehead AN, Russell B. *Principia Mathematics*. Cambridge, England:The University Press, 1960–1963.

[3] Brooke A, Kendrick D, Meeraus A. *GAMS: A User's Guide, Release 2.25*. South San Francisco (CA):The Scientific Press, 1992.

[4] Geoffrion AM. The SML language for structured modeling. *Operations Research*, 1992;40(1):38–75.

[5] Greenberg HJ. ANALYZE: A computer-assisted analysis system for linear programming models. *OR Letters*, 1987;6(5):249–259.

[6] Kimbrough SO, Pritchett CW, Bieber MP, Bhargava HK. The coast guard's KSS project. *Interfaces*, 1990;20(6):5–16.

[7] Kimbrough SO, Pritchett CW, Bieber MP, Bhargava HK. An overview of the coast guard's KSS project: DSS concepts and technology. In Volonino L, editor, *Transactions of DSS-90: Information Technology for Executives and Managers, Tenth International Conference on Decision Support Systems*, pages 63–77, Cambridge (MA):1990.

[8] Rand RH. *Computer Algebra in Applied Mathematics: An Introduction to MACSYMA*. Boston:Pitman, 1984.

[9] Wolfram S. *Mathematica: A System for Doing Mathematics by Computer*. Redwood City (CA):Addison-Wesley, 2nd edition, 1991.

[10] Char BW. *Maple V Language Reference Manual*. New York:Springer-Verlag, 1991.

[11] Keiper J, Wickham-Jones T. Designing tools for visualization and optimization. *ORSA Journal on Computing*, 1994;6(3):273–277.

[12] Reps TW. *Generating Language-Based Environments.* Cambridge (MA):The MIT Press, 1986.

[13] Holsapple CW, Park S, Whinston AB. Generating structure editor interfaces for OR procedures. Technical report, Lexington (KY):Center for Robotics and Manufacturing Systems, University of Kentucky, 1989.

[14] Vicuña F. *Semantic Formalization in Mathematical Modeling Languages.* [dissertation], Los Angeles (CA):Computer Science Department, UCLA, 1990.

[15] Scheifler RW, Gettys J. The X window system. *ACM Transactions on Graphics*, 1987;5(2):79–109.

[16] Knuth DE. *The TEXbook.* Reading (MA):Addison-Wesley, 1986.

[17] Lamport L. *LATEX a Document Preparation System.* Reading (MA):Addison-Wesley, 1986.

[18] Tufte ER. *The Visual Display of Quantitative Information.* Cheshire, Connecticut:Graphics Press, 1983.

Chapter 12

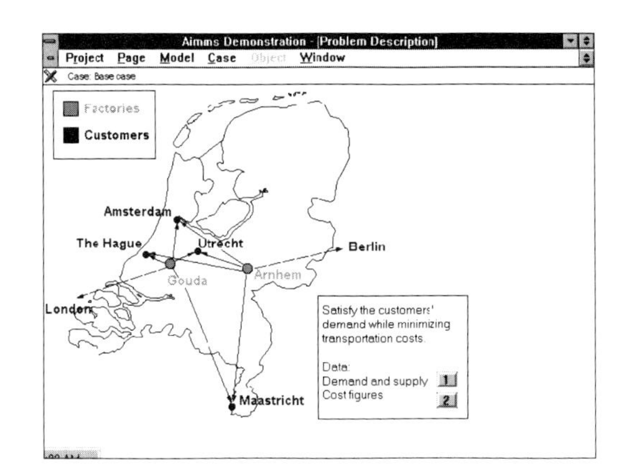

Hypertext

Optimization models attempt to represent complex interactions through constraints, decision variables and objectives. The components of an optimization model are tightly interlinked. Hypertext as discussed in Chapter 6 provides a useful navigation tool for such complex information. Hypertext has begun to be more widely used to support optimization modeling and solution analysis. We describe several systems that use hypermedia effectively for optimization, including Max (Section 12.1), Mentor (Section 12.2), AIMMS (Section 12.3) and gW (Section 12.4). We conclude with a discussion of how the World Wide Web is being used, or might be used for optimization (Section 12.5).

12.1 Max

Kimbrough *et al.* [1], [2] developed a hypertext-based system, Max, for organizing mathematical programming and other types of problem-solving projects. Essentially any important modeling construct, such as variable names and values, parameter names and values, and model names, can serve as a hypertext link (see Figures 11.4 and 12.1). Users need only point to the item of interest in order to obtain more information about that item. When the information is displayed, users can request additional information about items in the new window, and so on. In this fashion, users can navigate through a complicated optimization project.

In the example from Max below, when the user clicks on **asset**, the name of a model, more information about the model is displayed (Figure 11.4). User's can click on any bold-face item, which can represent variables, numeric values, models, among other entities in order to discover more information.

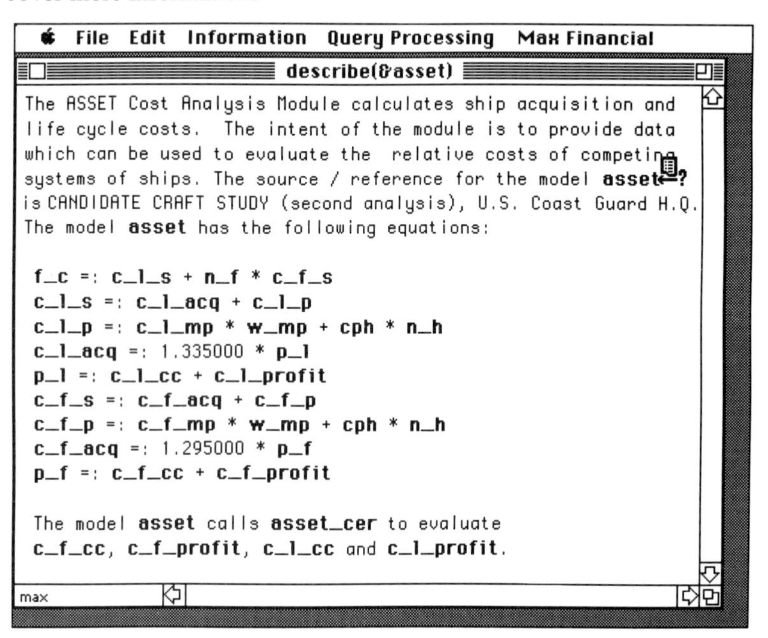

Figure 12.1

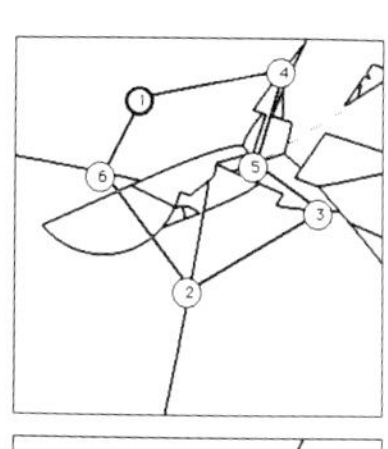

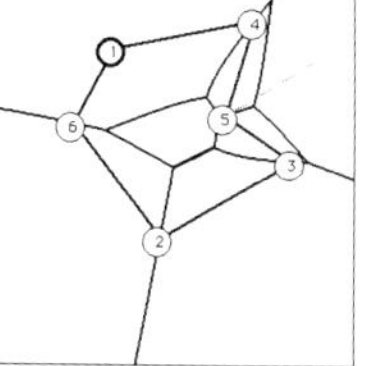

12.2 Mentor

The MENTOR project in the UK [3], [4], [5] has developed hypermedia for teaching mathematical programming, forecasting, simulation, and other traditional OR topics. Mentor provides text, graphics, and video illustrations of various concepts. Moreover, analytical tools (for example, VILP, discussed in Section 15) can be closely integrated.

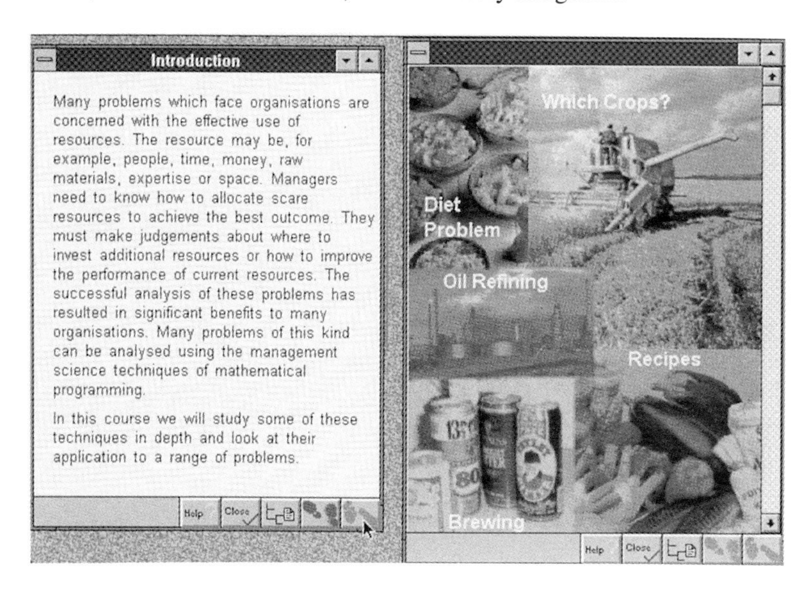

Figure 12.2

To teach the graphical solution technique for linear programming for example, MENTOR provides a tool that allows users to add one constraint at a time and see how the constraint affects the feasible region. Using the instructions in the left-hand window, users can explore a graphical illustration of a two-variable linear program.

In the image below, the user has clicked on the button labeled "constraint 1" on the right to draw the constraint that appears on the left.

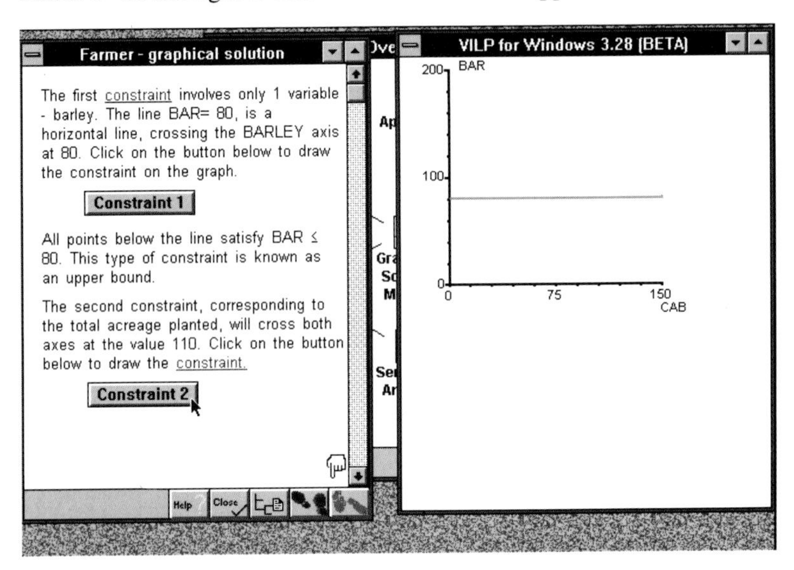

Figure 12.3

Continuing from Figure 12.3, the user has added an additional constraint.

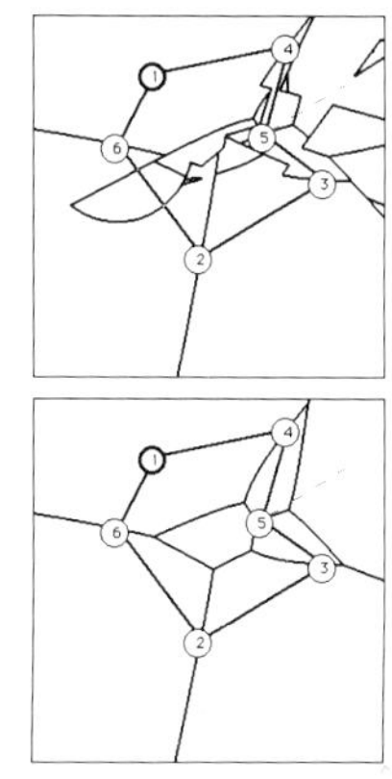

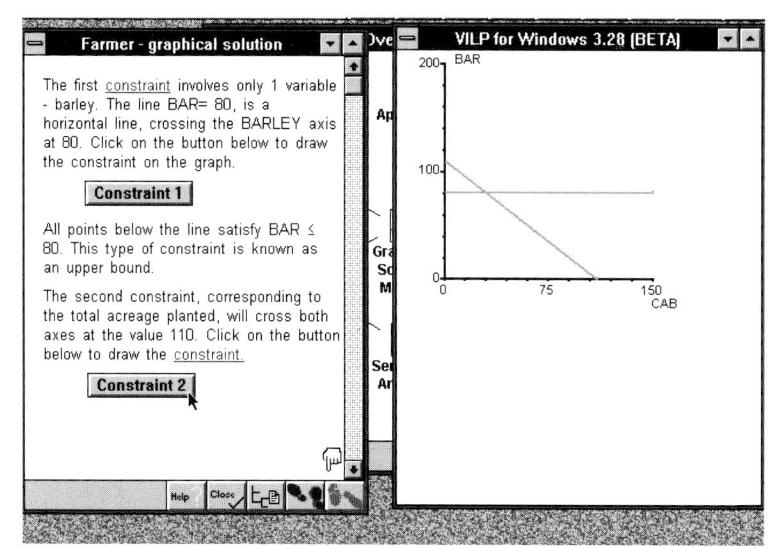

Figure 12.4

12.3 AIMMS

AIMMS [6] provides a development environment for mathematical programming models quite similar to hypermedia authoring environments such as Hypercard and Toolbook. Modelers can create graphical user interfaces by drawing push-buttons, text, input tables, graphics and and linking them together. Furthermore,the text can be linked to the underlying optimization model. The end user can enter new input values, press a push button, the model will be re-solved, with the results then automatically reported within the hypermedia environment. Underlying the hypertext interface is an algebraic modeling language modeled after GAMS.

For example, below is a sample screen from AIMMS for a Hitchcock-Koopmans transportation problem. The map of the Netherlands was copied into AIMMS and the map drawn. The pushbuttons in the lower right allow users to see various model inputs (Figure 12.6).

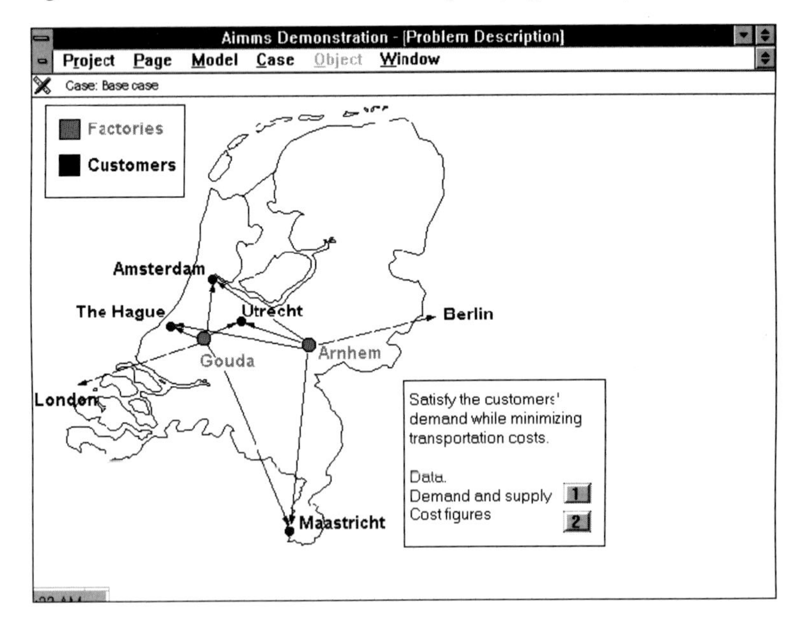

Figure 12.5

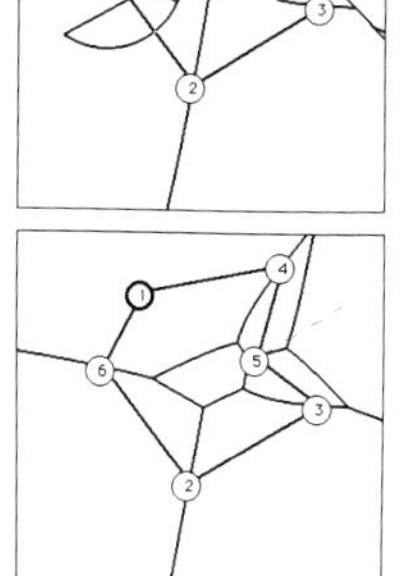

Below is a bar chart representing the transportation costs from Arnhem to the other locations in the problem. Using the "floating object index" on the left, users can scroll through the cities to see a bar chart of transportation costs for each city. Users can change a transportation cost by using the mouse to change the size of a bar in the bar chart.

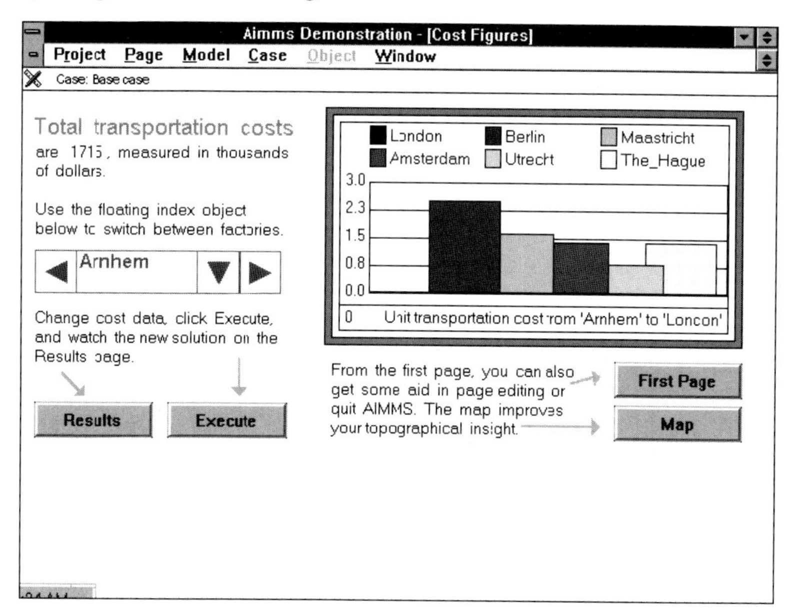

Figure 12.6

By pressing the "Execute" button in Figure 12.6, the model is solved to optimality. By pressing the "Results" button, the solution below is displayed.

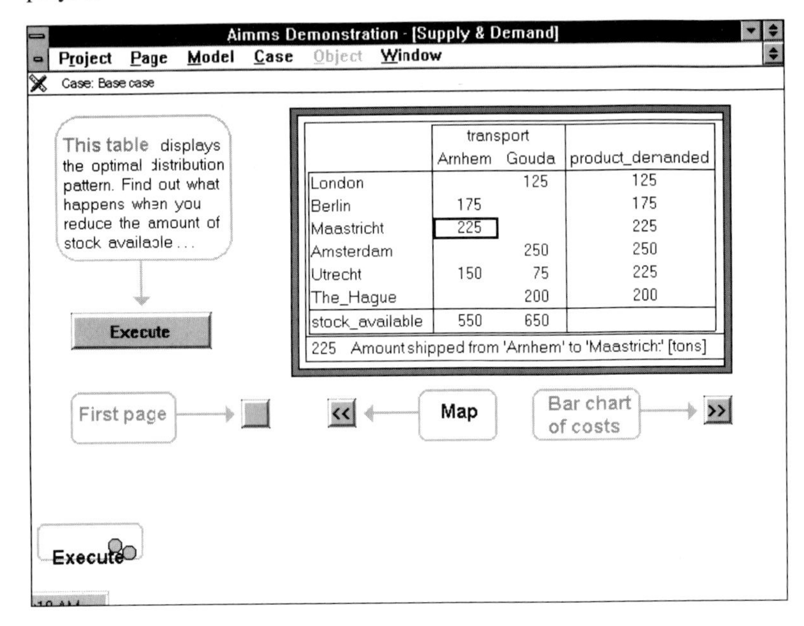

Figure 12.7

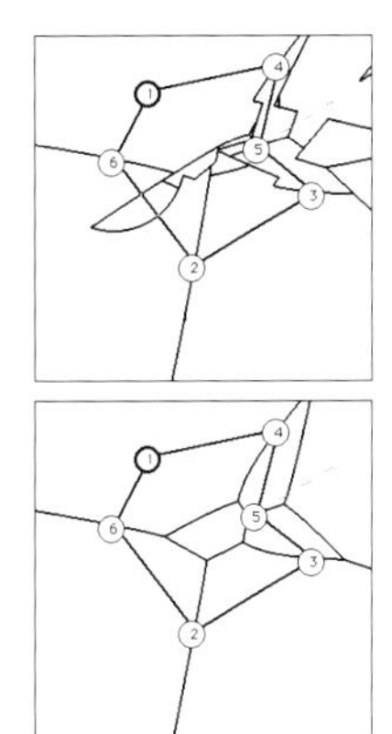

12.4 gW

gLPS [7] (see Chapter 8) has also been enhanced to support hypertext links [8]. Relying on the WEBs hypertext engine developed by Monnard and Pasquier [9], a user of gW (gLPS for WEBS) can create hypertext documents that link together textual model descriptions, graphical models, and solution results together. The figure below, for example, shows a graphical gLPS model, a textual description of the model, the algebraic representation of the model (in LPL) and the solution results. The dots in each window represent a link to another document. For example, when the user clicks on the dot next to the decision variable node in the gLPS model in the upper left, the window containing the optimal solution would be displayed. Similarly, if the user were to click on the dot in the algebraic formulation in the window on the upper right, the textual description of the decision variable would appear. As a modeler interacts with gW, she can add hypertext links to connect models (graphical and textual), textual descriptions of the models, and solutions.

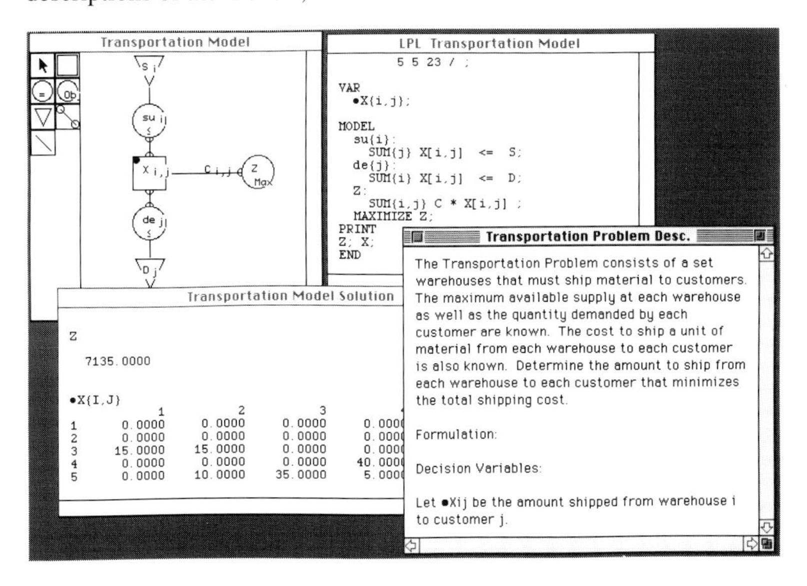

Figure 12.8

12.5 World Wide Web

Using the World Wide Web (Section 6.1.1) for optimization and other modeling is now beginning to develop. The basic idea is simple: user's would submit their optimization problems to a remote computer which would then solve the problem and then report the answer back to the user. For example, Bhargava and Krishnan have proposed Decision Net, a catalogue of modeling tools available on the net [10], [11], [12].

IBM's OSL now allows users to submit optimization problems using a Web page in order to obtain benchmark results on the speed of solution using the OSL solver on different computing platforms [13]. Below is an example of how one submits a problem to OSL over the World Wide Web. The answer is currently returned via e-mail, though it could also be returned directly as a Web page.

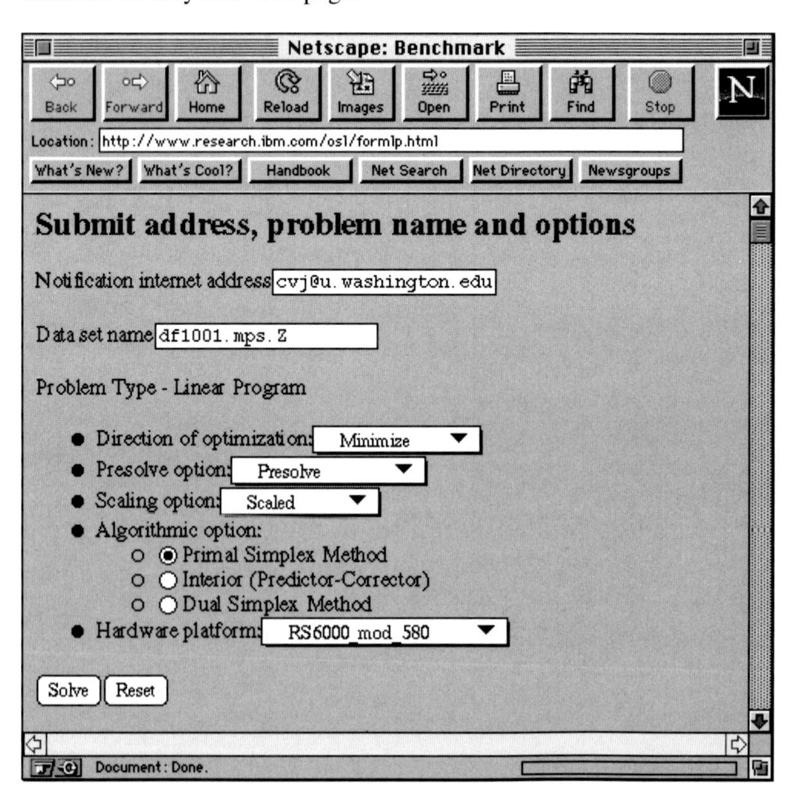

Figure 12.9

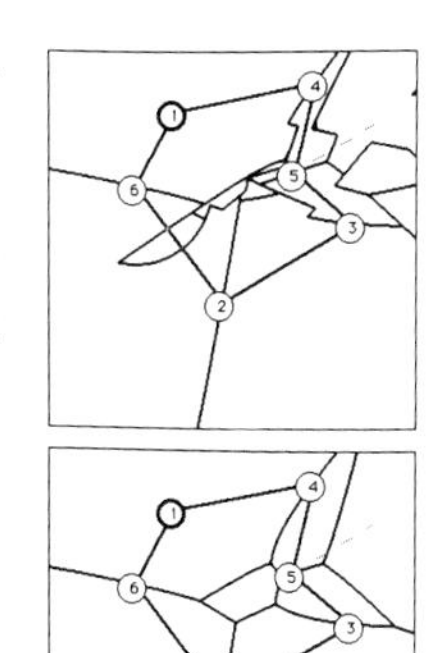

12.6 Summary

Hypermedia provides a natural, easy-to-use mechanism for navigating through complicated information. Since models are complex, hypertext should become a standard interface style for browsing through optimization models and solutions. The lure of clicking a mouse on a variable, column, or constraint to obtain more information is irresistible. The explosive growth in the World Wide Web makes that particular form of hypertext a standard.

Bibliography

[1] Kimbrough SO, Pritchett CW, Bieber MP, Bhargava HK. The coast guard's KSS project. *Interfaces*, 1990;20(6):5–16.

[2] Kimbrough SO, Pritchett CW, Bieber MP, Bhargava HK. An overview of the coast guard's KSS project: DSS concepts and technology. In Volonino L, editor, *Transactions of DSS-90: Information Technology for Executives and Managers, Tenth International Conference on Decision Support Systems*, pages 63–77, Cambridge (MA):1990.

[3] Belton V, Elder MD, Thornbury HE, Candy E. MENTOR–a teacher and friend? Technical Report 93/17, Strathclyde, Scotland:University of Strathclyde, 1993.

[4] Belton V, Elder MD, THornbury HE. Early experiences of MENTORING: Design and use of multimedia materials for teaching or/ms. Technical Report 95/5, Strathclyde, Scotland:University of Strathclyde, 1995.

[5] Mentor Project. The Mentor project. Technical report, Glasgow, Scotland:University of Strathclyde, 1995.
URL: *http://www.strath.ac.uk/Departments/MgtSci/mentor.html*

[6] Bisschop J, Entriken R. *AIMMS The Modeling System.* PO Box 3277, 2001 DG Haarlem The Netherlands:Paragon Decision Technology, 1993.

[7] Collaud G, Pasquier-Boltuck J. gLPS: A graphical tool for the definition and manipulation of linear problems. *European Journal of Operations Research*, 1994;72:277–286.

[8] Collaud G, Pasquier J. gLPS for WEBSs: A scriptable object oriented hypertext system for learning linear optimisation. Technical report, CH-1700, Switzerland:Institute of Informatics, University of Fribourg, 1995.

[9] Monnard J, Pasquier J. WEBSs: an electronic book shell with an object-oriented scripting environment. In Thalmann NM, Thalmann D, editors, *Virtual Worlds and Multimedia*, pages 69–83:John Wiley and Sons, 1992.

[10] Bhargava HK, Krishnan R. Welcome to DecisionNet. Technical report, Monterey (CA):Naval Postgraduate School, 1995.
URL: *http://bhargava.as.nps.navy.mil/dNethome.html*

[11] Bhargava H, Krishnan R, Kaplan D. On generalized access to a WWW-based network of decision support services. In *Proceedings of the Third ISDSS Conference*, June 1995.

[12] Bhargava H, King A, McQuay D. DecisionNet: Modeling and decision support over the World Wide Web. In *Proceedings of the Third ISDSS Conference*, June 1995.

[13] IBM Corporation. The OSL home page. Technical report, Almaden (CA):IBM Corporation, 1995.
URL: *http://www.research.ibm.com/osl/osl*

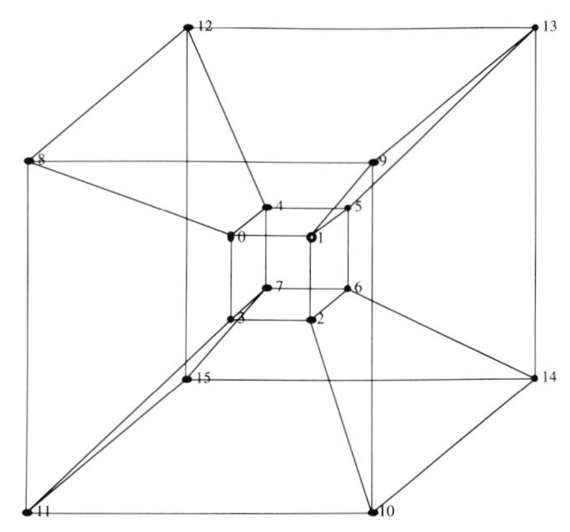

Chapter 13

Networks and Graphs

Although not as ubiquitous as algebra or tables, as discussed in Chapter 8, networks and graphs have often been used for optimization. In addition to graph-based languages for optimization, many systems have been implemented for specific types of problems, including project management software and interactive vehicle routing systems. Vehicle routing systems make extensive use of different types of graphs and networks [1], [2], [3], [4], [5] since a road map is just a graph. With respect to graph-based representations, this chapter concentrates on two topics: general systems for representing and manipulating graphs (Section 13.2), and techniques for drawing graphs in a clear, coherent fashion (Section 13.3).

13.1 Definitions

First, however, it will be useful to define a graph more precisely:

Definition 2 *An (undirected)* graph *G, is an ordered pair* $G = \langle N, E\rangle$ *where N is a set of* nodes, *and* $E \subseteq N \times N$ *is a set edges, consisting of a set of unordered pairs of nodes.*

In a graph, nodes are usually drawn as circles or squares and edges are drawn as lines connecting its associated nodes. A *network* (or *directed graph* or *digraph*) orders the nodes that comprise each edge. For simplicity, though, unless otherwise specified this chapter uses the term "graph" to mean either a graph or a network. Note further that the edges in a network are sometimes called *arcs*, though this chapter shall generally use the term "edge" to refer to either edges in a graph or arcs in a network, unless otherwise specified.

13.2 General Graph-Based Modeling Systems

Some problems such as project management, production planning and vehicle routing use very specific types of graphs. All the visual languages for optimization modeling that have been proposed (Chapter 8) can be modeled as a graph.

Numerous interactive computer graphics systems have been built for problems that rely on a graph-based representation. Each of these systems provides features specifically targeted to its application area. Given the variety of graph-based representations that people use to model and represent their problems, instead of building specialized systems for each different type of graph, perhaps one can construct a general graph-based modeling system for representing and manipulating any type of graph.

Just as spreadsheets provide a general, easy-to-use interface for working with tables, perhaps one could provide a general, easy-to-use interface for working with graphs. If one wanted to create project management models, such a *graph-based modeling system* should make such a task tractable. Similarly, if one wanted to work with graphical representations of vehicle routing problems, the same tool should be able to help. Such a system must provide at least the following capabilities:

- Ability to specify the different types of nodes and edges in the graph. For example, in a project management application, if nodes are used to represent tasks, values such as the early start, early finish, late start, late finish and slack should be associated with each node.
- Ability to input the graph-based model by drawing with a mouse.
- Ability to create a pictorial representation of the graph.
- Ability to restrict the allowable graphical representations.
- Ability to limit the structure of the graph, depending on the application. For example, project management graphs cannot have cycles; common vehicle routing graphs must have cycles since a vehicle often is constrained to return to its point of origin.

A variety of systems have been proposed, some originating from the field of graph theory and combinatorics, some arising from computer science, some from the field of optimization, and some arising from formal language theory. We discuss each in turn.

13.2.1 Combinatorics

Several efforts arose from the field of combinatorics and graph theory. Typically these systems are designed to explore questions important to

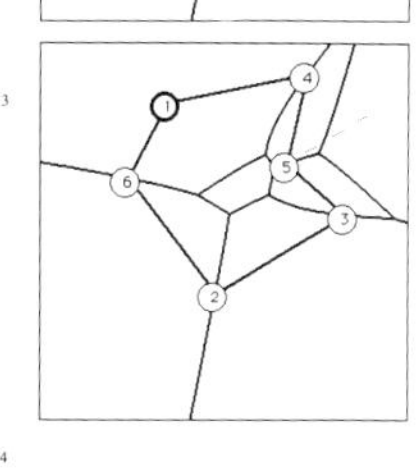

the field. These systems provide a variety of built-in techniques for identifying classic properties of graphs such as finding hamiltonian cycles, strongly connected components, and minimum weight spanning trees.

Among these systems is Cabri [6]. Cabri, which runs on Apple Macintosh, provides a variety of facilities for creating graphs and examining classic graph-theoretic properties of graphs.

For example, this graph from Cabri [6] illustrates the adjacency relationships of a four-dimensional cube.

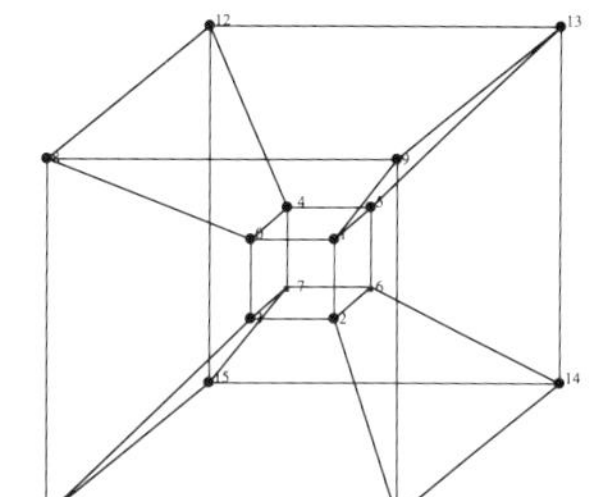

Figure 13.1

Here is a hamiltonian cycle identified by Cabri [6] for the previous graph (Figure 13.1)

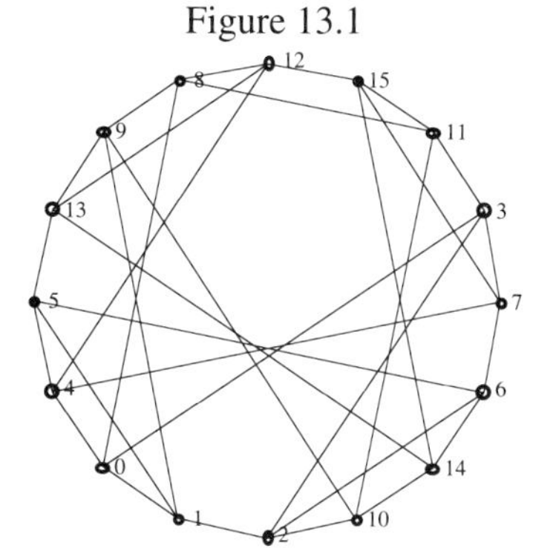

Figure 13.2

Combinatorica [7], a package provided with Mathematica, provides a standard package of functions for graph theory. Consider the graph at right.

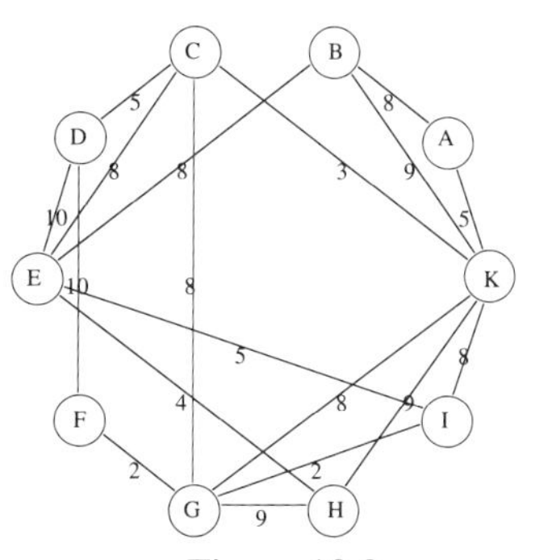

Figure 13.3

At right is its minimum weight spanning tree.

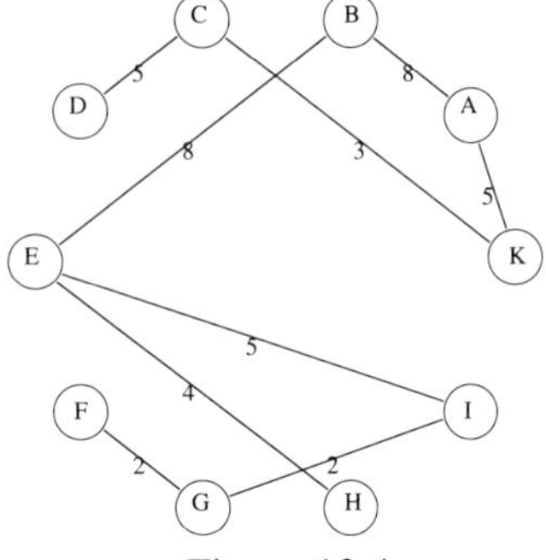

Figure 13.4

13.2.1.1 Optimization

MIMI, a system for production planning and scheduling, supports block-structured representations of mathematical programs. Actually, MIMI supported only a simple, but efficient, database of lists (called sets) and two-dimensional tables. A recent extension [8], MIMI/G, provides a visual modeling language based on a graph (network) representation. The idea is quite simple: tables are interpreted as adjacency matrices for purposes of constructing the graph. This provides an immediate translation of a block-structured representation to a graphical representation. For example, consider the block structured formulation below.

GENERIC MODEL STRUCTURE

ROW BY COL				A	MAT PLA A	P	POP PLA POPINC	B	BOP PLA BOPINC	T PLA	FIN DMC T	D	FIN DMC D
B	PRD	PLA	E	+1		YIELD		BLEND		-1			
B	FIN	DMC	E							+1		-1	
K	UNI	PLA	L			1		1					

Figure 13.5

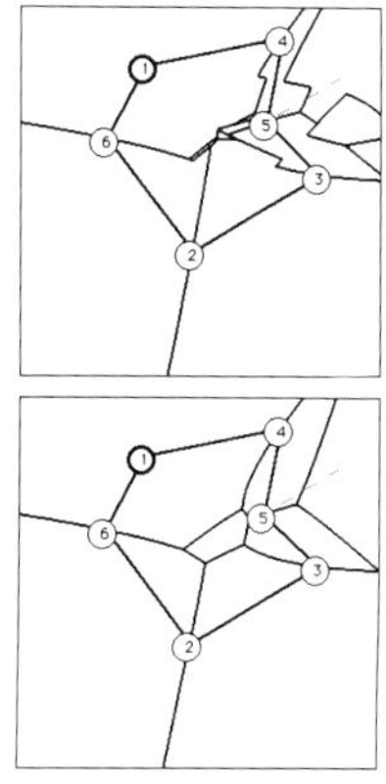

The formulation can also be represented by the graph-based representation below.

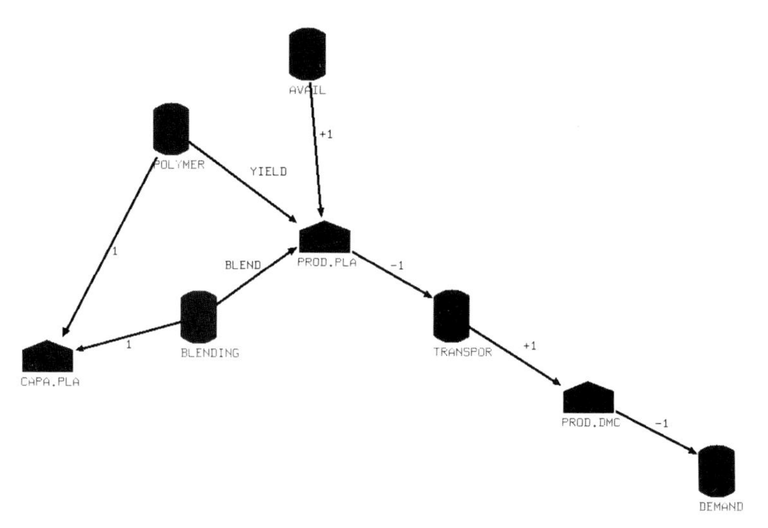

Figure 13.6

Note that Figure 13.6 is essentially an activity-constraint diagram, in that constraints and activities are represented by the node types in the table at right.

constraints	
activities	

Figure 13.7

Interpreting tables as the adjacency matrix of a graph is certainly not a new idea. Greenberg [9] was the first to exploit the idea. Much of the analysis provided by his ANALYZE system [10], [11] arose out of considering the simplex tableau as an adjacency matrix. Since all data in MIMI is stored in tables, any table can be displayed as a graph. For example, below is a graph-based representation of how different crude streams (for example, LH1..LH3; CN1..CN2) can be mixed to produce a particular grade of finished product (PG).

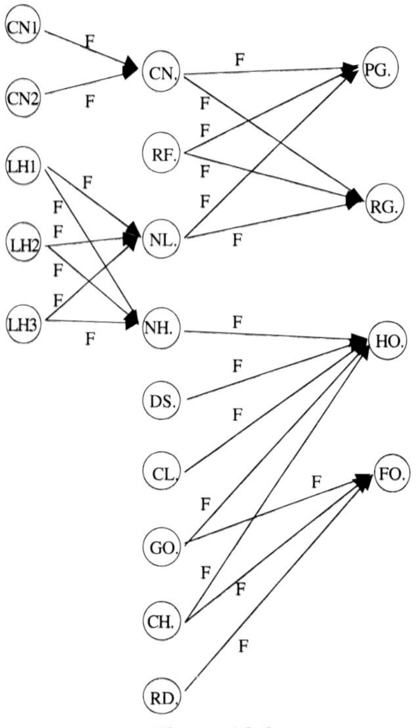

Figure 13.8

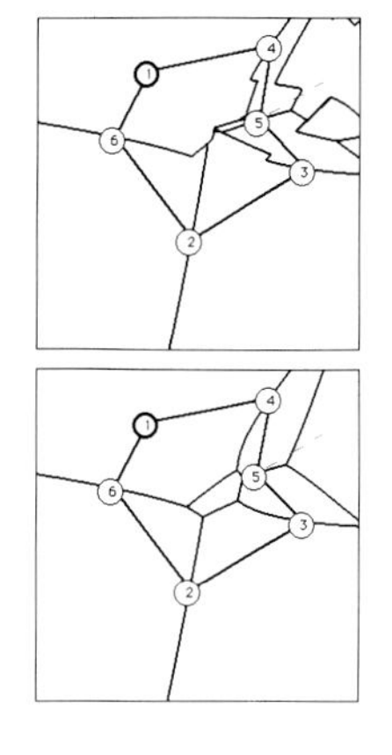

It was based on the following MIMI table.

STR BY PLL									
	NL.	NH	DS	RF	CN.	PG.	RG.	HO	FO
NL.						F	F		
LH1	F	F							
LH2	F	F							
LH3	F	F							
NH.								F	
DS.								F	
GO.								F	F
RD.									F
RF.						F	F		
CN1					F				
CN2					F				
CL.								F	
CH.								F	F
CN.						F	F		

Figure 13.9

13.2.2 Computer Science

Researchers from computer science have also developed systems for working with graphs. These systems usually are intended to support graphs and networks typically found in computer science, such as representations of data structures, flow charts of various forms, among others.

A variety of commercial systems, subroutine libraries, and actual code are now available to create, store, draw and manipulate graphs. Knuth [12] developed a set of standard data structures and a subroutine library called the Stanford Graphbase to allow programmers to easily construct graph algorithms. Although the Stanford Graphbase does not provide built-in procedures for building a user interface, it does provide necessary machinery for the underlying manipulation of graphs that a general user interface requires. North [13] developed `dotty` and `lefty`, two interacting systems for building editors for working with different types of graphs. GraphEd [14], a graph-grammar based graph-based modeling system, also provides layout algorithms. daVinci [15] provides a variety of algorithms for laying out a graph. Netpad [16] was developed as an extensible front-end for graph-theoretic algorithms. It runs under Unix and X Window. Netpad facilitates linking in external algorithms through the use of Unix interprocess communication mechanisms. A carefully defined interface allows external algorithms and Netpad to communicate.

A commercial graph drawing system is available from Tom Sawyer Software [17]. Below is an image showing a layout of a not small graph from their system.

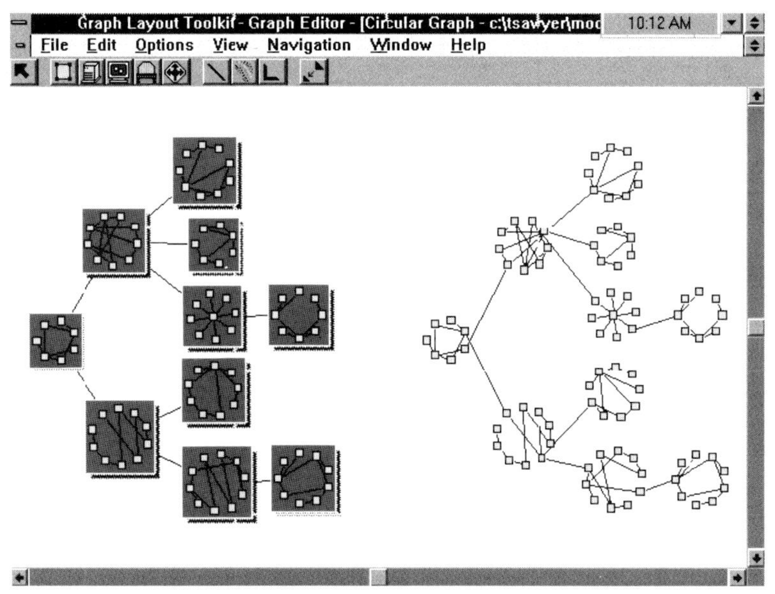

Figure 13.10

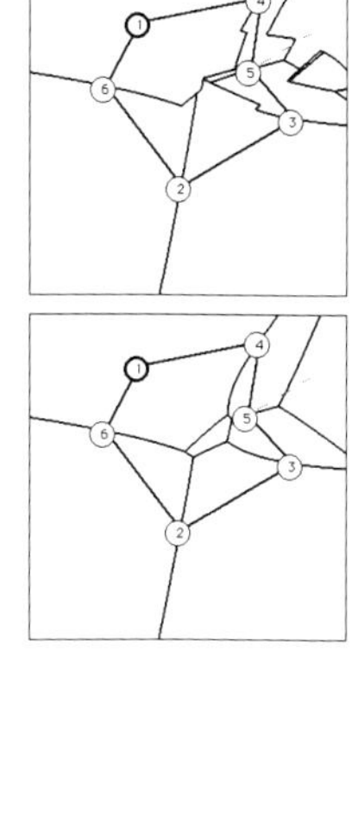

13.2.3 Graph-Grammars

A particular type of general graph-based modeling system relies on a subset of formal language theory called *graph-grammars*. Just as string grammars (discussed in Section 3.3.1) provide formal machinery for defining a textual language, graph-grammars provide formal machinery for defining a graph language, that is, a class of allowed graphs.

Graph-grammars have been used for a variety of applications, but perhaps most notably, for modeling biological systems. In the figure below, for example, are a computer-generated (left) and actual (right) image of a type of moss (Physcomitrella patens) from [18]. The growth of the moss was modeled using a graph-grammar (image courtesy F. D. Fracchia).

Figure 13.11

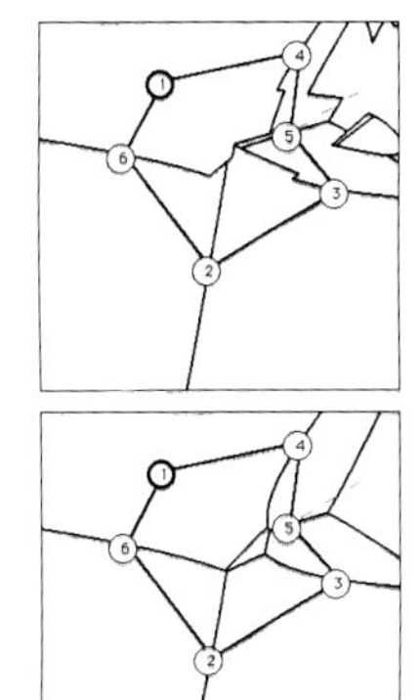

There is no single graph-grammar approach that dominates the field. Several different approaches, each with many different variations have been proposed. All the approaches, however, more or less, build from the basic ideas developed for a string grammar (see Section 3.3.1).

A graph-grammar, like a string grammar, defines a set of *productions* that can be used to construct a particular type of graph. In a graph-based modeling system, the productions form a series of procedures that an end-user can invoke to edit a graph.

Each production consists of a left and right-hand side, G^L and G^R, respectively. Since this is a graph-grammar, however, G^L and G^R are graphs, not strings. Analogously to a string grammar, a graph-grammar production when applied to a graph G will replace G^L by G^R.

In contrast to string grammars, however, G^R must be attached to $G \setminus G^L$. That is, edges (and perhaps nodes) will need to be added in order to connect the nodes in G^R to G. In a string grammar, where the characters are just arranged in a linear sequence, inserting G^R into the string is obvious. In a graph-grammar, performing such a connection is not as obvious. Graph-grammars, therefore, must provide some mechanism for *embedding* G^R into G.

Systems based on graph-grammars include GraphEd [14], Progres [19] and Networks [20], [21]. Consider the graph-grammar used in Networks [20], [21]. Assume that we wish to develop a series of productions for editing graphs representing solutions to simple vehicle routing problems. The graph will consist of a set of warehouse nodes and customer nodes. Edges will represent trips among warehouses and customers. Solutions to this problem must consist solely of cycles (routes) containing one warehouse node and one or more customer nodes. Furthermore, each customer node must appear on exactly one route. One desirable operation is to reroute a customer. For example, in the picture below, we wish to move customer a from its existing route to a location between customers e and d on another route.

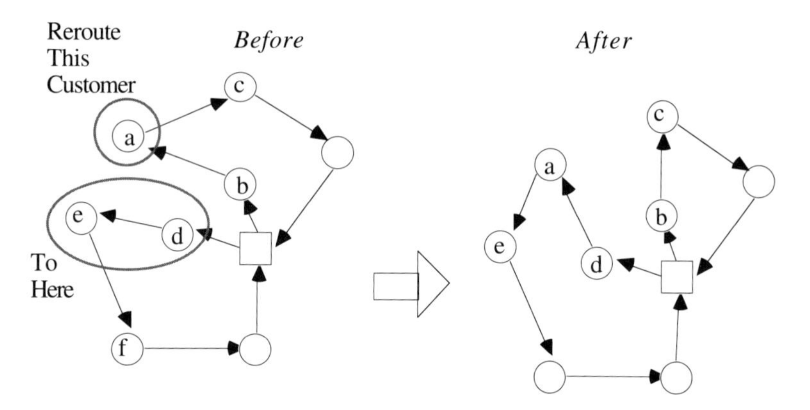

Figure 13.12

This operation involves deleting three edges $((a, c), (d, e), (e, f))$ and adding three new edges $((a, e), (d, e), (b, c))$. The graph consisting of the the three deleted edges and the three added edges, as shown at right, essentially forms the graph-grammar production used in [21].

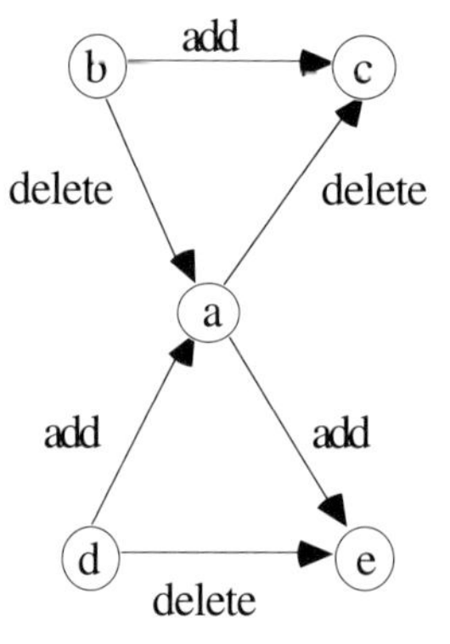

Figure 13.13

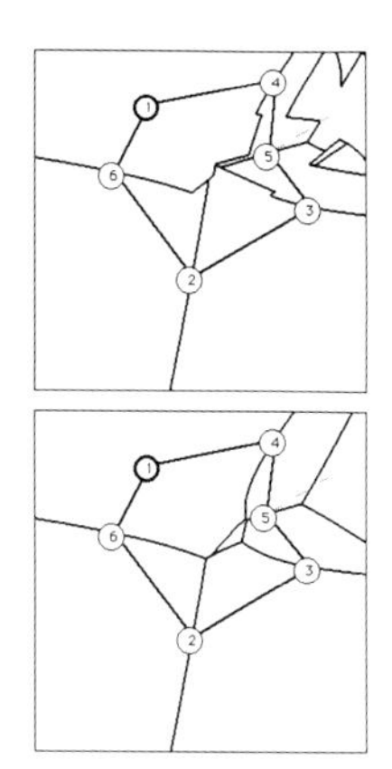

The actual production uses more complicated labels. Note that the production is itself a graph. Because of this, one should theoretically be able to use the graph-based modeling system to allow one to create the actual productions. That is exactly what is done in Networks [21].

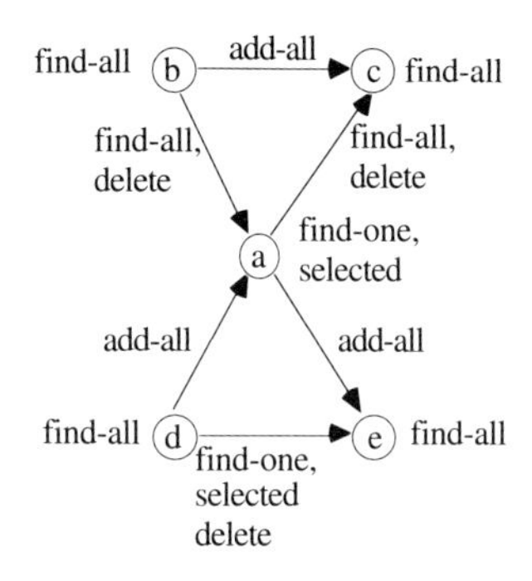

Figure 13.14

The figure below shows some of the graph-based modeling systems created using Networks including graphical syntax-directed editors for bar charts (upper left), decision trees (upper right), project management networks (lower left), and linear programs (lower right) from [22]. Note that the height of the bars in the bar chart is determined by the values of the probabilistic nodes (smaller circles) in the decision tree. Similarly, the values of the leaves of the decision tree are determined from other graphs. For example, the value of 120.0 for the leaf node labeled `lp1` in the decision tree is calculated by the linear program. Finally, the project management network is used to manage the other models. Editors for Structured Modeling [23] and PM* [24] have also been developed.

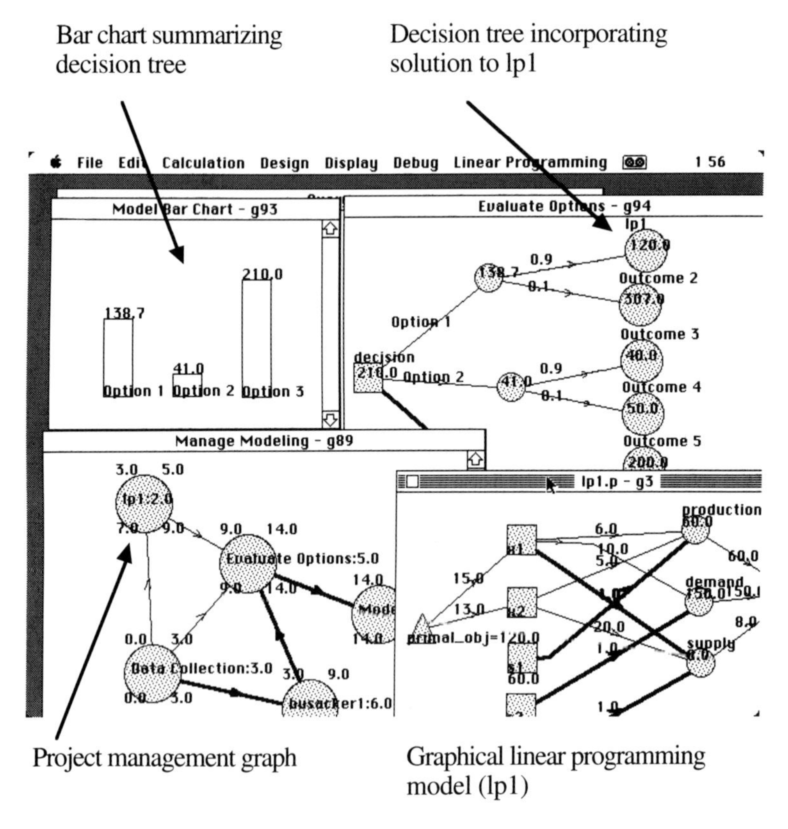

Figure 13.15

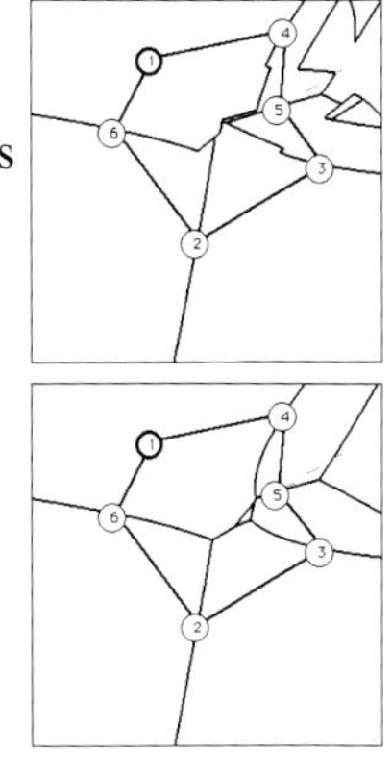

Different graph-grammar formulations have different levels of theoretical and practical expressive power. Theoretical expressive power means the ability of a particular graph-grammar scheme to describe a particular class of graphs. For example, can a particular graph-grammar formalism describe a particular class of graphs (e.g., vehicle routing, project management). Practical expressive power refers to the ease with which one can specify a set of productions to accomplish a particular task. A Turing machine, for example, is theoretically a very powerful computer. Yet Turing machines are not very convenient for real-world programming.

The work described in [21], for example, was based on the graph-grammar scheme proposed by Nagl [25], [26], [27] and Göttler [28], [29], [30], which is theoretically as powerful as a Turing machine. In the original graph-grammar formulation, however, the production to reroute a customer (below) is significantly more complicated than the one shown in Figure 13.14.

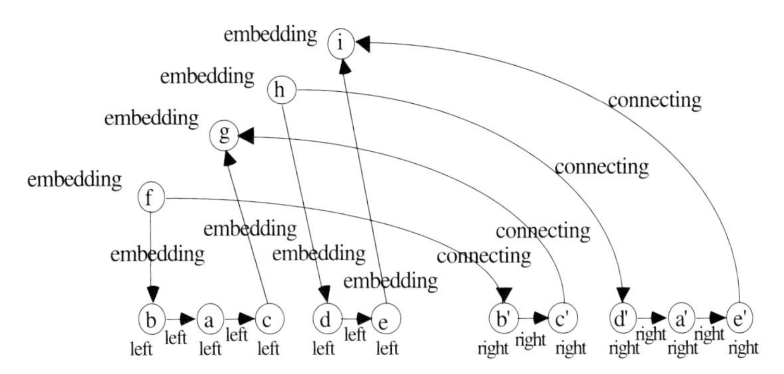

Figure 13.16

13.2.4 Summary

Given the fundamental nature of graph and network representations for optimization modeling, systems that can edit and display a variety of different types of network representations are potentially very powerful. Although many such systems have been developed, only a few are available commercially. Most are academic prototypes, with all the accompanying problems. It seems likely, however, that such systems will become more readily available in the near future.

13.3 Graph Drawing

Large graphs can grow into a tangle of spaghetti. Text and tables at least provide a clear linear or rectilinear representation. This section discusses several approaches that have been proposed for presenting graphs and networks clearly. *Graph layout* or *embedding* has been the subject of intensive investigation by researchers from a number of fields including integrated circuit designers who are interested in finding feasible layouts of VLSI chips [31] and computer scientists interested in algorithms to produce readable views of graphs [32], [33], [34], [35], [36], [37], [38], [39]. The discussion in this section is taken largely from [38], [40], [41] and [7], though other sources are also used. Di Battista, Eades, Tamassia and Tollis [41] provide a general bibliography on graph drawing.

Of course a graph representation is not always the most effective way to present a particular set of information. For example, the complete bipartite graph $K_{10,10}$ with 20 nodes and 100 edges is shown here.

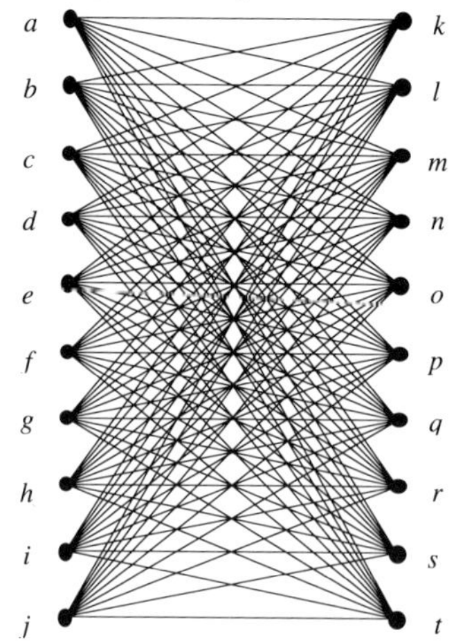

Figure 13.17

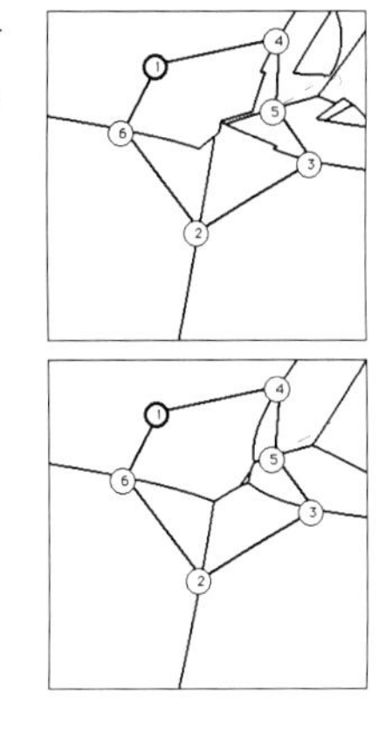

$K_{10,10}$ is probably more effectively represented by the two-dimensional table (adjacency matrix) below.

	k	l	m	n	o	p	q	r	s	t
a	X	X	X	X	X	X	X	X	X	X
b	X	X	X	X	X	X	X	X	X	X
c	X	X	X	X	X	X	X	X	X	X
d	X	X	X	X	X	X	X	X	X	X
e	X	X	X	X	X	X	X	X	X	X
f	X	X	X	X	X	X	X	X	X	X
g	X	X	X	X	X	X	X	X	X	X
h	X	X	X	X	X	X	X	X	X	X
i	X	X	X	X	X	X	X	X	X	X
j	X	X	X	X	X	X	X	X	X	X

Figure 13.18

If one edge were to be removed from $K_{10,10}$, say edge (h, m), then the effectiveness of the tabular representation over the graphical representation becomes even more clear. Here is the table,

	k	l	m	n	o	p	q	r	s	t
a	X	X	X	X	X	X	X	X	X	X
b	X	X	X	X	X	X	X	X	X	X
c	X	X	X	X	X	X	X	X	X	X
d	X	X	X	X	X	X	X	X	X	X
e	X	X	X	X	X	X	X	X	X	X
f	X	X	X	X	X	X	X	X	X	X
g	X	X	X	X	X	X	X	X	X	X
h	X	X		X	X	X	X	X	X	X
i	X	X	X	X	X	X	X	X	X	X
j	X	X	X	X	X	X	X	X	X	X

Figure 13.19

and here is the corresponding graph.

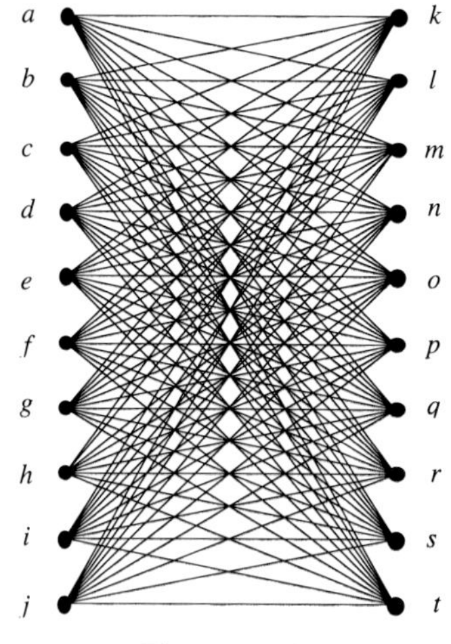

Figure 13.20

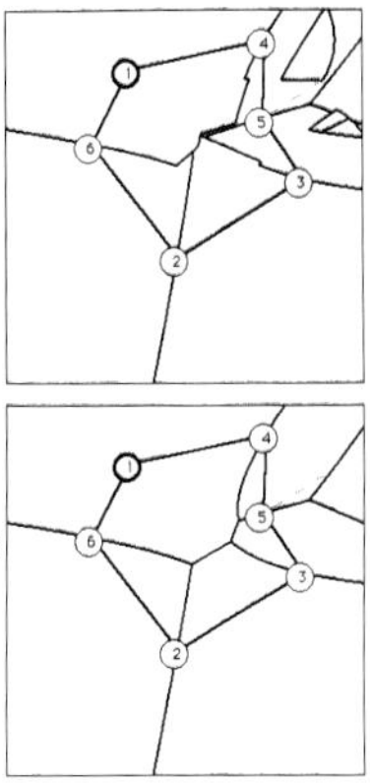

Of course this may be an unfair example. One could attempt to create a different layout of the nodes so that the graph becomes more readable.

A circular layout, for example, has the advantage that no edges can overlap. For $K_{10,10}$, a circular might be considered to be slightly clearer, but that opinion is certainly debatable

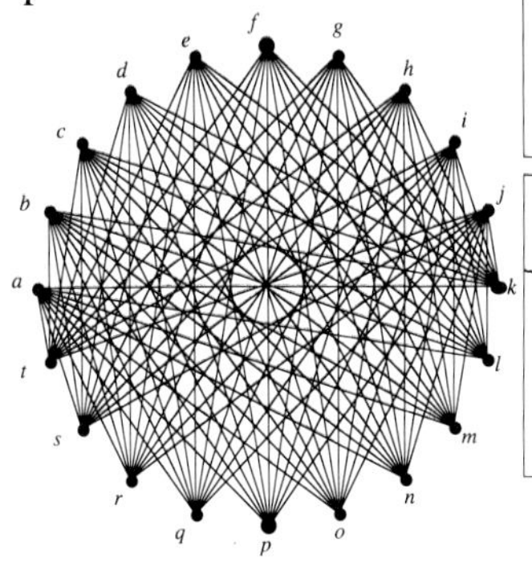

Figure 13.21

A circular layout of $K_{10,10} \setminus (h, m)$ makes the deleted edge somewhat more prominent, but not tremendously.

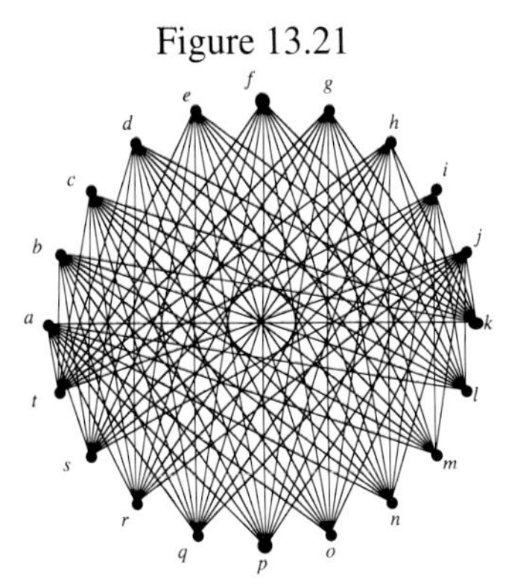

Figure 13.22

One could also try other layout algorithms. For example, a spring equilibrium algorithm, discussed in Section 13.3.5 produced the following horrible embedding of $K_{10,10} \setminus (h, m)$. As horrible as this embedding is, at least the nodes were located so as to partition the two underlying sets in the graph.

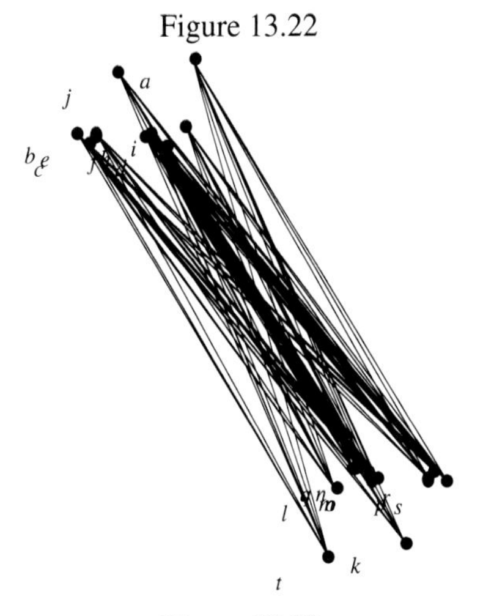

Figure 13.23

No matter the layout algorithm, for $K_{10,10}$, a tabular representation seems far clearer than a graphical representation. If a graph is sparse, however, a graphical representation generally dominates the tabular representation. Exactly at what level of sparsity a graphical representation dominates a tabular representation is an interesting research question. The problem remains, however, of how to create a readable, graphical drawing of a graph.

13.3.1 Layout Criteria

Graph *layout* algorithms seek to calculate positions for nodes and edges that produce a clear, readable representation or embedding of a graph. One must of course define what one means by a clear, readable embedding. Since beauty is in the eye of the beholder, different, often conflicting aesthetic criteria may be imposed. Listed below are some typical qualities sought in graph embedding (taken from [38] and [40]):

- Minimize the number of edges that cross.
- All nodes and edges should be seen, that is, that no node or edge should completely obscure any other node or edge.
- The area covered by the graph should be as small as possible.
- The total length of all edges should be as small as possible.
- If edges are allowed to bend, the total number of bends introduced into edges should be minimized.
- The vertices should be evenly distributed throughout the image.
- The diagram should be balanced both horizontally and vertically.
- Ensure that the angles between edges attached to a node are larger than a certain minimum.

Further constraints can be imposed to complicate the drawing operation. For example, node positions may be constrained to a fixed, rectilinear grid. Edges might be constrained to consist of exclusively horizontal and vertical line segments. When this grid standard is imposed, minimizing the number of edge bends may become important.

Usually the above criteria are in conflict. For example, the embedding that minimizes the number of edge bends may not minimize the number of crossing edges [38]. It seems unlikely that one will be able to develop a single algorithm that will produce an acceptable layout for all purposes.

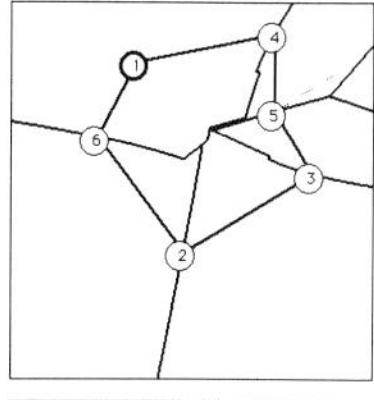

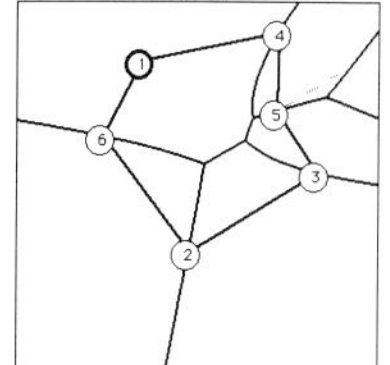

Most layout algorithms are specialized to graphs having particular characteristics such as trees (Section 13.3.2), planar graphs (Section 13.3.3), acyclic graphs (Section 13.3.4), or general graphs (Section 13.3.5). Similarly, different users, even for the same graph, may express different preferences for layouts.

Sometimes, just a few simple layout algorithms can help. For example, if edges overlap, one can draw the graph as a circular embedding (Figure 13.21), or one can perturb the node locations to reveal overlapping edges.

Consider an example from [7]. The graph at right is $K_{2,2,2}$.

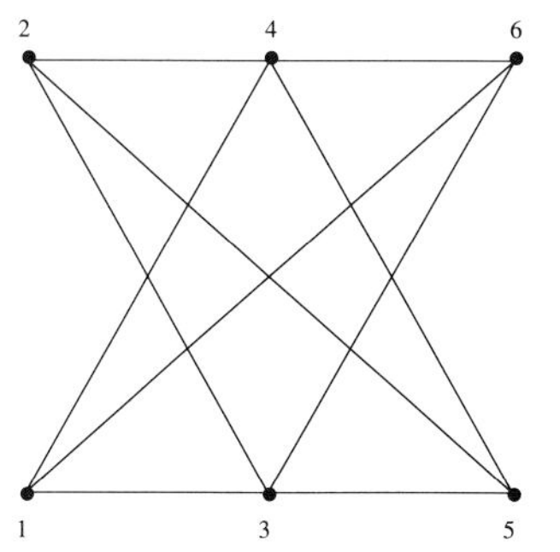

Figure 13.24

After a random perturbation of node positions, the overlapping edges are revealed.

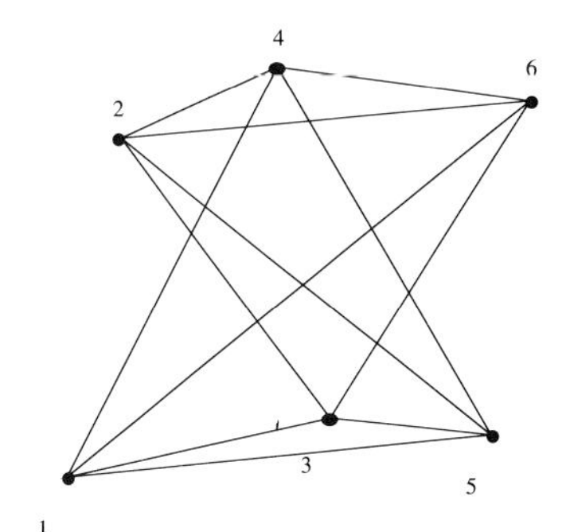

Figure 13.25

We now discuss different approaches for drawing different types of graphs, including trees, planar graphs, acyclic directed graphs, and general graphs.

13.3.2 Tree Layout

Layout algorithms for trees usually assume either that the layout will orient the tree in a linear direction (top to bottom or left to right).

Another layout (constructed using Mathematica) style places the root of the tree at the center of a circle, with descendants oriented around the center, i.e., a *radial* layout.

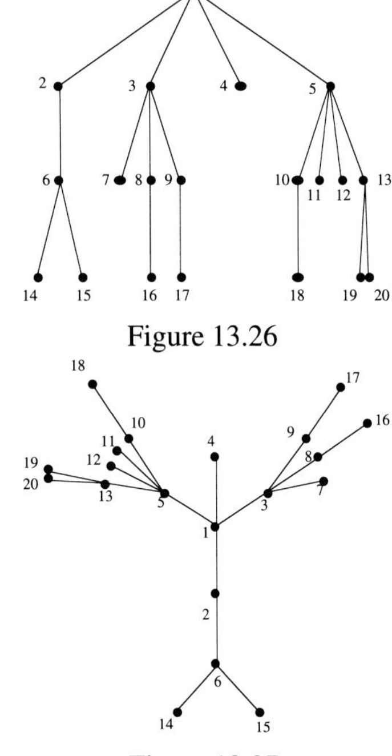

Figure 13.26

Figure 13.27

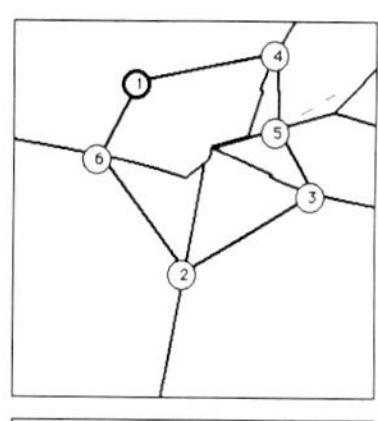

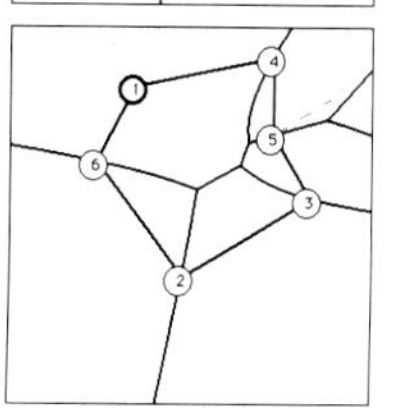

Layout algorithms for trees generally try minimize the area covered by the tree while at the same time forbidding any crossing edges. For a linear orientation of a tree, minimizing area is equivalent to minimizing the maximum width of the tree.

One basic algorithm for drawing a tree oriented vertically (Wetherell and Shannon [42], Vaucher [39]) positions each node as far to the left as possible (assuming the tree is being drawn vertically), subject to constraints on its position relative to its parent and children. This basic algorithm, however does not draw isomorphic subtrees identically.

Below is a tree drawn using the algorithm of Wetherell and Shannon. This figure comes from [43]. Although the two subtrees of the root are isomorphic, they are drawn quite differently.

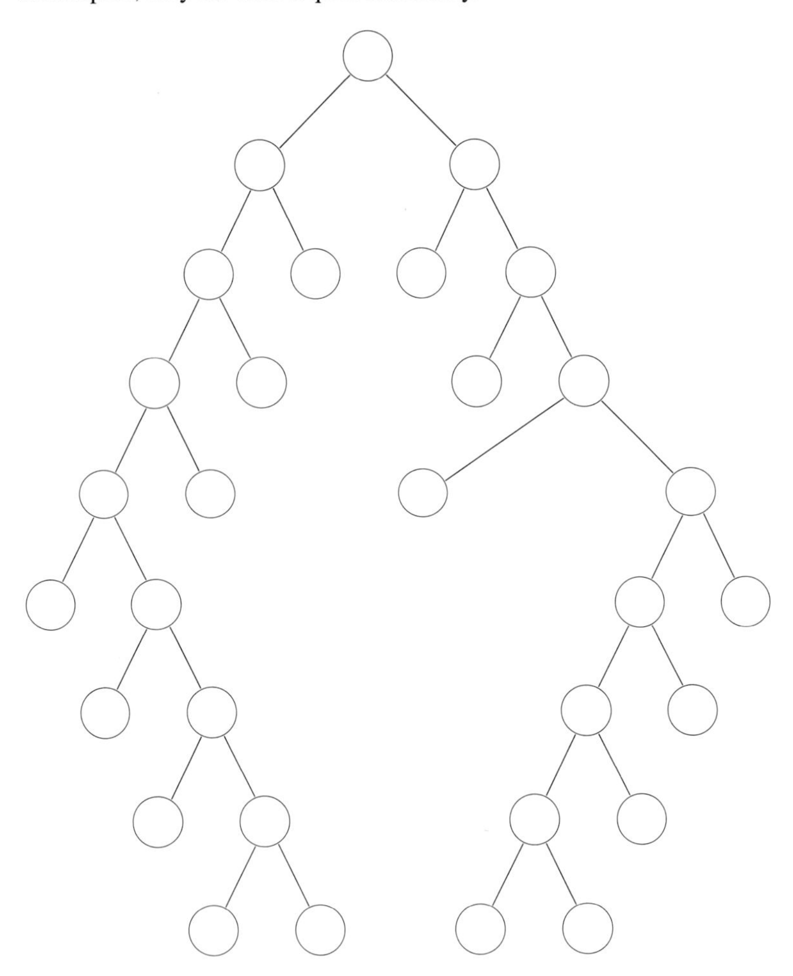

Figure 13.28

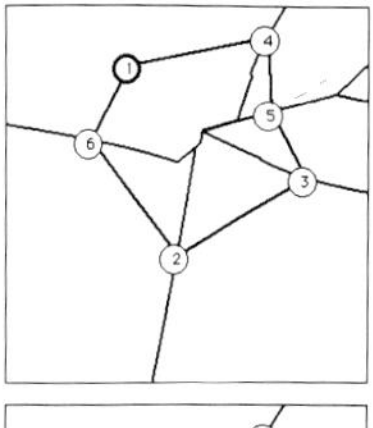

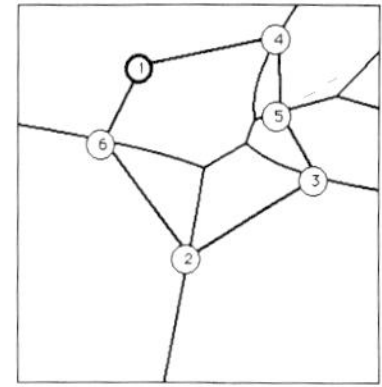

Drawing the same tree using the algorithm of Reingold and Tilford [43] draws both subtrees in a symmetric fashion. Their algorithm follows a simple recursive algorithm: for each node, draw each subtree, separating each subtree by a small distance and then draw the node halfway between each subtree. This simple algorithm draws isomorphic subtrees similarly.

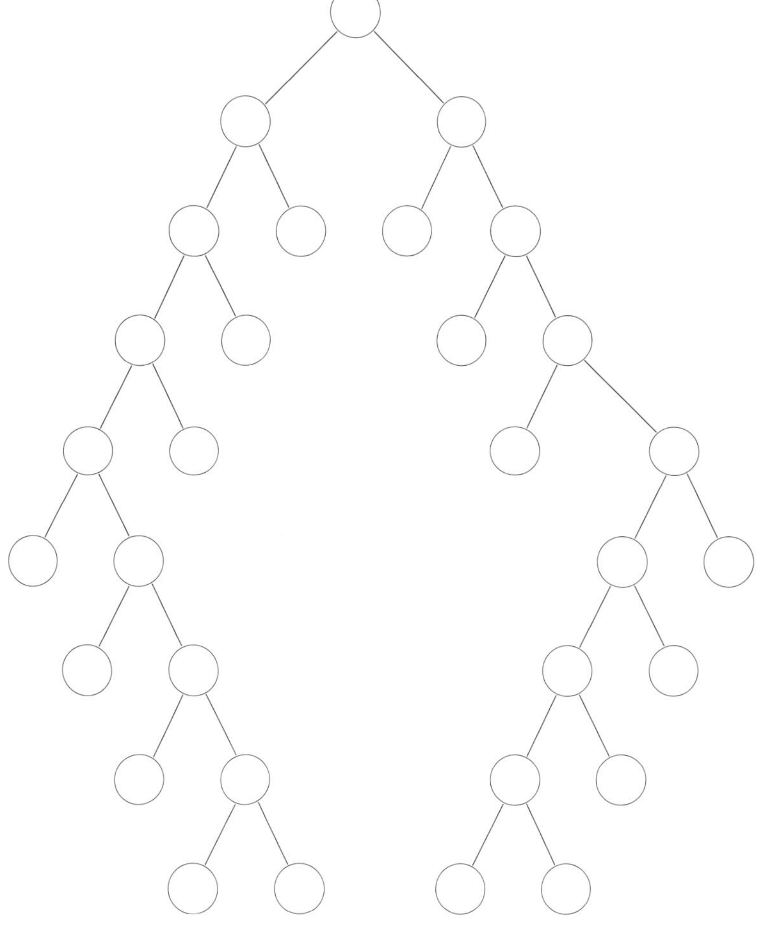

Figure 13.29

Supowit and Reingold [44] showed that polynomial algorithms exist for drawing trees to minimize the number of crossing edges, ensure symmetry and to minimize area. The proof reduces the problem to linear programming. If the nodes must be positioned on a grid, however, then the problem becomes *NP*-hard.

Robertson, Mackinlay and Card [45] explored the use of three-dimensional representations of trees (Figure 13.30) called *cone trees.* In a cone tree, the children of any node are arranged in a circle with the parent node positioned above the center of the circle. The resulting representation consists of a series of cones, the tip of each cone representing the root of a subtree. Their implementation of cone-trees combined three-dimensional representation with animation. When users point to a node in the tree, the cones are smoothly rotated so that the selected node is moved closest to the user.

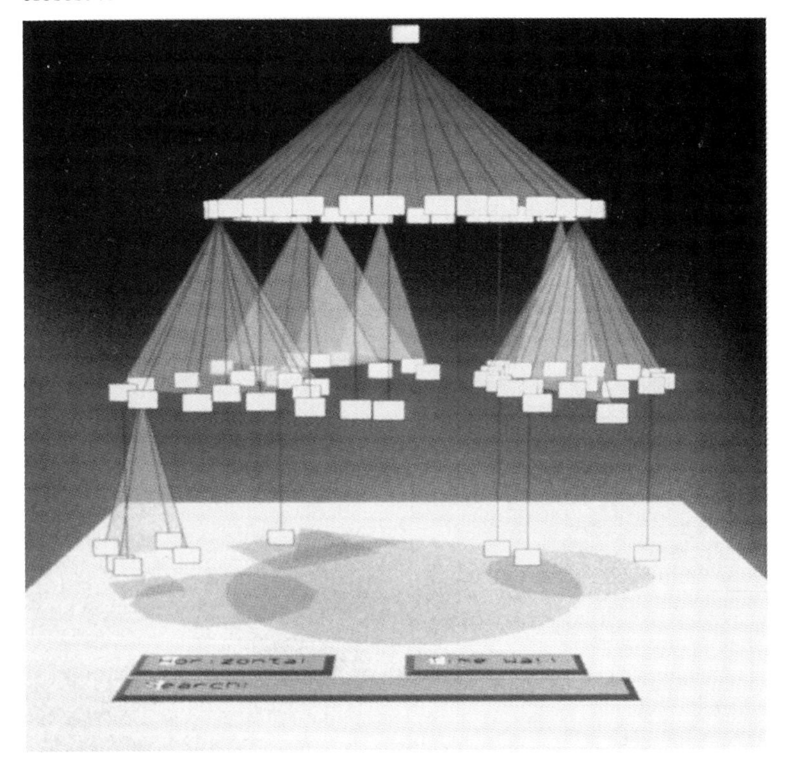

Figure 13.30

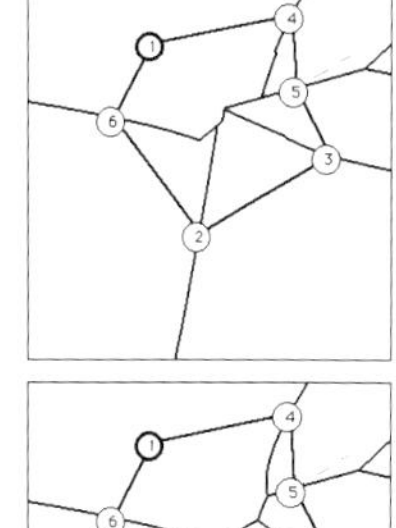

Shneiderman [37] proposed a representation for trees called *treemaps*. We assume that associated with each node in the tree is a numeric value. For example, if one is representing a tree of directories and files as found on a typical computer, the numeric value could be the size of a computer file (leaf node) or directory (all other nodes) .

Consider the tree displayed at right consisting of 6 nodes. The number to the left of each node label gives the value associated with the node.

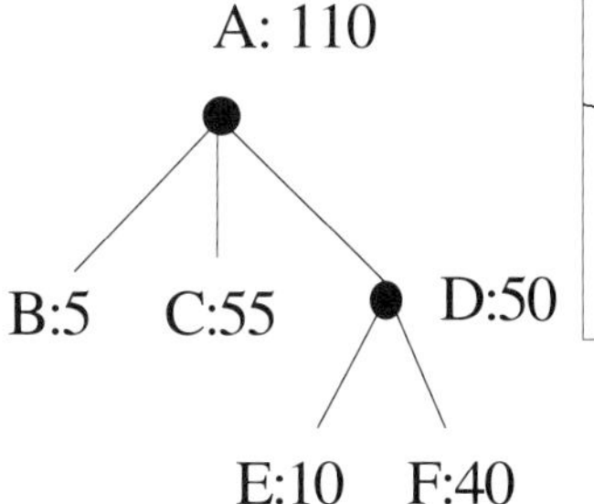

Figure 13.31

The treemap representation of Figure 13.31 is drawn in a rectangular area below. The entire rectangle represents the root node of the tree. The child nodes of the root are drawn as a set of rectangles arranged vertically in the window. The width of each rectangle is proportional to the numeric value of the node. Within each rectangle, its child nodes are drawn as a set of horizontal rectangles. The height of those rectangles is made proportional to the numerical value of the associated node. If any of those nodes have children, then they are represented as a set of vertical rectangles whose height is proportional to their numeric value, and so on.

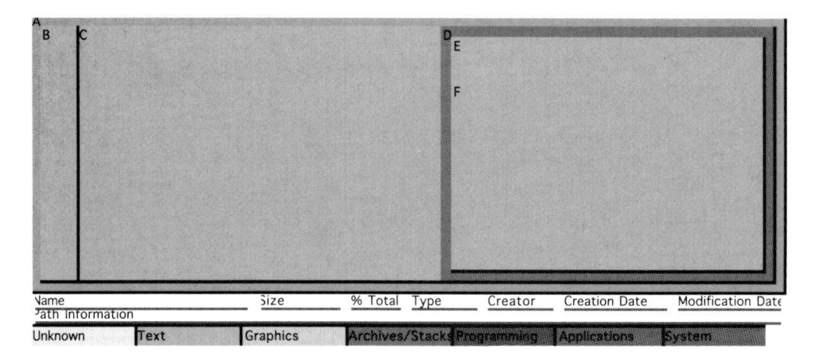

Figure 13.32

Shneiderman calls the process of creating the alternating horizontal and vertical rectangles "slice and dice." Treemaps are clearly related to classic Venn diagrams [46] used in set theory.

Although the previous example illustrates how treemaps are created, the real value of treemaps lies in their ability to display extremely large trees. Treemaps allow for a very high information density, or as Tufte [47] (see Section 4.1.2) calls it, a very high data to ink ratio.

For example, the tree map shown below illustrates the contents of the "Applications" directory of the author's hard disk. This treemap consists of several hundred individual files arranged into a directory tree. Note the large directories (represented as vertical stripes) for Mathematica and Microsoft Excel. Treemap software courtesy of B. Shneiderman.

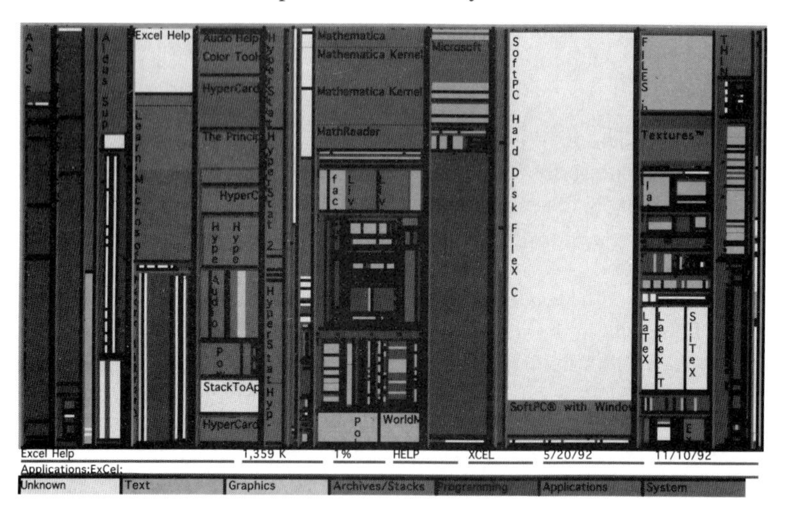

Figure 13.33

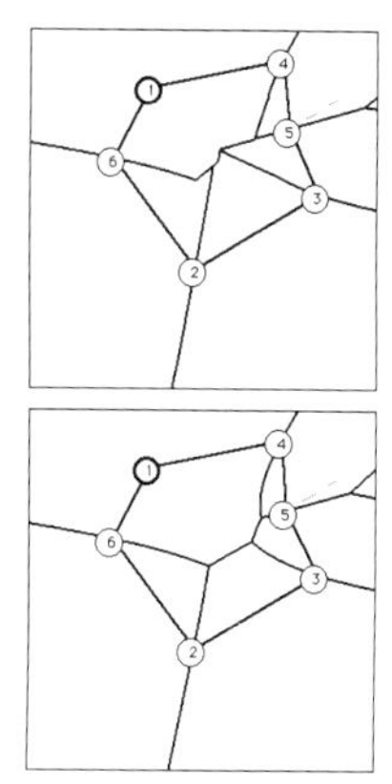

If one "zooms in" on the Mathematica directory, the treemap appears as shown below

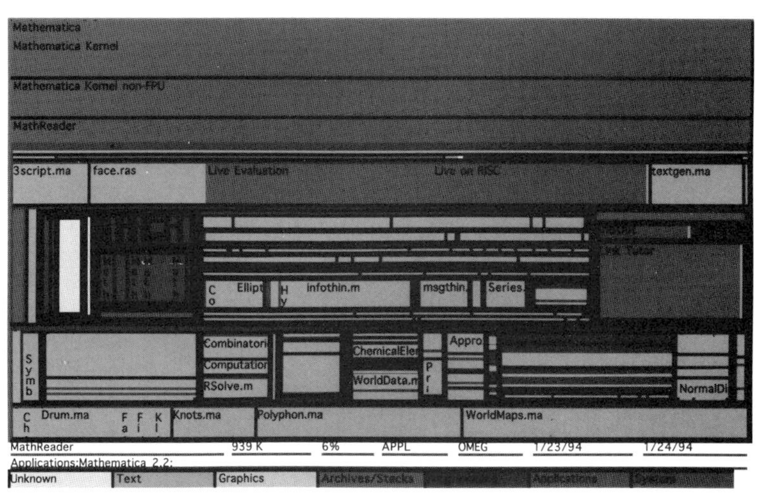

Figure 13.34

If one "zooms in" on the Packages directory, the treemap appears as shown below. Note the "Combinatorica" package [7], which was used to draw many of the graphs shown in this chapter.

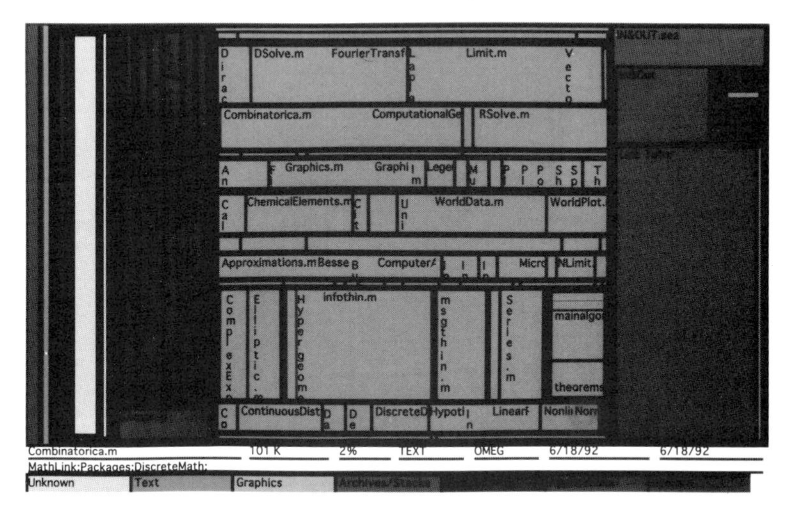

Figure 13.35

13.3.3 Planar Graphs

Efficient, though complex, algorithms exist to detect whether or not a graph is *planar*, that is, can be drawn without any crossing edges at all [48]. Similarly efficient ($O(n)$) algorithms have been developed to actually draw clear representations of planar graphs [49].

Even if a graph is planar, though, many possible embeddings are possible since an internal face in one embedding can be made the external face in another embedding. For example, compare Figures 13.36a and 13.36b, where in (b) is derived from (a) by moving face 2 to the outside.

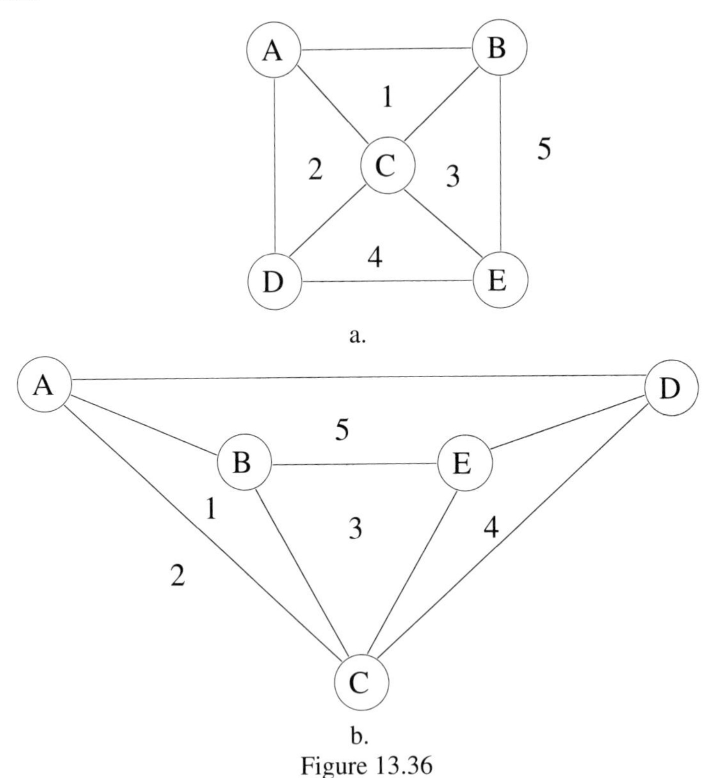

Figure 13.36

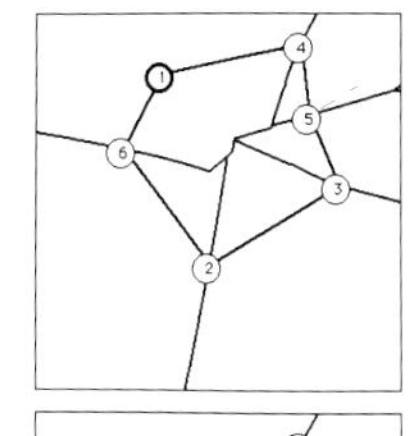

Even if a graph is planar, it may be sometimes preferable to draw the graph with edges that cross. For example, compare the two different layouts of the graph of a three-dimensional cube below. Although the graph is planar, the planar layout in (a) is rarer than the layout in (b) which has edges that cross. This example comes from [41].

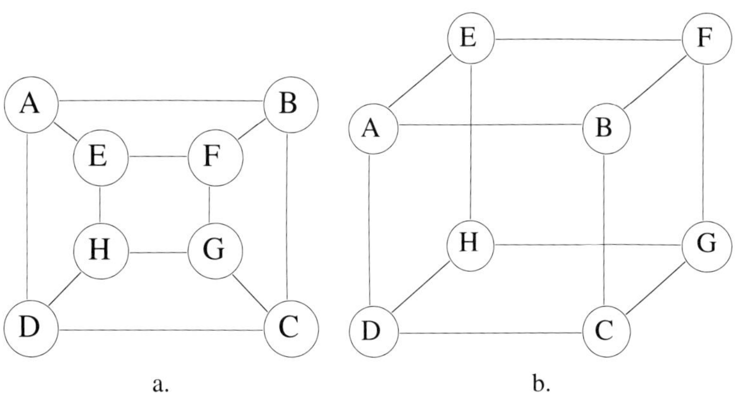

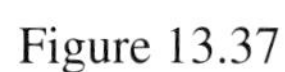
Figure 13.37

Fáry showed that it is possible draw a planar graph without having to bend any edges [50]. The embedding that results from his algorithm, however, frequently produces very tiny faces. Consider the Fáry embedding of a 10,000 node graph below [51], [52].

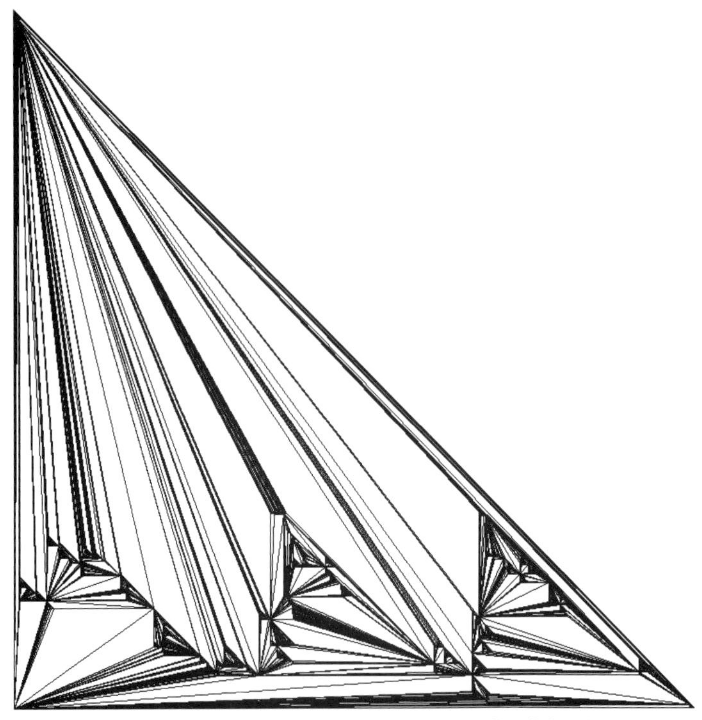

Figure 13.38

A detailed view of the lower right corner of the embedding is shown below. Although the graph is planar, it is not necessarily a particularly clear embedding.

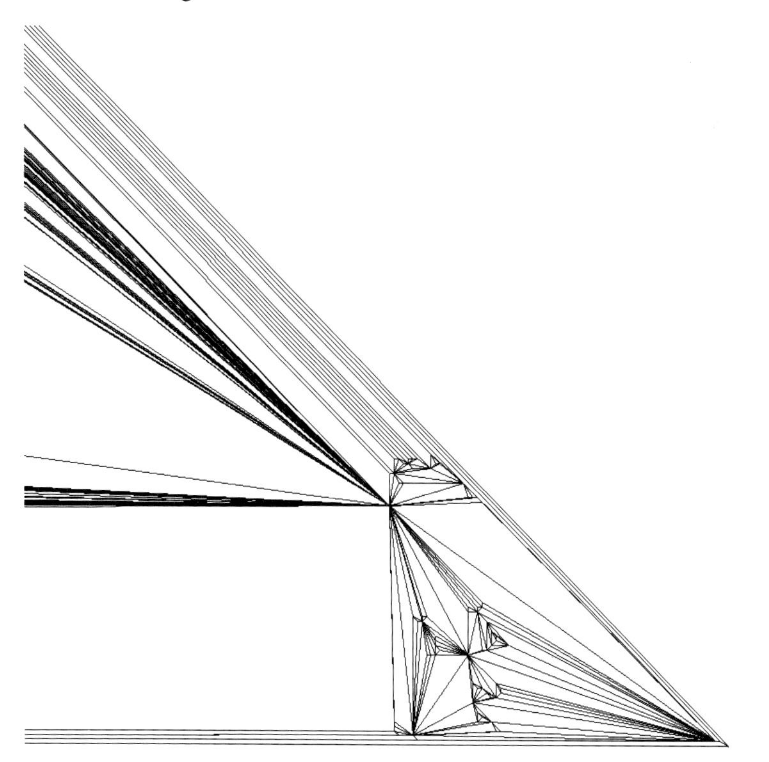

Figure 13.39

Building on Fary's work, Tutte [53] showed that it is always possible to draw a three-connected planar graph so that the resulting faces are convex, that is, a *convex drawing*. For example, consider the planar embedding of a clique of 4 nodes, K_4, at right.

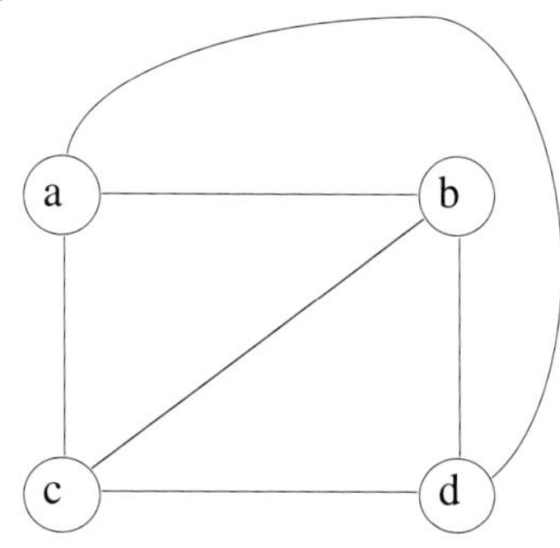

Figure 13.40

Tutte's algorithm is $O(n^{1.5})$ where n is the number of nodes in the graph. Chiba, Onoguchi and Nishizeki [49], [49] developed an $O(n)$ algorithm for producing such a convex drawing. The graph of Figure 13.40 can be drawn without any bent edges as shown at right.

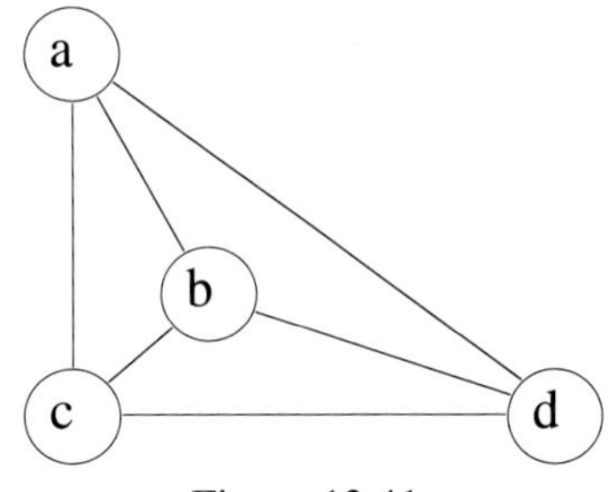

Figure 13.41

In an *orthogonal* embedding, in contrast, edges are allowed to bend but only at 90° angles. More precisely, nodes can only be located at positions on a rectilinear grid, and edges can only consist of horizontal and vertical line segments of the grid. To the right is an orthogonal layout of K_4. Note that an orthogonal layout is possible only for planar graphs whose nodes have at most four attached edges.

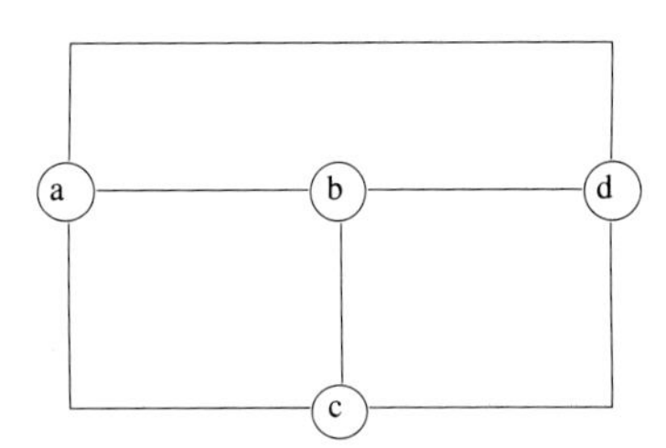

Figure 13.42

An obvious goal for an orthogonal embedding is to minimize the number of edge bends. Tamassia [54] developed an $O(n^2 \log n)$ algorithm based on a minimum cost network flow problem. The formulation considers the relationship between the nodes, N, edges, E, and faces F in a given planar graph. In the figure at right, the faces are numbered. Note that face 4 is an external face of the graph.

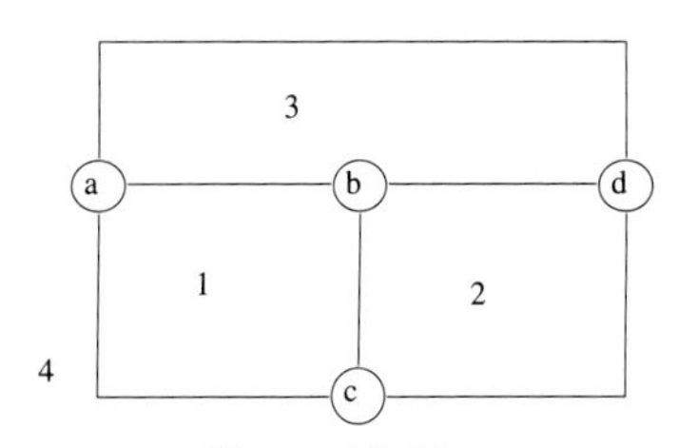

Figure 13.43

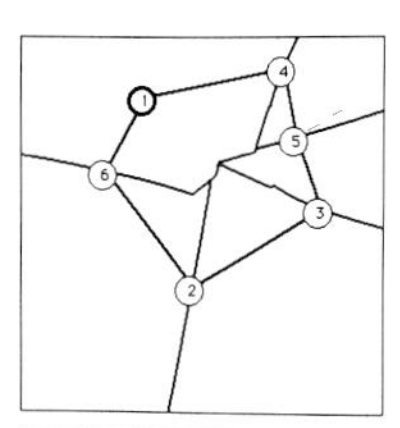

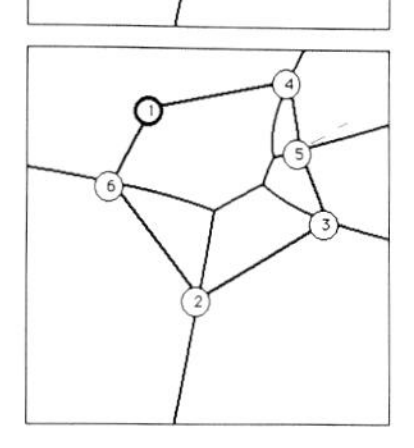

Given a planar graph $G = (N, E)$ with nodes N and edges E, the network flow formulation, $G' = (N', A')$ where:

- $N' = N \cup F$ is the set of nodes in the minimum cost network flow formulation. The formulation includes one node for each node and face in the original graph.
- $A' = E_N \cup E_F$ is the set of arcs in the minimum cost network flow formulation. The arcs, defined more precisely below, are partitioned into two types, those (A_N) associated with the nodes and those (A_F) associated with the faces of the original graph.

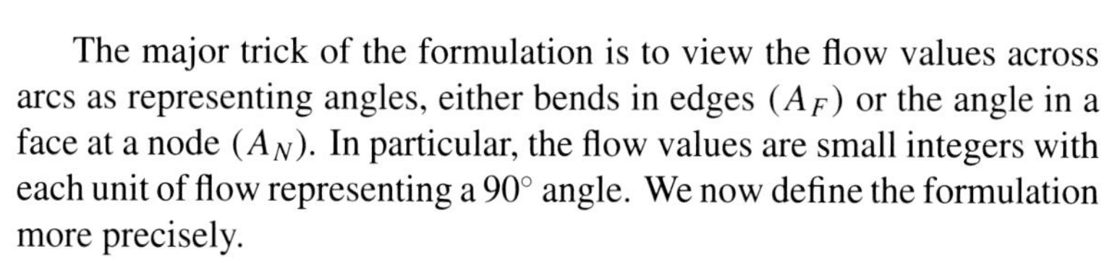

The major trick of the formulation is to view the flow values across arcs as representing angles, either bends in edges (A_F) or the angle in a face at a node (A_N). In particular, the flow values are small integers with each unit of flow representing a 90° angle. We now define the formulation more precisely.

The supply $s_{n'}$ at each node $n' \in N'$ is defined as follows:

$$s_{n'} = \begin{cases} 4 & \text{if } n' \in N \\ 4 - 2|n'| - 8\,\text{ext}(n') & \text{if } n' \in F \end{cases} \tag{13.1}$$

where $|f|$ gives the number of edges adjacent to face f and $\text{ext}(f) = 1$ if f is an external face, and 0, otherwise.

The intuition for setting $s_{n'} = 4$, $n' \in N$ stems from the fact that the sum of the angles incident to n' must be 360° which is represented by a flow value of 4. The supply at face nodes ensures that each face is a rectilinear polygon.

The supply values at each node can be shown to be closely related to Euler's formula for planar graphs. Euler's formula relates the number of nodes, edges and faces (including the external face) in a planar graph. It can be stated as:

$$|N| + |F| - |E| = 2 \tag{13.2}$$

With respect to the supply values in Tamassia's formulation, note that the sum of the supply values must equal 0, i.e.,

$$\sum_{n' \in N} s_{n'} + \sum_{n' \in F} s_{n'} = 0 \tag{13.3}$$

Expanding the previous equation yields,

$$\sum_{n' \in N} 4 + \sum_{n' \in F} 4 - 2|n'| - 8\,\text{ext}(n') = 0 \tag{13.4}$$

The above can be rewritten as

$$4|N| + 4|F| - 8 - 2\sum_{n' \in F} |n'| = 0 \tag{13.5}$$

The value $\sum_{n' \in F} |n'|$ counts each edge twice, so equation 13.5 can be written:

$$4|N| + 4|F| - 4|E| = 8 \tag{13.6}$$

Dividing both sides of the above equation by 4 yields Euler's formula.

Edges E_N consist of pairs (n, f) for each $n \in N$ and f adjacent to n. The flow across these arcs represents the angle made at node n in face f. That is, the angle at node n in face f is given by $90x_{nf}$.

Arcs A_F consists of all ordered pairs (f_1, f_2), where $f_1, f_2 \in F$ and f_1 and f_2 share a common edge. A positive flow $x_{f_1 f_2}$ across arc (f_1, f_2) represents a bend of $90x_{f_1 f_2}$ degrees in their common edge. The angle of $90x_{f_1 f_2}$ degrees is the angle internal to f_1.

The lower bound, l_{nf}, upper bound, u_{nf}, and unit cost c_{nf} for arc $(n, f) \in A_N$ is defined as follows:

$$\begin{aligned} l_{nf} &= 1 \\ u_{nf} &= 5 - \deg(n) \\ c_{nf} &= 0 \end{aligned} \tag{13.7}$$

where $\deg(n)$, the *degree* of n, is defined as the number of edges adjacent to N. The lower bound reflects the fact that each node makes at least a 90° angle in each of its faces. The upper bound reflects the fact that the sum of the angles emanating from the face can be no more than 360°. Since the objective is to minimize the number of edge bends, the cost coefficient is 0, since the angles represented by these arcs do not represent bends in original edges.

The lower bound, $l_{f_1 f_2}$, upper bound, $u_{f_1 f_2}$, and unit cost $c_{f_1 f_2}$ for arc $(f_1, f_2) \in A_F$ is defined as follows:

$$\begin{aligned} l_{f_1 f_2} &= 0 \\ u_{f_1 f_2} &= m \\ c_{f_1 f_2} &= 1 \end{aligned} \tag{13.8}$$

where m is the maximum number of bends allowed in a single edge. Note that if one sets m to a suitably large number, then the formulation would find the absolute minimum number of total edge bends possible for a particular graph. The unit cost of 1 ensures that the objective function counts the total number of edge bends.

Tamassia [54] proves the following propositions:

Proposition 1 *Each feasible integer circulation in G' represents a computable orthogonal embedding of G.*

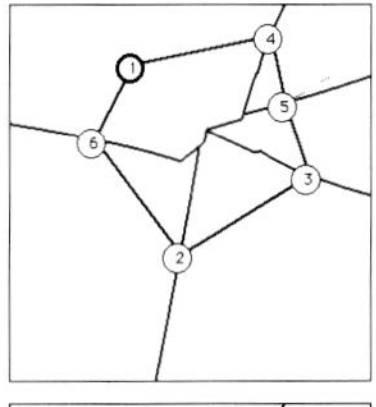

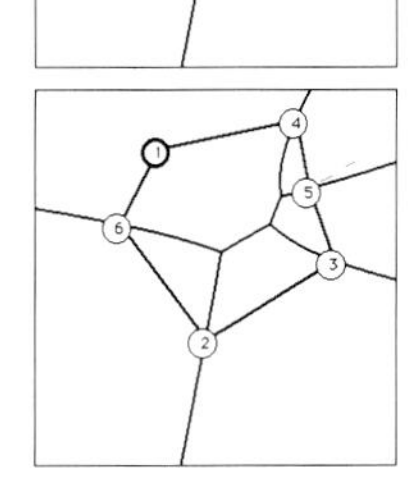

Proposition 2 *Each orthogonal embedding of G observing the bounds on the maximum number of bends in a single edge* (m) *corresponds to a feasible integer circulation in* G' *whose cost is equal to the number of bends in the edges of G.*

Since the input parameters to the network flow formulation are all integer, an optimal solution with all integer flow values exists, and be the previous propositions, an optimal solution to the minimum cost network flow formulation produces an orthogonal embedding that minimizes the number of edge bends.

In the example at hand, a piece of the network flow formulation is shown below. This piece shows the edges attached to node a and face node 1. Other edges are hidden in order to improve clarity.

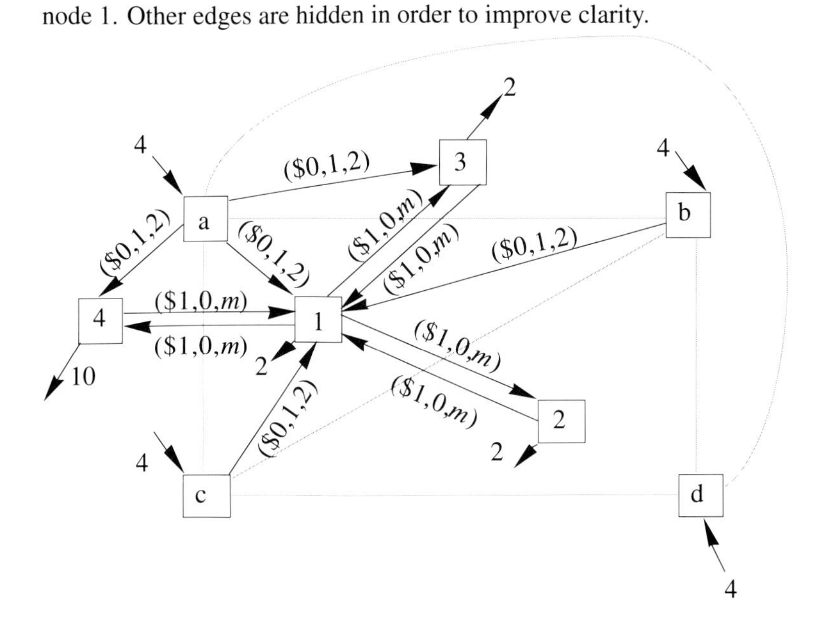

Figure 13.44

When this problem is solved, an optimal solution with objective function value of 4 is shown below. The objective function value indicates that an orthogonal planar embedding having only 4 edge bends can be produced. Figure 13.42 illustrates such an embedding.

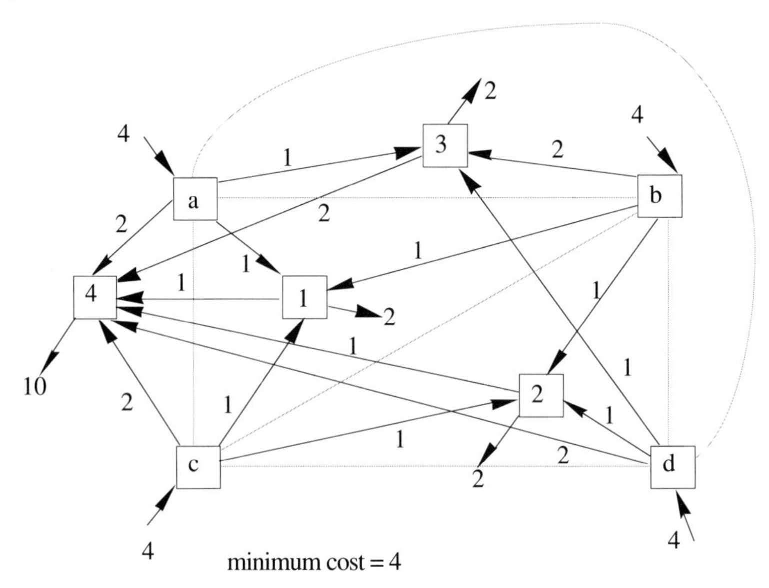

Figure 13.45

The orthogonal layout problem for planar graphs had been conjectured to be *NP*-hard by Storer [55] in 1984, but Tamassia's [54] formulation in 1987 as a minimum cost network flow algorithm disproved this conjecture.

To the right is an example from [54] of a larger graph with a non-optimal orthogonal layout having 5 edge bends.

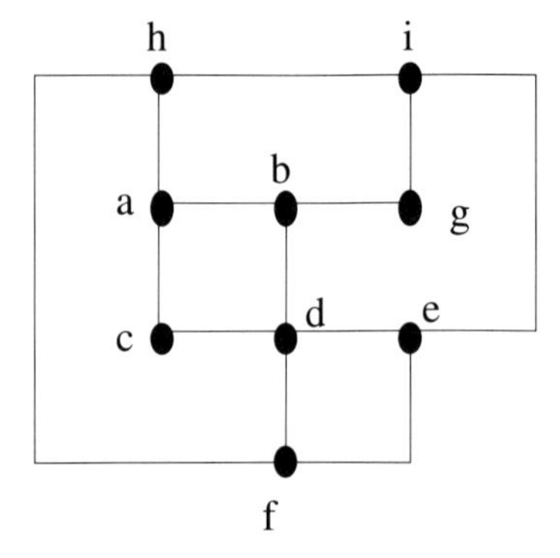

Figure 13.46

To the right is the optimal layout of the previous graph with just 4 edge bends.

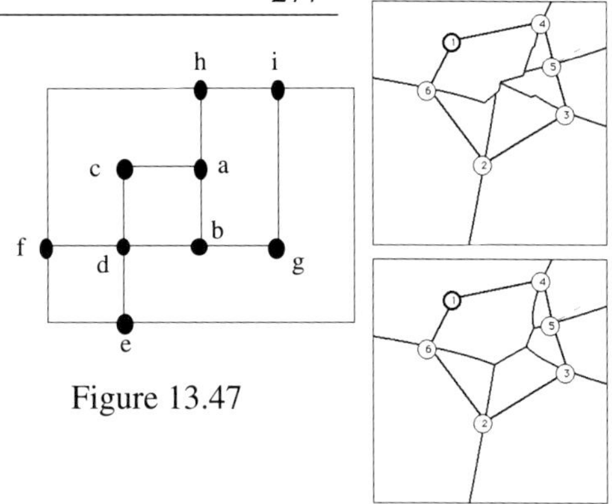

Figure 13.47

13.3.4 Acyclic Directed Graphs

More general than the planar graphs of the previous section are acyclic directed graphs, which are often used for modeling. Perhaps the most widely used example is project management (PERT/CPM). A common type of algorithm for drawing this class of graphs first partitions the nodes into a series of k levels numbered $1, \ldots, k$ such that for all edges (n, m) the level for n is less than the level for m. Nodes at the same level are drawn in the same horizontal row.

The next step is to order or sequence the nodes at each level i from left to right. Different sequences will produce different numbers of edges that cross. Eades *et al.* [56] showed that minimizing the number of edges that cross for a k-level graph is *NP*-hard even for graphs having just two levels (i.e., bipartite graphs), so most practical algorithms will be heuristics. Several heuristic algorithms have been proposed for this type of graph ([57], [58], [59], Messinger [35], [60] and [61]). Cruz and Tamassia [40] call this a *layering* algorithm.

13.3.5 General Graphs

Algorithms for general graphs have been less studied than for specific types of graphs. Note that algorithms for drawing planar graphs can be applied to non-planar graphs if fictitious nodes are added at every edge crossing point. Similarly, layout algorithms for directed acyclic graphs can also be applied to general directed graphs if the nodes are partitioned into a hierarchy. Edges that point up can temporarily change direction until the layout is actually complete.

For general graphs, several heuristics have been proposed that rely on a "spring-equilibrium" model [62], [63]. Edges are modeled as springs that pull adjacent nodes together. To counteract the force exerted by edges, nodes also exert a repulsive force on each other. The algorithm then seeks a local minimum energy for the entire system. For sparse, two-connected graphs, this type of algorithm can work reasonably well [38]. For graphs containing dense subgraphs, however, the drawings are generally poor.

Attempts have also been made to draw general graphs in three dimensions [64], [65]. An attempt to use three-dimensional representations for trees was discussed in Section 13.3.2. SemNet [64] was developed to explore relationships in large Prolog knowledge bases. One layout algorithm provided in SemNet essentially performed the spring-equilibrium algorithm in three dimensions. Ware [65] showed that visualization of three dimensional graphs was enhanced by the use of motion to enhance parallax and by the use of stereo glasses to view the graph.

13.3.6 Hierarchical Graphs

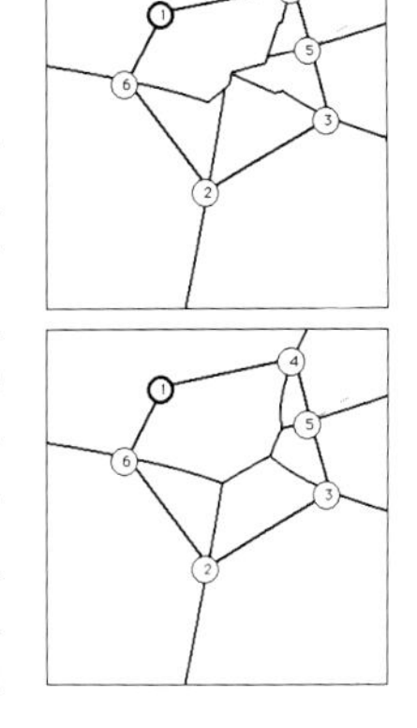

Particularly for large graphs, it often makes sense to group together a subgraph and then represent that subgraph as a single node. For example, a map of a city might include both a detailed street map and an overview of the major routes into and out of the city. In the overview, the detail is suppressed, and the city is represented essentially as a single node. We shall call this style of graph a *hierarchical* graph. Of course graphs themselves are often used to represent hierarchies— a tree, for example. A hierarchical graph groups a subgraph in a given graph into single node. A variety of extensions to the basic definition of a graph that could model a hierarchical graph have been proposed including hypergraphs [66] and higraphs [67].

A *hypergraph* [66] extends the definition of a graph to allow edges (now *hyperedges*) to contain any number of nodes (not just two). Although Berge [66] and others have developed an extensive theory of hypergraphs, little work has been done to consider how to draw hypergraphs, though Harel [67] briefly discusses the topic.

One possibility, shown at right, would simply draw a boundary around the nodes contained in a hyperedge. Unfortunately, boundaries for different edges can readily overlap, which can make it difficult to comprehend.

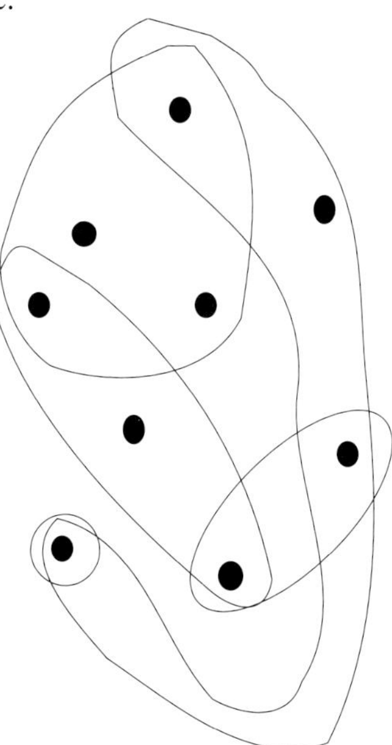

Figure 13.48

Another possibility, shown at right, would connect the nodes in a hyperedge by a connected series of lines. This representation of a hypergraph would then be amenable to the graph layout techniques discussed in this chapter.

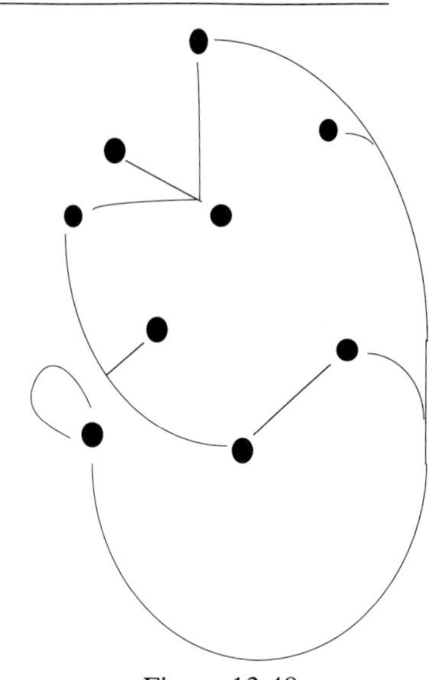

Figure 13.49

A *higraph* [67] represents another approach for representing hierarchical graphs. Higraphs provide their hierarchical capabilities by using ideas from Venn diagrams.

A Venn diagram [46] is a graphical representation of the relationships among sets. In a Venn diagram, sets are represented by closed figures. Set unions and intersection are represented by overlapping the figures representing the appropriate sets.

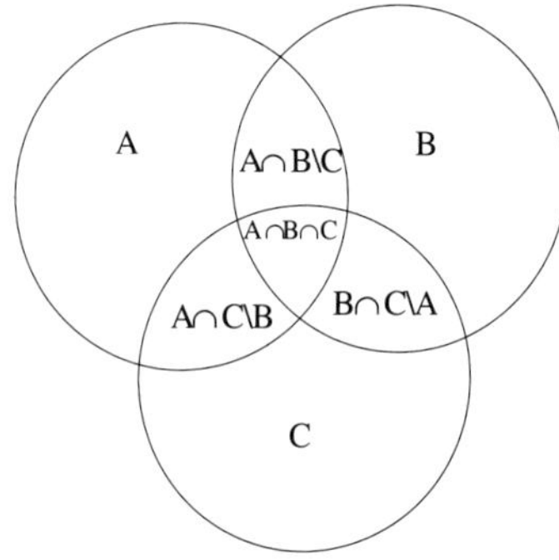

Figure 13.50

The Venn diagram of Figure 13.50 is somewhat ambiguous. The label "A," for example, refers to set A represented by its enclosing circle. The label "A," however, is located in the region of the diagram representing $A \setminus B \setminus C$. A higraph extends a Venn diagram by including a single rounded rectangle in each separate region defined by the Venn diagram, as shown at right. Harel calls the area they enclose *blobs*. In other words, in a drawing of a higraph, if two blobs overlap, there must be another blob wholly contained in the overlapping region. So, in the diagram at right, set $A = D \cup H \cup I \cup K$, $H = A \cap B \setminus C$, $K = A \cap B \cap C$.

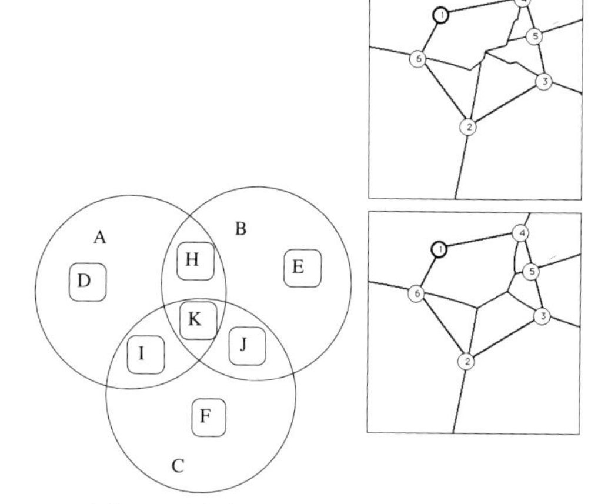

Figure 13.51

One can now connect blobs by directed or undirected edges or both. Edges can connect any blob to any other. That means that an edge can connect a blob contained in one blob to a blob that contains many other blobs.

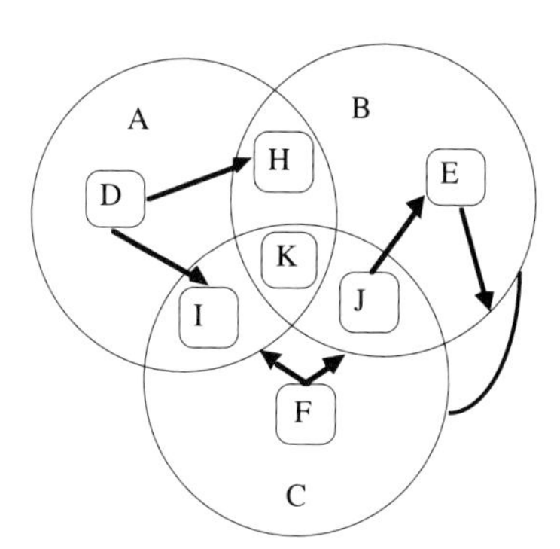

Figure 13.52

Finally, higraphs provide an additional construct for representing bipartite graphs. A vertical line drawn down the middle of a blob indicates that each blob on the left is to be connected to each blob on the right. $K_{10,10}$ (Figure 14.17) would therefore be represented by the higraph at right.

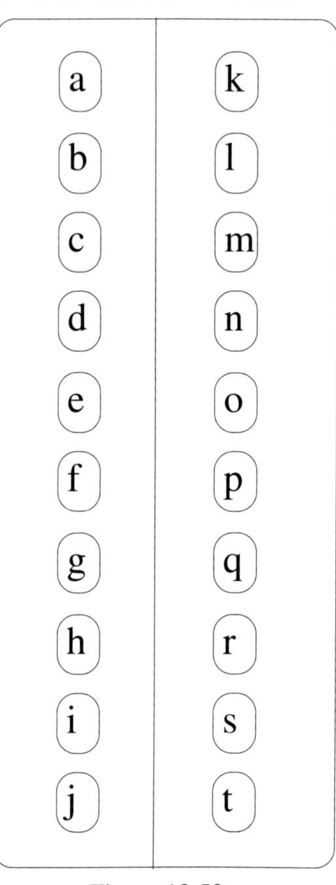

Figure 13.53

Higraphs can provide quite economical representations of many graphs that require a very dense representation if drawn normally. For example, at right is a remarkably simple representation of K_{10}, a 10-clique, as a higraph. This representation, of course, depends on interpreting the single edge as indicating a connection between each pair of nodes contained in the outer blob.

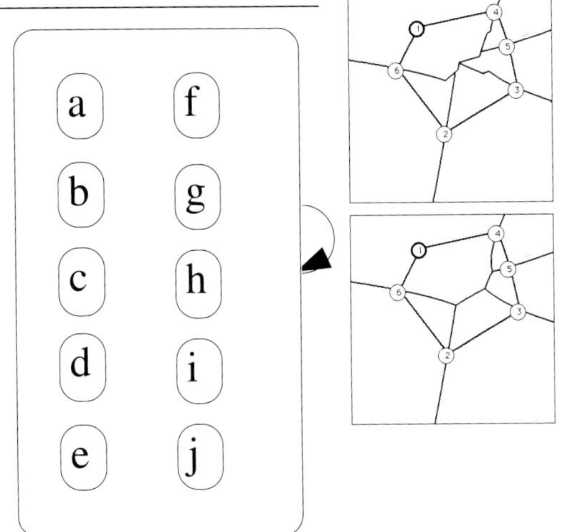

Figure 13.54

13.3.7 Interactive Layout

All the algorithms to this point take a graph as input and output a series of positions for all the nodes and edges. Several authors have explored the concept of *interactive* layout. Interactive layout systems [68], [69], [64], [33], [36], [70] do not assume that a computer algorithm will be able to generate an acceptable layout all by itself. Rather, they provide tools for the user to help with the layout process. As Henry and Hudson state [33]:

> ...even when traditional layout algorithms do a good job minimizing such properties, they do not always produce the most readable graphs or make the best use of available resources such as screen space. The reason for this is that the "best" layout depends on what information the user is currently focused upon. Consequently, a single canonical layout algorithm cannot always produce the best results. Users must be able to interact with the layout process so they can customize it to meet their current needs and interests.

Interactive techniques useful for displaying a graph include:

- *Move a node or edge.* By using this capability, in the worst case, a user can create an acceptable layout.
- *Geometric zoom and pan.* As graphs grow large, they usually cannot fit on a single computer display and hope to be comprehensible. Users should be able to focus in on just the portion of the graph desired.
- *Hide nodes and edges.* Users should be able to temporarily hide nodes and edges. This allows a user to remove clutter and focus in on those nodes and edges of particular interest.
- *Hierarchical graphs.* Users should be able to group together selected nodes and edges into aggregate nodes. This allows users to hide complexity. Algorithms for creating such groupings automatically could also be helpful.
- *Multiple views.* Users should be able to see different views of the graph (from panning or zooming, for example) simultaneously. One should be able to see both an overview of the entire graph and simultaneously see details of the graph.
- *Database-style queries.* In addition to being able to point to nodes and edges with a mouse (to move or hide them, for example), users

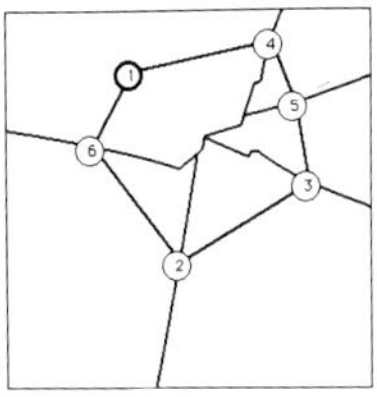

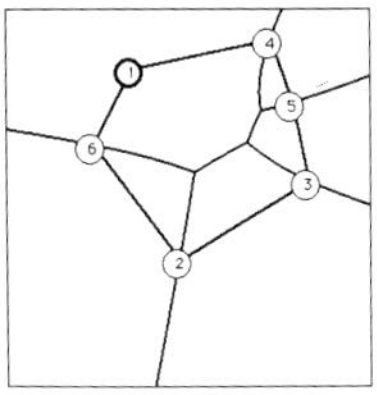

should be able to select groups of nodes and edges using database style queries. In a project management application, for example, user's might wish to see all those tasks that have a slack of 2 days or less. Although not all possible queries are easily implemented (find the minimum cost hamiltonian tour on a given graph), many simple queries are easily implemented and can can actually help the user.

Interactive layout stands in direct contrast to the algorithmic approaches so far discussed. In interactive layout, a user would want to provide suggestions to the layout system in order to provide guidance as to what constitutes a useful layout.

For example, users could identify subgraphs and request that a particular layout style be applied to that subgraph. The layout algorithm of Messinger [35] and the interactive system of Henry and Hudson [33] apply such a "divide and conquer" approach. Henry and Hudson give the example of interactively deriving a graph layout to highlight the two shortest paths between a pair of nodes in a 116 node graph. The shortest paths were laid out horizontally using a horizontal layout algorithm. Then 30 more invocations of layout algorithms were applied to other portions of the graph in order to achieve a readable layout.

Note that these suggestions concern local properties of the graph such as "make these nodes close together ," or "draw these nodes with a particular shape." These suggestions are essentially *constraints* on the allowed layout of the graph. In contrast, the algorithmic approaches discussed so far determine a layout globally. provide great flexibility to the user.

but it seems difficult to use constraints to define certain aesthetic criteria, for example, planarity.

Typical constraint based approaches, e.g., Marks [71] rely on expert system style constraint resolution. His system, ANDD was written in Prolog. Constraint-based approaches, however, cannot readily define all useful constraints. For example, it would be difficult to specify that a graph should be drawn with a planar embedding using just a set of local constraints.

Another challenge of interactive layout involves ensuring a stable layout. If nodes or edges are added or removed from a graph, the output of a traditional layout algorithm may be radically different from the original layout. Because a user's short-term memory is limited, such a radical change would detract from their comprehension of the structure of the graph. Instead, one might desire an algorithm that makes only minimal changes in the layout in response to changes in the structure of the graph. Böhringer and Paulisch [69] used simple linear constraints to achieve such dynamic stability using Sugiyama's [60] spring equilibrium layout algorithm.

13.3.8 Distorted/Fish-eye Views

An interactive layout technique called the *distorted* or *fish-eye* view [72] has been the subject of intensive investigation. A fish-eye view provides a special kind of zoom capability. In particular, a user can point to part of a graph to make it larger. The rest of the graph does not disappear off the edge of the computer display (or off the edge of a window), however; rather, the rest of the graph becomes physically smaller as the part of the graph of interest becomes larger. In other words, the image is distorted to make certain areas larger and more detailed at the expense of other areas of the image. The Perspective Wall, presented in Section 10.3 is one example of a fish-eye view. For a graph-based example, consider the figure (from [36]) below showing a network of cities in North America

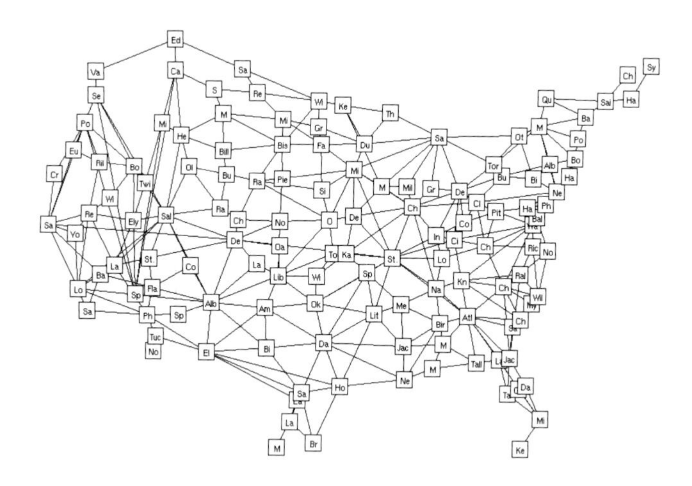

Figure 13.55

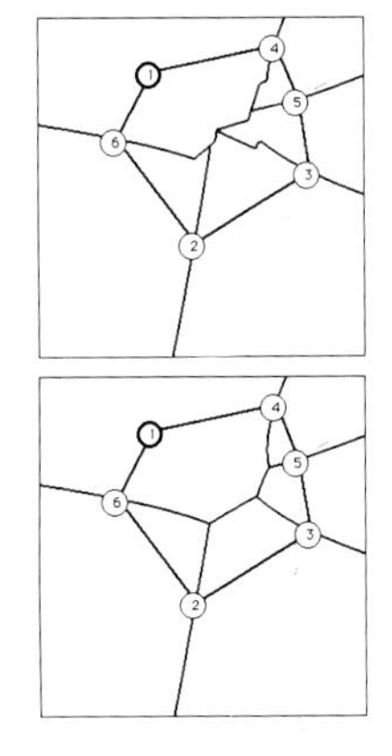

Consider the fish-eye view of this graph below wherein St. Louis and nearby cities are made the focus. One can see the details of the network surrounding St. Louis while still seeing the global context.

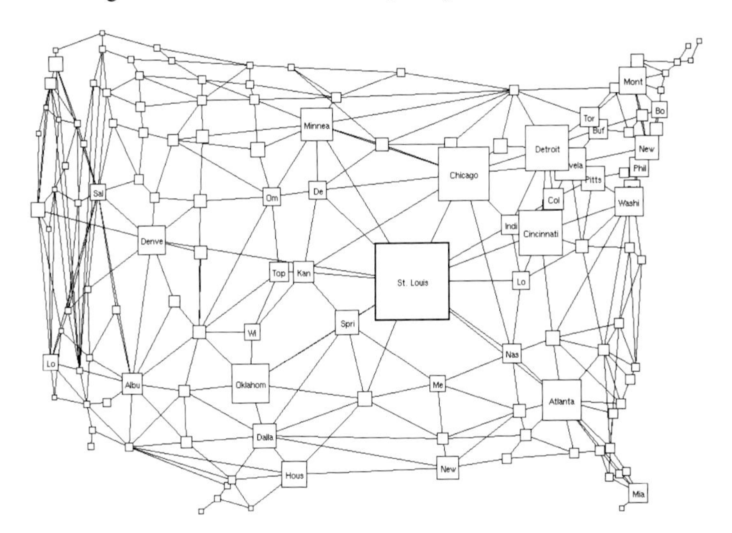

Figure 13.56

The user can change the focus of the view smoothly, yielding an interactive animation. The image below, for example, was produced as the user changed the focus from St. Louis to Salt Lake City.

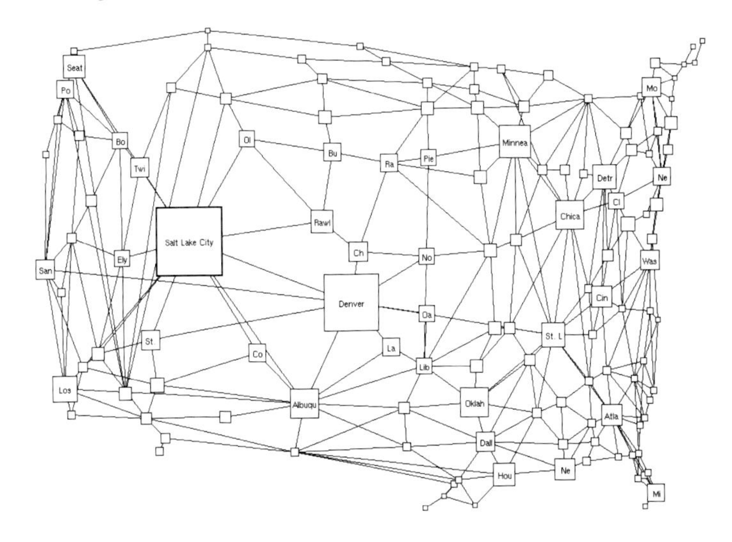

Figure 13.57

Bartram, *et al.* [68], extended the fish-eye view for graphs to include hierarchical nodes and edges. In particular, nodes can be grouped together into hierarchical nodes. Moreover, as the focus point changes, or as the nodes are increased or decreased in size, the level of detail shown changes appropriately. For example, below is the starting view of a hierarchical graph.

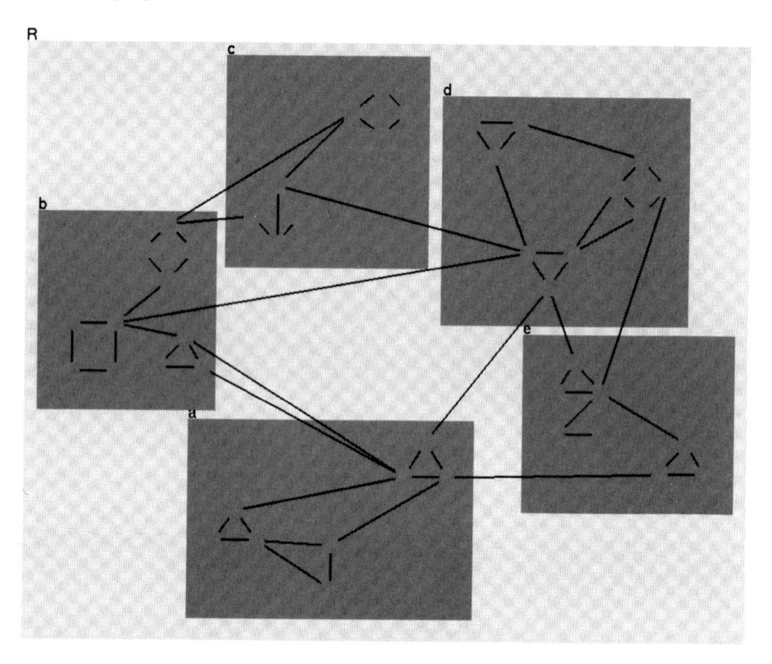

Figure 13.58

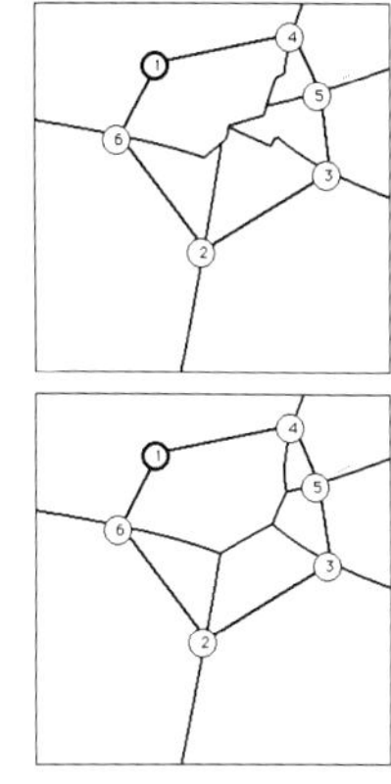

If the user increases the size of the lower node (using the mouse), more detail inside the node is displayed, and the other nodes become smaller.

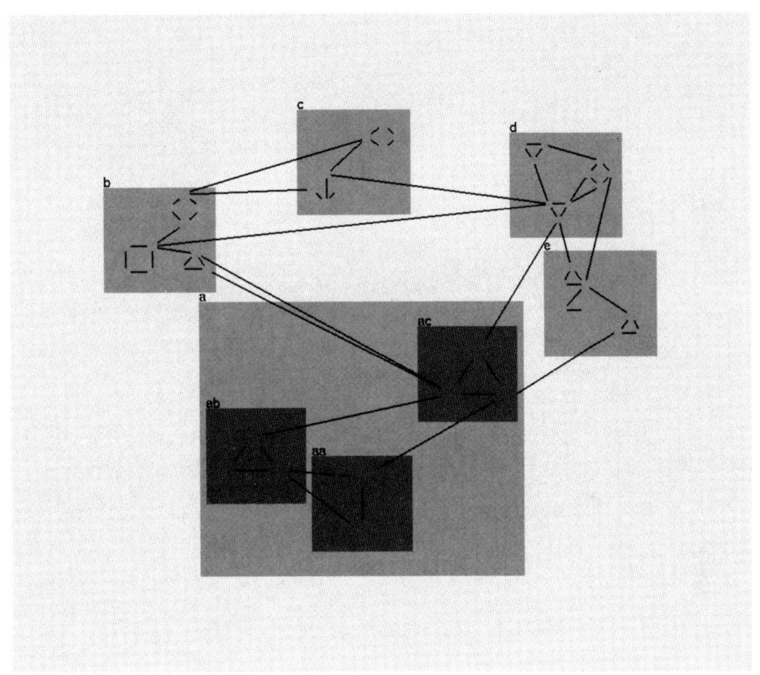

Figure 13.59

If the user increases the size of the lower node, more detail inside the node is displayed, and the other nodes become smaller.

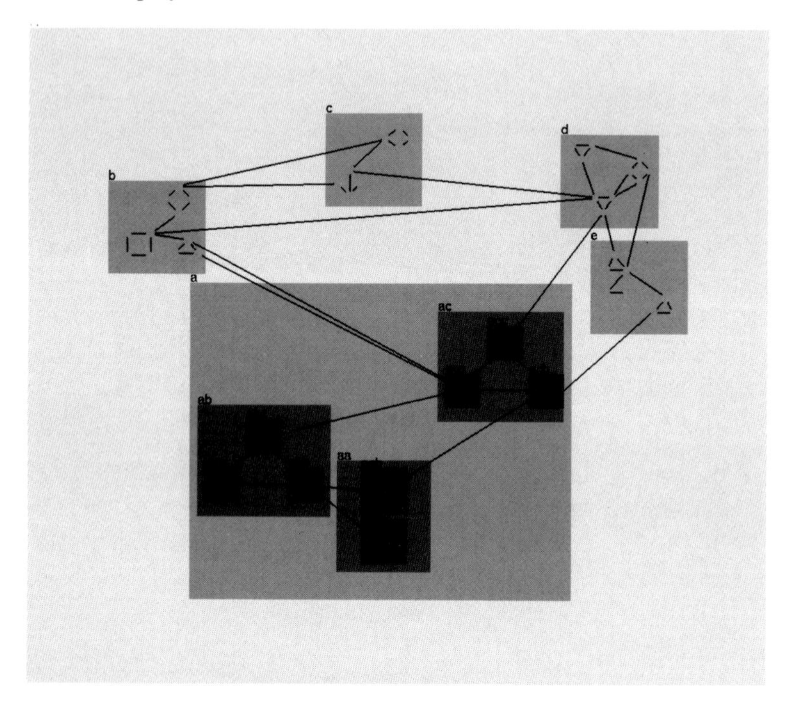

Figure 13.60

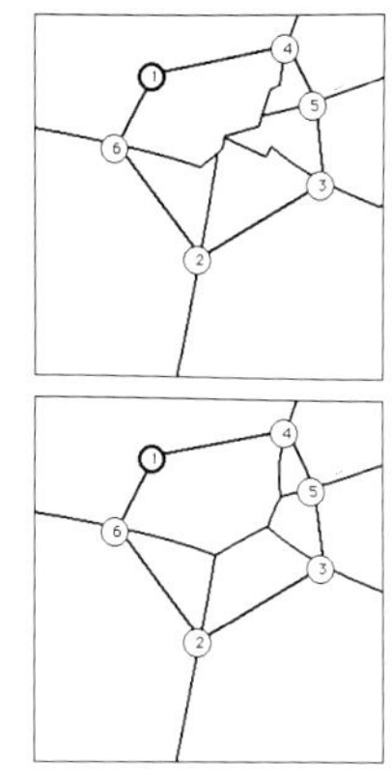

The user can then expand the inner nodes further, resulting in the image below:

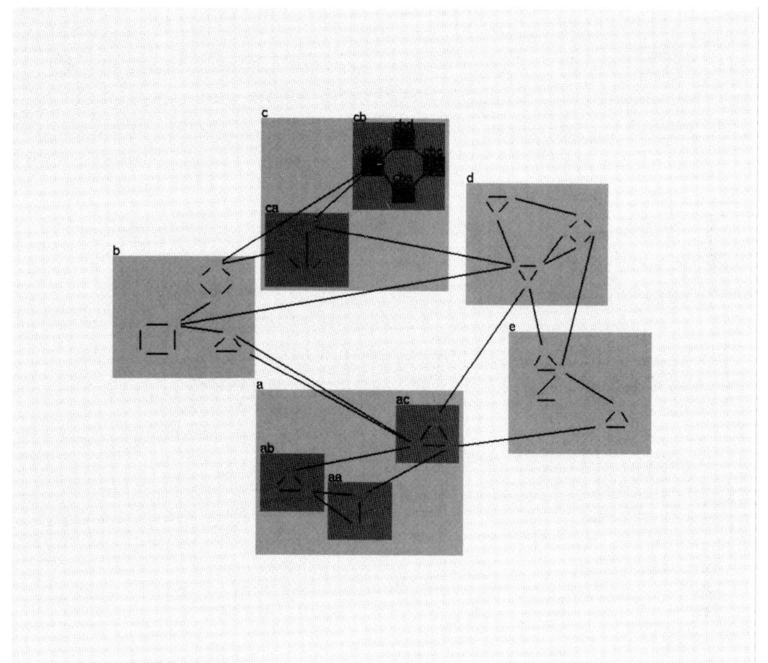

Figure 13.61

Schaffer, *et al.* [70] conducted an experiment to test the effectiveness of the fish-eye zoom compared to the traditional geometric zoom. Subjects had to diagnose and repair a problem in a telecommunications network. A problem in the network is disrupting a telephone connection between a pair of locations, though the source of the problem is buried in the detailed network connecting the locations. Their study showed that subjects were able to diagnose and repair a breakdown in a telephone network faster using the fish-eye zoom compared to a traditional geometric zoom.

Carpendale, Cowperthwaite and Fracchia [73], [74] developed a distorted view technique that considers the surface on which a graph (or other image) is drawn to be a rubber sheet or pliable surface. The user can pull *focus* points of the pliable surface towards their eye, with the corresponding part of the image increased in size. Below, for example, is an undistorted view of h_8, an eight-dimensional hypercube.

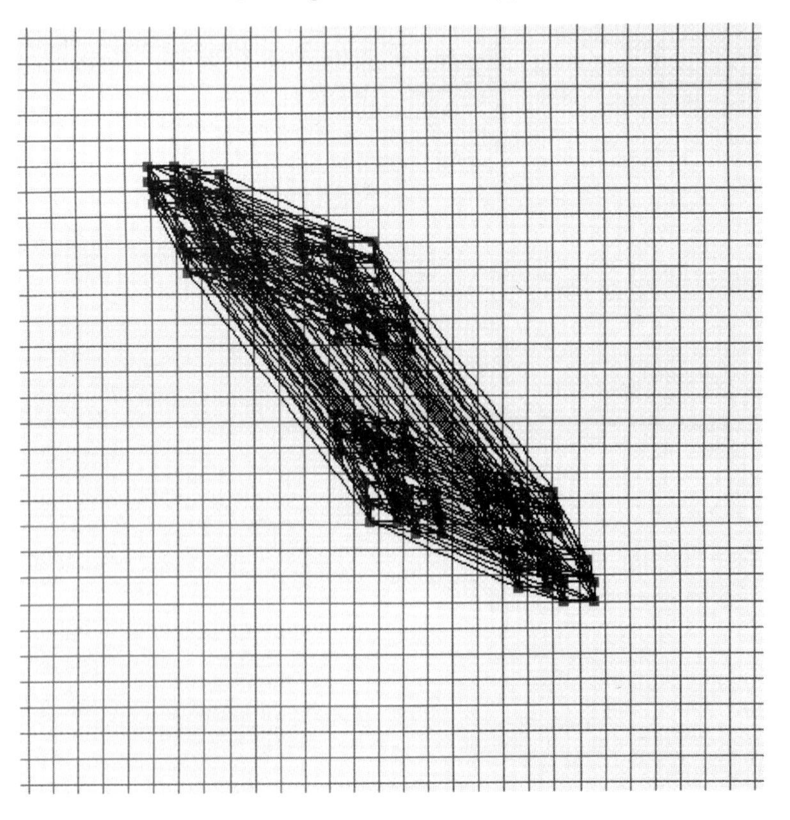

Figure 13.62

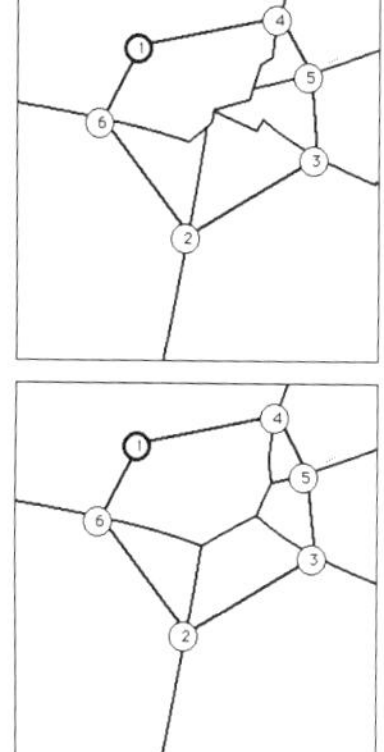

In the figure below, one point of the rubber sheet is drawn towards the viewer's eye. Note that the display of the distortion in the underlying rectilinear grid gives the viewer a sense of the location and magnitude of the distortion applied.

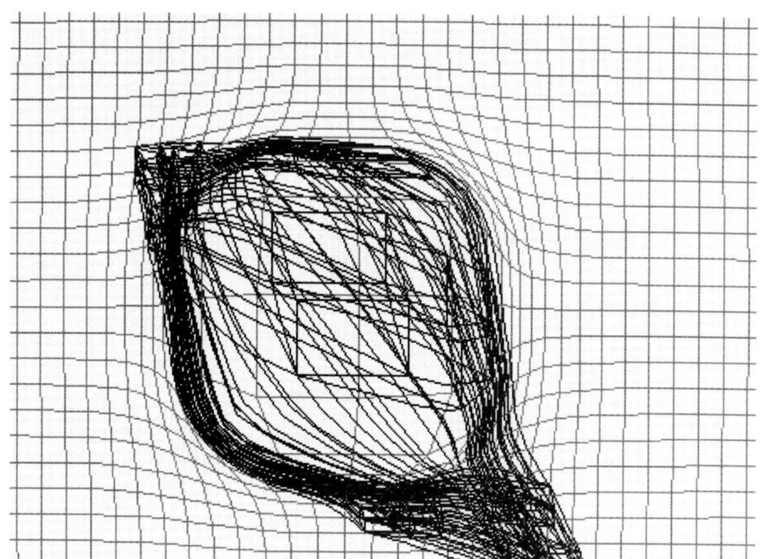

Figure 13.63

This is shown more clearly in the side view, below.

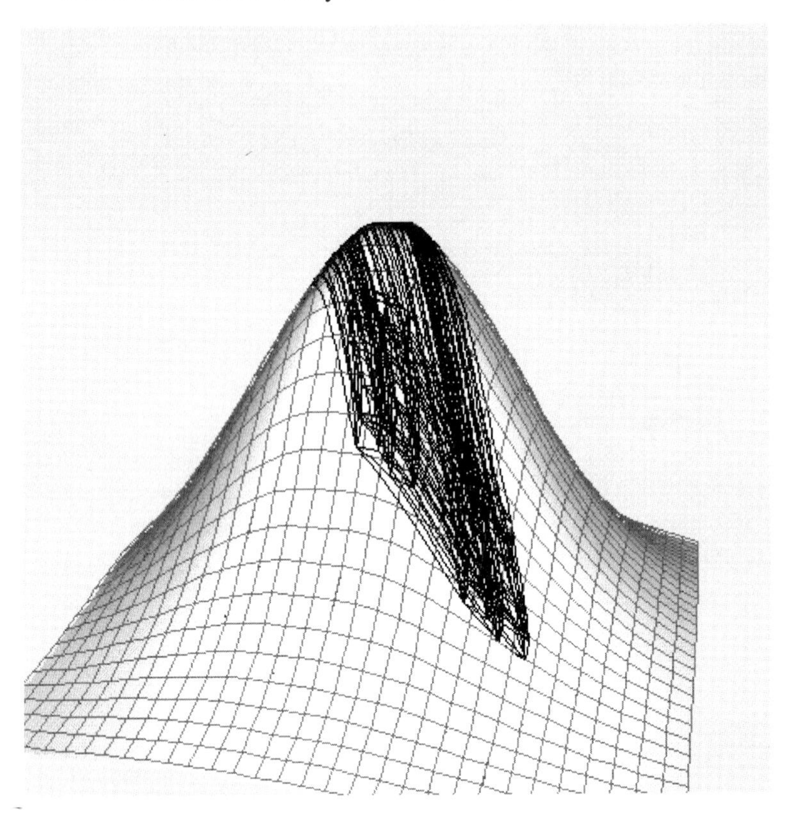

Figure 13.64

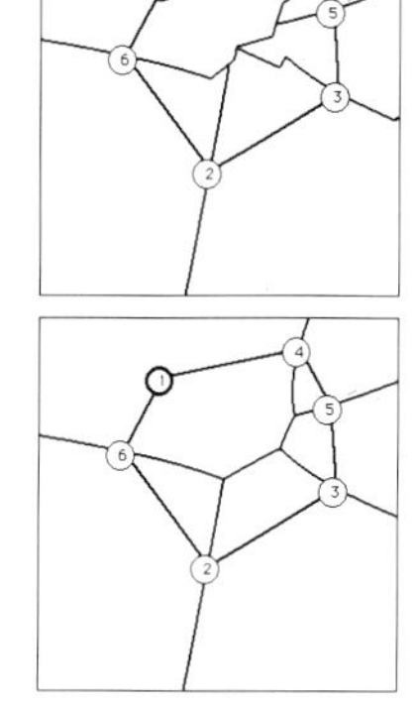

Multiple foci are also allowed. In the view below, two different foci have been pulled towards the viewer. In addition, the underlying surface is shaded to give the viewer even more indication as to the location and magnitude of the distortion applied.

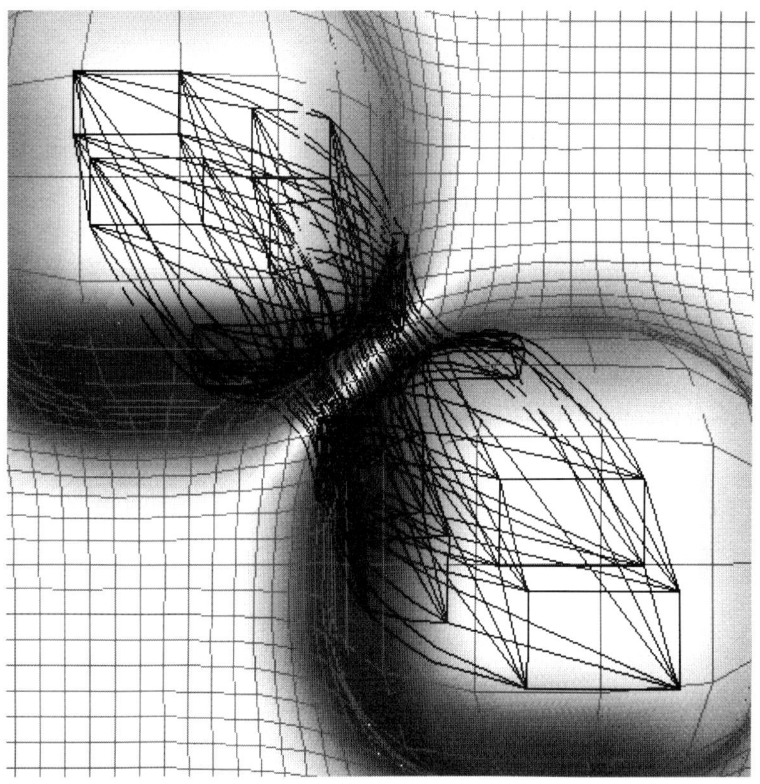

Figure 13.65

The pliable surface technique provides a simple metaphor for constructing fish-eye views. It allows for multiple foci, and provides a variety of visual clues to help the viewer perceive the location and magnitude of the distortion. These cues are especially important for abstract images such as graphs (as opposed to maps), wherein the underlying geometry may not be familiar to the viewer. Note further that the pliable surface technique is not limited to graphs, but can be applied to any two-dimensional image.

13.3.9 Summary of Distorted Views

All of the distorted view techniques assume a fixed graph that has been previously embedded in a plane. Given the fixed layout of the fixed graph, the distorted view allows the user to focus in on one or more detailed sections of a network. An interesting research problem involves the issue of combining graph editing, graph layout algorithms that reposition nodes and edges to achieve some desired characteristics with distorted view techniques. As a user edits a graph, the layout of the graph should probably be adjusted to reflect the changes. The graph should not be altered so radically, however, that the user loses all understanding of the graph. Moreover, users would seem to benefit from some indication of the nature of the distortion. The distorted grid and smooth shading used in the work in pliable surfaces [73], [74] may help. Smooth transitions while view of the graph is being distorted should help the viewer maintain their mental map of the graph under study.

13.4 Summary

Graphs and networks are ubiquitous in optimization, for modeling, embedding in algorithms, and for present the results of solutions. We currently lack ubiquitous tools for working with graphs and networks; this stands in sharp contrast to tabular representations, which are manipulable using commonly available spreadsheets and relational databases. Although a variety of systems have been proposed for allowing people to manipulate graphs and networks, none has yet predominated. Perhaps a generic graph-based modeling system will emerge out of Geographic Information Systems. One can always use a standard graphics drawing package to draw a network, but that package does not understand the definition of a graph (nodes connected by edges).

One of the major hurdles to using graph-based representations concerns the spaghetti produced when large graphs are drawn. Although this is a subject of active investigation, a variety of useful layout algorithms

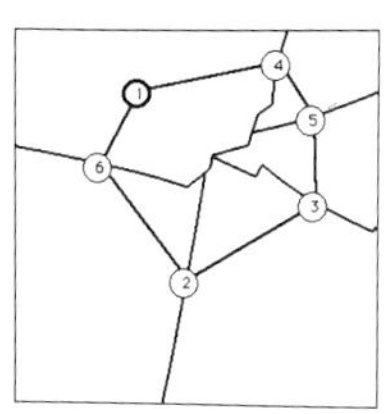

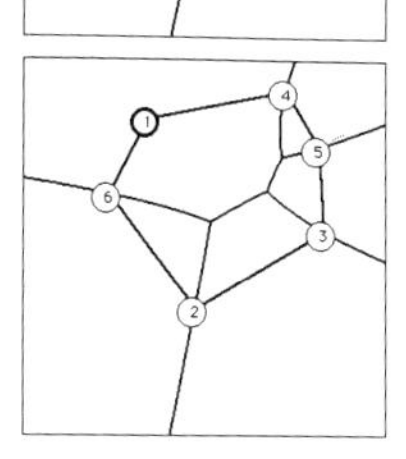

and techniques have been developed that can be applied now. To be truly useful, though, these techniques must allow the user to interactively modify the layout of the graph in a predictable, stable manner.

Bibliography

[1] Babin A, Florian M, James-Lefebvre M, Spiess H. EMME/2: An interactive graphic method for road and transit planning. Technical Report 204, Montréal, Canada:Centre de Recherche sur les Transports, Université de Montréal, 1982.

[2] Bodin L, Fagan G, Levy L. Vehicle routing and scheduling problems over street networks. Technical report, College Park (MD):College of Business and Management, University of Maryland, 1993.

[3] Fisher ML, Greenfield A, Jaikumar R. Vergin: A decision support system for vehicle scheduling. Technical Report 82-06-02, Philadelphia (PA):Department of Decision Sciences, The Wharton School, University of Pennsylvania, 1982.

[4] Mak KT, Srikanth K, Morton A. Visualization of routing problems. Technical Report CRIM 90-03, Chicago:Department of Information and Decision Sciences, College of Business Administration, University of Illinois at Chicago, 1990.

[5] Savelsbergh MWP. Computer aided routing. Technical report, Amsterdam, The Netherlands:Centrum voor Wiskunde en Informatica, 1988.

[6] Dao M, Habib M, Richard JP, Tallot D. Cabri, an interactive system for graph manipulation. In Tinhofer G, Schmidt G, editors, *Graph-Theoretic Concepts in Computer Science*, pages 58–67. Berlin:Springer-Verlag, 1986.

[7] Skiena S. *Implementing Discrete Mathematics: Combinatorics and Graph Theory with Mathematica*. Redwood City (CA):Addison-Wesley, 1990.

[8] Jones CV. MIMI/G: A graphical environment for mathematical programming and modeling. Technical report, Burnaby (BC), V5A 1S6 CANADA:Faculty of Business Administration, Simon Fraser University, 1995. forthcoming, *Interfaces*.

[9] Greenberg HJ. A new approach to analyze information in a model. In Gass SI, editor, *Energy Models: Validation and Assessment*, number NBS Pub No. 569 in National Bureau of Standards Reports, pages 517–524. Gaithersburg, Maryland:National Bureau of Standards, 1978.

[10] Greenberg HJ. ANALYZE: A computer-assisted analysis system for linear programming models. *OR Letters*, 1987;6(5):249–259.

[11] Greenberg HJ. *A Computer-Assisted Analysis System for Mathematical Programming Models and Solutions: A User's Guide for ANALYZE.* Boston (MA):Kluwer, 1993.

[12] Knuth DE. *The Stanford GraphBase: A Platform for Combinatorial Computing.* Reading (MA):Addison-Wesley, 1993.

[13] North SC, Koutsofios E. Applications of graph visualization. In Davis and Joe [75].

[14] Himsolt M. GraphEd: An interactive graph editor. In *Proceedings Symposium on Theoretical Aspects of Computer Science (STACS) 89*, volume 349 of *Lecture Notes in Computer Science.* Berlin:Springer-Verlag, 1989.
URL: *ftp://ftp.uni.passau.de/pub/local/graphed*

[15] Frölich M, Werner M. daVinci v1.2 user manual. Technical report, Postfach 330 440, D28334 Bremen, Germany:Department of Computer Science, Universität Bremen, 1993.
URL: *mailto:daVinci@Informatik.Uni-Bremen.DE*

[16] Dean N, Mevenkamp M, Monma CL. Netpad: An interactive graphics system for network modeling and optimization. In Balci O, Sharda R, Zenios SA, editors, *Computer Science and Operations Research: New Developments in their Interfaces*, pages 231–243. Oxford:Pergamon Press, 1992.

[17] Tom Sawyer Software. *Graph Layout Toolkit, 1.08 edition.* 1824B Fourth Street, Berkeley (CA) 94710:Tom Sawyer Software, 1995.

[18] Ashton NW, Zacharias TA, Fracchia FD. Genetic analysis of phytohormone action and morphogenesis in *physcomitrella patens.* In Basile DV, Mankiewicz PS, editors, *Memoirs of the Torrey Botanical Club*, volume 25. Bronx (NY):Sheridan Press, September 1993.

[19] Schürr A. PROGRES, a visual language and environment for PROgramming with Graph REwrite Systems. Technical Report AIB 94-11, Aachen, Germany:RWTH, 1994.
URL: *http://www-i3.informatik.rwth-aachen.de/research/progres/*

[20] Jones CV. An introduction to graph-based modeling systems, part I: Overview. *ORSA Journal on Computing*, 1990;2(2):136–151.

[21] Jones CV. An introduction to graph-based modeling systems, part II: Graph-grammars and the implementation. *ORSA Journal on Computing*, 1991;3(3):180–206.

[22] Jones CV. Attributed graphs, graph-grammars, and structured modeling. *Annals of OR*, 1992;38:281–324.

[23] Jones CV. Animated sensitivity analysis. In Balci O, Sharda R, Zenios SA, editors, *Computer Science and Operations Research: New Developments in their Interfaces*, pages 177–196. Oxford, United Kingdom:Pergamon Press, 1992.

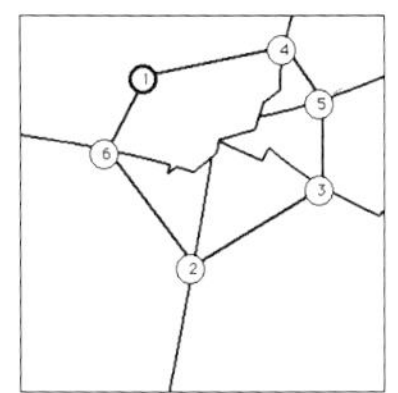

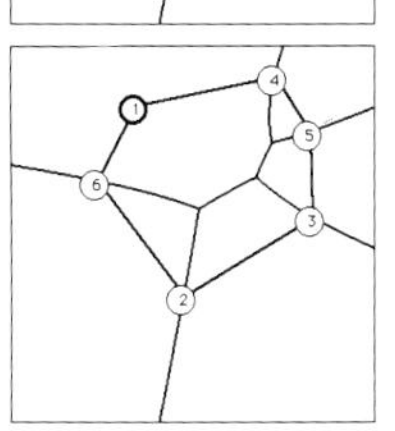

[24] Jones CV, D'Souza K. Graph-grammars for minimum cost network flow modeling. Technical report, Burnaby (BC), V5A 1S6 CANADA:Faculty of Business Administration, Simon Fraser University, 1992.

[25] Nagl M. Formal languages of labeled graphs. *Computing*, 1976;16.

[26] Nagl M. A tutorial and bibliographical survey on graph-grammars. In Claus V, Ehrig H, Rozenberg G, editors, *Graph-Grammars and their Application to Computer Science and Biology*, pages 70–126. Berlin:Springer-Verlag, 1979.

[27] Nagl M. Set theoretic approaches to graph grammars. In Ehrig H, Nagl M, Rozenberg G, Rosenfeld A, editors, *Graph-Grammars and their Application to Computer Science*, pages 41–54. Berlin:Springer-Verlag, 1987.

[28] Göttler H. Attributed graph-grammars for graphics. In Ehrig H, Nagl M, Rozenberg G, editors, *Graph-Grammars and their Application to Computer Science*, pages 130–142. Berlin:Springer-Verlag, 1983.

[29] Göttler H. Graph-grammars and diagram editing. In Ehrig H, Nagl M, Rozenberg G, Rosenfeld A, editors, *Graph-Grammars and their Application to Computer Science*, pages 216–231. Berlin:Springer-Verlag, 1987.

[30] Göttler H. Graph grammars, a new paradigm for implementing visual languages. In Dershowitz N, editor, *Rewriting Techniques and Applications*, pages 152–166. Berlin:Springer-Verlag, 1989.

[31] Lengauer T. *Combinatorial Algorithms for Integrated Circuit Layout*. New York:Wiley, 1990.

[32] Ganser ER, North SC, Vo KP. Dag—a program that draws directed graphs. *Software–Practice and Experience*, 1988;18(11).

[33] Henry TR, Hudson SE. Interactive graph layout. In *Proceedings of the ACM Symposium on User Interface Software and Technology*, Snowbird (UT):October 1991. Association for Computing Machinery.

[34] Knuth DE. Computer-drawn flowcharts. *Communications of the ACM*, 1963;6.

[35] Messinger EB, Rowe LA, Henry TR. A divide-and-conquer algorithm for the automatic layout of large directed graphs. *IEEE Transactions on Systems, Man and Cybernetics*, 1991;21(1).

[36] Sarkar M, Brown MH. Graphical fisheye views of graphs. In *Proceedings of ACM CHI '92 Conference on Human Factors in Computing Systems*, pages 83–92, Monterey (CA):1992.
URL: *http://gatekeeper.dec.com/pub/DEC/SRC/research-reports /abstracts/src-rr-084a.html*

[37] Shneiderman B. Tree visualization with tree-maps: 2-d space-filling approach. *ACM Transactions on Graphics*, 1992;11(1):92–99.

[38] Tamassia R, Battista GD, Batini C. Automatic graph drawing and readability of diagrams. *IEEE Transactions on Systems, Man and Cybernetics*, 1988;18(1):61–79.

[39] Vaucher JG. Pretty-printing of trees. *Software–Practice and Experience*, 1980;10:553–561.

[40] Cruz IF, Tamassia R. How to visualize a graph: Specification and algorithms. Technical report:Tufts University, 1994.

[41] Battista GD, Eades P, Tamassia R, Tollis IG. Algorithms for drawing graphs: An annotated bibliography. Technical report, Providence (RI):Department of Computer Science, Brown University, 1994.
URL: *ftp://ftp.cs.brown.edu/pub/papers/compgeo/gdbiblio.ps.gz*

[42] Wetherell C, Shannon A. Tidy drawing of trees. *IEEE Transactions on Software Engineering*, 1979;SE-5(5):514–520.

[43] Reingold E, Tilford J. Tidier drawing of trees. *IEEE Transactions on Software Engineering*, 1981;SE-7:223–228.

[44] Supowit K, Reingold E. The complexity of drawing trees nicely. *Acta Informatica*, 1983;18:377–392.

[45] Robertson GG, Mackinlay JD, Card SK. Cone trees: Animated 3d visualizations of hierarchical information. In Robertson SP, Olson GM, , Olson JS, editors, *Proc. ACM Computer-Human Interaction '91 Conference on Human Factors in Computing Systems*, pages 189–194, New York:1991. ACM Press.

[46] Venn J. *Symbolic Logic*. Bronx (NY):Chelsea Publishing Company, 2 edition, 1971. reprint of 1894 edition.

[47] Tufte ER. *The Visual Display of Quantitative Information*. Cheshire, Connecticut:Graphics Press, 1983.

[48] Hopcroft J, Tarjan R. Efficient planarity testing. *Journal of the ACM*, 1974;21:549–569.

[49] Chiba N, Onoguchi K, Nishizeki T. Drawing planar graphs nicely. *Acta Informatica*, 1985;22.

[50] Fáry I. On straight line representations of planar graphs. *Acta Scientiarum Mathematicarum Szegediensis*, 1948;11.

[51] Fraysseix HD, Pach J, Pollack R. Small sets supporting Fáry embeddings of planar graphs. In *Proceedings of the 20th Annual ACM Symposium on Theory of Computing*, pages 426–433, 1988.

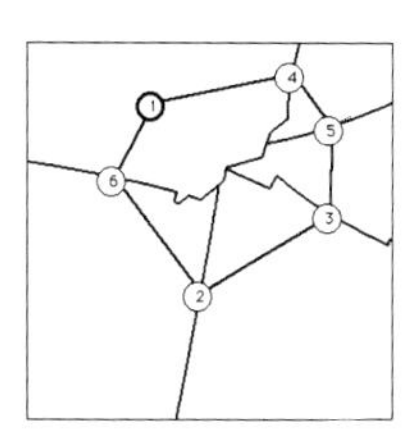

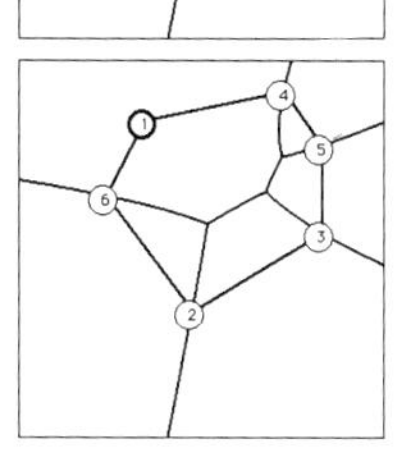

[52] Schnyder W. Embedding planar graphs on the grid. In *Proceedings of the ACM-SIAM Symposium on Discrete Algorithms*, pages 138–148, 1990.

[53] Tutte WT. Convex representations of graphs. *Proceedings of the London Mathematical Society*, 1960;10:304–320.

[54] Tamassia R. On embedding a graph in the grid with the minimum number of bends. *SIAM Journal of Computing*, 1987;16(3):421–444.

[55] Storer JA. On minimal node-cost planar embeddings. *Networks*, 1984;14:181–212.

[56] Eades P, McKay B, Wormald N. An NP-hard crossing number problem for bipartite graphs. Technical Report 60:Department of Computer Science, University of Queensland, 1985.

[57] Carpano MJ. Automatic display of hierarchized graphs for computer aided decision analysis. *IEEE Transactions on Systems, Man and Cybernetics*, 1980;SMC-10(11):705–715.

[58] Eades P, Wormald N. The median heuristic for drawing 2-layered networks. Technical Report 69:Department of Computer Science, University of Queensland, 1986.

[59] Eades P, Kelly D. Heuristics for reducing crossings in 2-layered networks. *Ars Combinatorica*, 1986;21.A:89–98.

[60] Sugiyama K, Tagawa S, Toda M. Methods for visual understanding of hierarchical systems. *IEEE Transactions on Systems, Man and Cybernetics*, 1981;SMC-11(2):109–125.

[61] Warfield J. Crossing theory and hierarchy mapping. *IEEE Transactions on Systems, Man and Cybernetics*, 1977;SMC-7(7):502–523.

[62] Eades P. A heuristic for graph drawing. *Congressus Numerantium*, 1984;42.

[63] Kamada T, Kawai S. An algorithm for drawing general undirected graphs. *Information Processing Letters*, 1989;31.

[64] Fairchild KM, Poltrock SE, Furnas GW. Semnet: Three-dimensional graphic representations of large knowledge bases. In Guindon R, editor, *Cognitive Science and its Applications for Human-Computer Interaction*, pages 201–233. Hillsdale (NJ):Lawrence Erlbaum Associates, 1988.

[65] Ware C, Hui D, Franck G. Visualizing object oriented software in three dimensions. In *Proceedings of CASCON '93 Conference*. IBM Center for Advanced Studies, 1993.

[66] Berge C. *Graphs and Hypergraphs*. Amsterdam:North-Holland, 1976. translated by E. Minieka.

[67] Harel D. On visual formalisms. *Communications of the ACM*, 1988;31(5):514–530.

[68] Bartram L, Ovans R, Dill J, Dyck M, Ho A, Havens WS. Intelligent graphical user interfaces to complex time-critical systems: The intelligent zoom. In Davis and Joe [75], pages 216–224.

[69] Böringer KF, Paulisch FN. Using constraints to achieve stability in automatic graph layout algorithms. In Chew JC, Whiteside J, editors, *Human Factors in Computing Systems, Empowering People, CHI '90 Conference Proceedings*, pages 43–51, Reading (MA):1990. Addison-Wesley.

[70] Schaffer D, Zuo Z, Bartram L, Dill J. Comparing fisheye and full-zoom techniques for navigation of hierarchically clustered networks. In *Proceedings of Graphics Interface '93*, pages 87–96, New York:1993. ACM Press.

[71] Marks J. A syntax and semantics for network diagrams. In *Proceedings of the 1990 Workshop on Visual Languages*, pages 104–110, Los Alamitos (CA):1990. IEEE Computer Society.

[72] Furnas GW. Generalized fisheye views. In *ACM CHI '86*, pages 16–34, 1986.

[73] Carpendale MST, Cowperthwaite DJ, Fracchia FD. 3-dimensional pliable surfaces: For the effective presentation of visual information. Technical report, Burnaby (BC), V5A 1S6 CANADA:School of Computing Science, Simon Fraser University, 1995.

[74] Carpendale MST, Cowperthwaite DJ, Fracchia FD. Graph folding: Extending detail and context viewing into a tool for subgraph comparisons. Technical report, Burnaby (BC), V5A 1S6 CANADA:School of Computing Science, Simon Fraser University, 1995.

[75] Davis WA, Joe B, editors. *Proceedings of Graphics Interface '94*, Toronto, 1994. Canadian Information Processing Society.

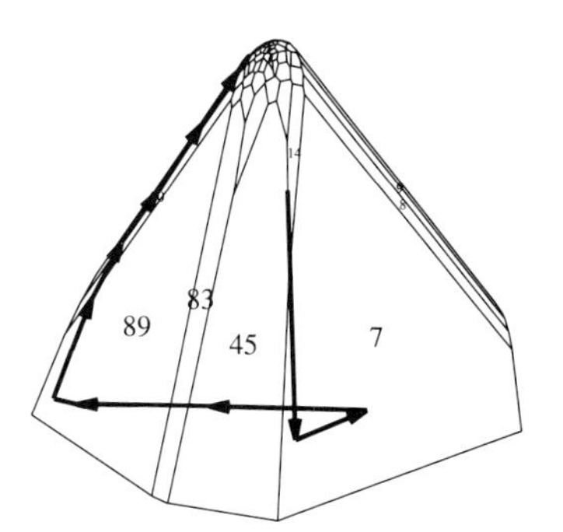

Chapter 14

Multiple Dimensions

A mathematical program identifies a subset of n-dimensional space that constitutes the set of possible or feasible solutions. Therefore, it is important to consider how to visualize n-dimensional spaces. This chapter discusses a variety of approaches that have been proposed for visualizing n-dimensional information. Throughout the chapter we consider the problem of visualizing a n-dimensional polyhedron, which is a core object used in optimization. We consider a variety of traditional techniques such as simple mathematical projection and techniques from statistics as well as relatively new techniques such as Parallel coordinates.

14.1 Projection

Real optimization problems involve polyhedra of very high dimension. One could simply project higher dimensional spaces into lower dimensional spaces, just as one projects three-dimensional computer graphics onto the two-dimensional surface of a display. Banchoff [1], [2] pioneered this technique to visualize higher dimensional objects such as a four dimensional sphere. Rucker [3] provides a variety of helpful suggestions to help one visualize four dimensional space. The images produced require much study before their behavior can be readily understood. To explore this possibility, we consider perhaps the simplest n-dimensional polyhedron, a n-dimensional hypercube. As will be seen, even visualizing a n-dimensional hypercube is challenging.

Definition 3 *An n-dimensional* hypercube *consists of the set of points $x = (x_1, \ldots, x_n) \in \Re^n$ such that $0 \le x_i \le 1$ for all i.*

The common representation of a three-dimensional cube at right is actually a set of two-dimensional lines. The three-dimensional cube can be constructed by connecting appropriate vertices of two squares (two-dimensional cubes). Figures 14.1-14.4 were created using Cabri [4].

A projection of a four-dimensional hypercube into two dimensions appears at right. It is still not too complicated, and one can, for example, see how the four-dimensional hypercube can be constructed from an inner and and outer (three-dimensional) cube whose vertices are attached by edges.

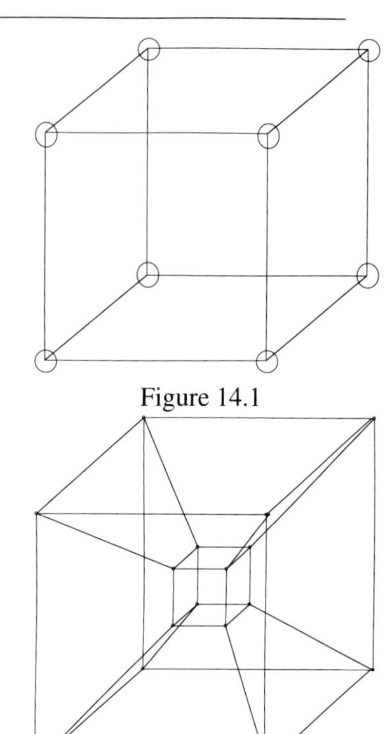

Figure 14.1

Figure 14.2

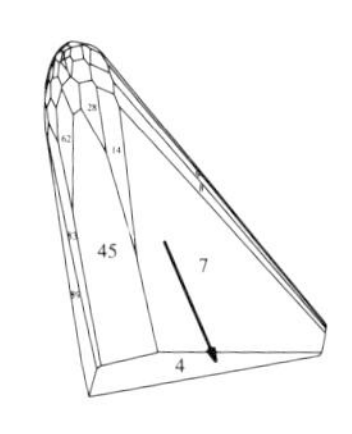

Even the projection of a five-dimensional hypercube into two dimensions is comprehensible. Again, the five-dimensional cube is constructed by connecting two four-dimensional cubes together, though this conclusion is not so obvious as in the four-dimensional case.

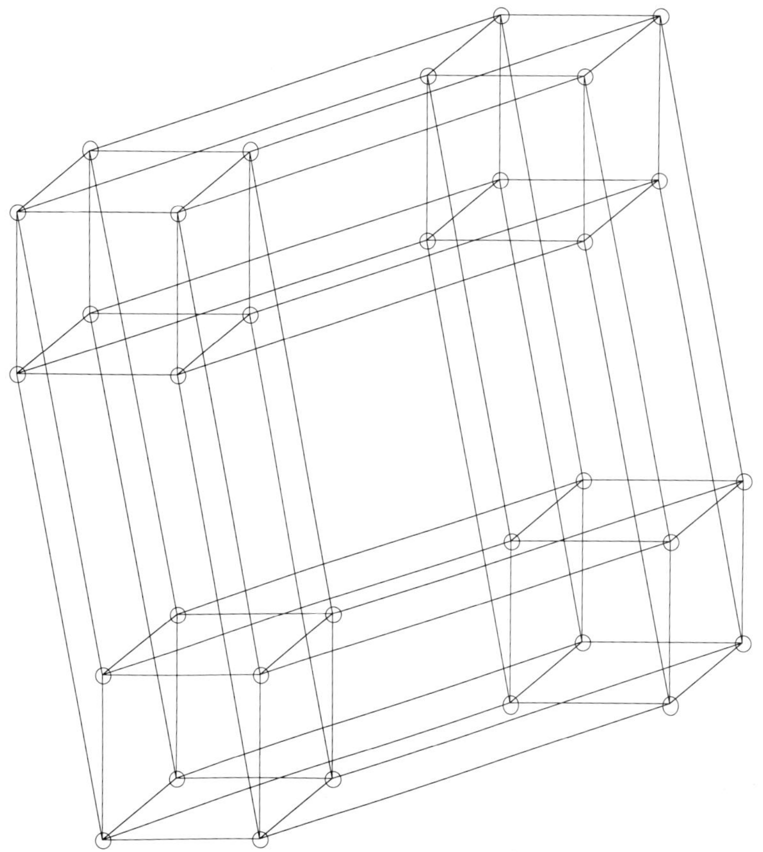

Figure 14.3

As the number of dimensions increases, however, the complexity eventually overwhelms human perception. Below, for example, is a nine-dimensional hypercube projected into two dimensions. This representation has little value except perhaps as abstract art.

Figure 14.4

Simple projection of higher-dimensional objects into two and three dimensions seems to reach a limit at around five or six dimensions, though there is no formal experimentation to support this claim. Typical optimization problems, however, deal with much, much higher dimensional spaces, so simple projection seems to be of limited use.

14.1.1 Graph Layout

Given the research on graph layout algorithms (see Section 13.3), one might consider drawing n-dimensional polytopes as graphs. Such a representation at least would show the adjacency relationships among vertices, though their orthogonal geometry would be distorted by projecting from n dimensions into 2.

We explored this possibility using two different commercially available systems (Mathematica and Maple) that provide a variety of graph layout algorithms. Consider Figure 14.1.1, however, which shows alternative representations of n dimensional hypercubes, created using Mathematica.

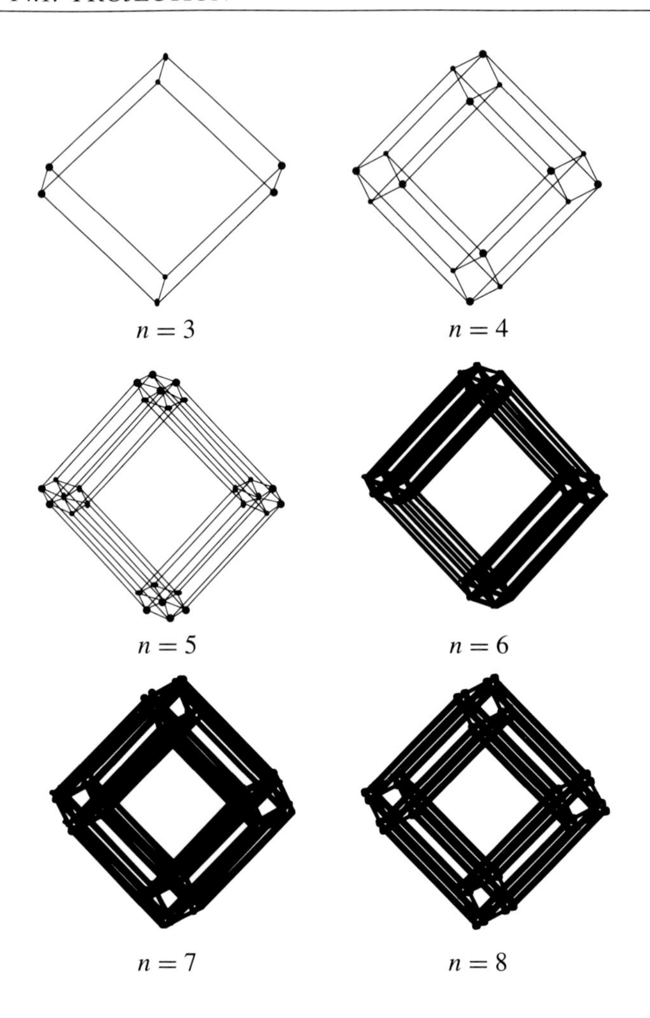

Figure 14.5 Hypercubes $h_3, \ldots, h_8$ created using Mathematica.

Using the standard graph layout provided by Mathematica, h_5 is legible, but h_6 is not. Using a different layout algorithm (ranked embedding algorithm from [5]) provided in Mathematica (below), h_6 is legible.

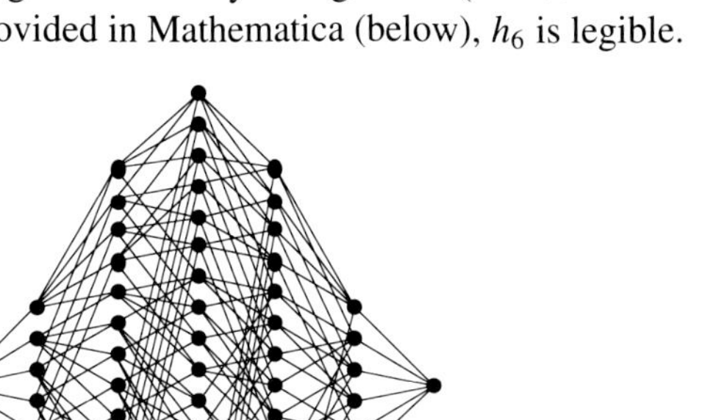

Figure 14.6

Even h_7 is almost legible.

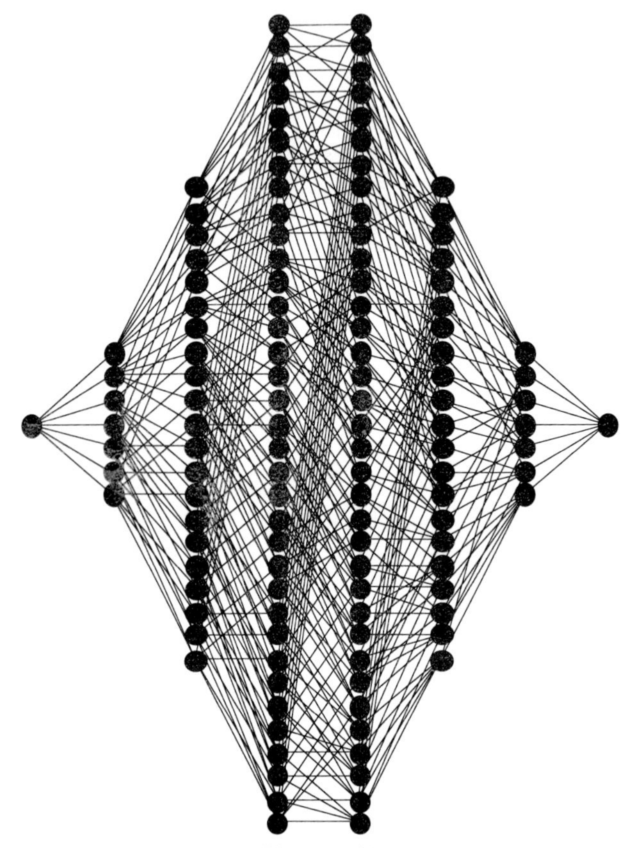

Figure 14.7

But h_8 remains a tangled mess.

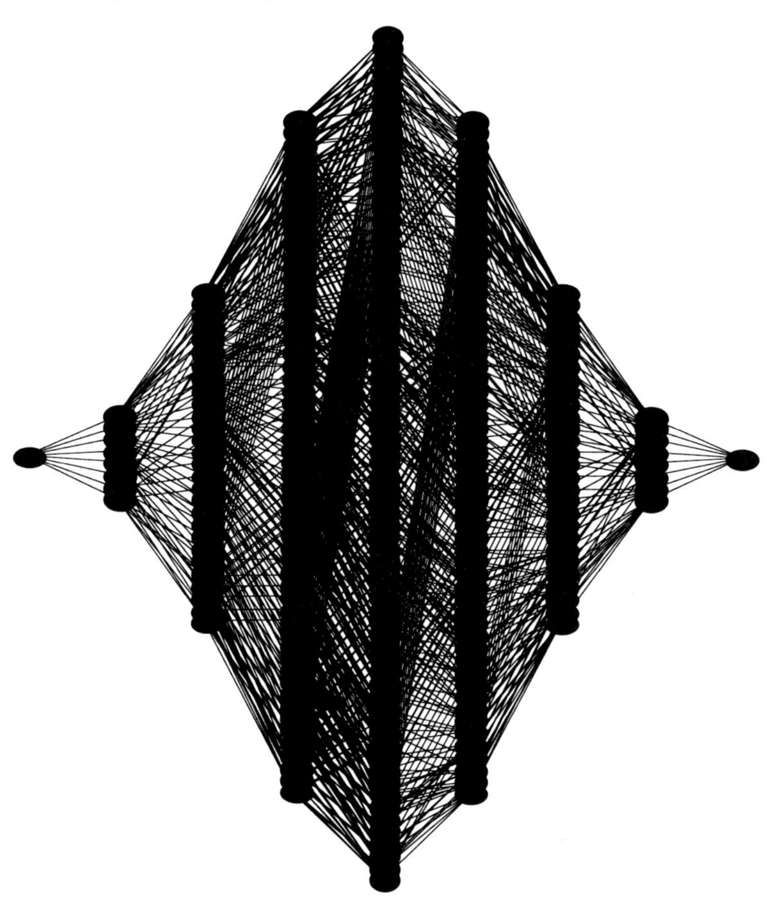

Figure 14.8

By carefully positioning the nodes of h_6 and h_7 along concentric circles (using Maple), the layouts are perhaps more legible. It is doubtful, however, if this would scale well to even larger dimensions since the nodes in the center become quite crowded.

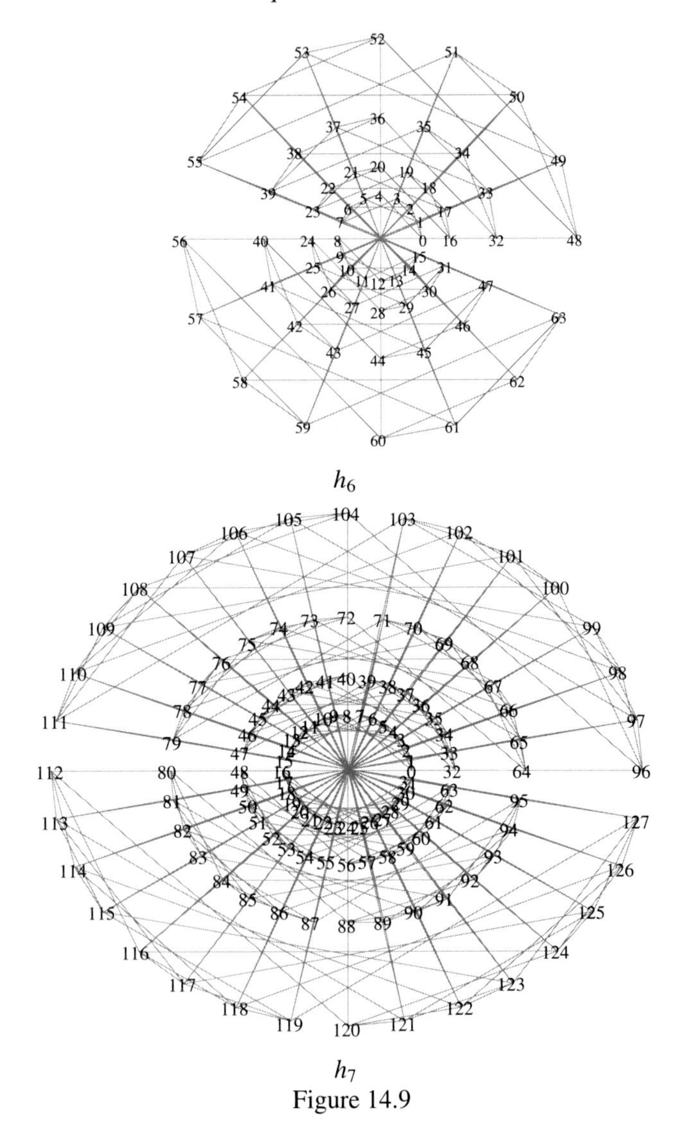

h_6

h_7

Figure 14.9

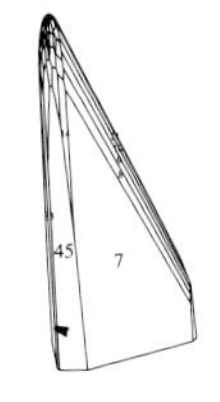

But again, h_8 becomes illegible.

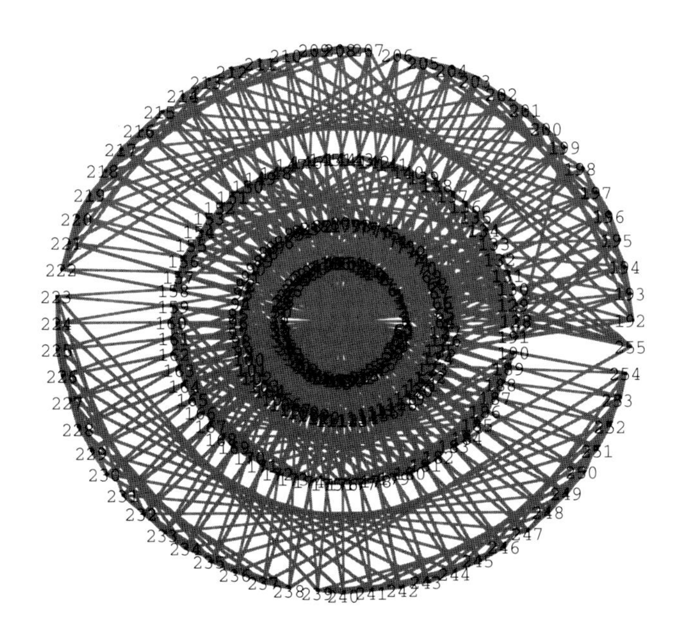

h_8

Figure 14.10

14.1.2 Visualizing Three-Dimensional Polyhedra

If one limits oneself to visualizing three-dimensional polyhedra, several notable attempts have been made [6], [7].

The image below shows the path taken by an interior point algorithm in finding the optimal solution to a three-variable linear programming problem (Figure 75 of [6]).

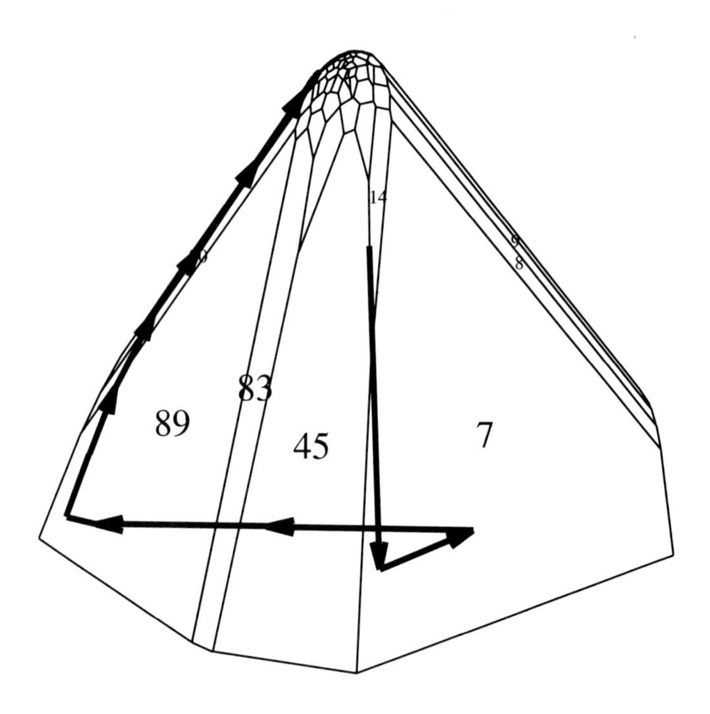

Figure 14.11

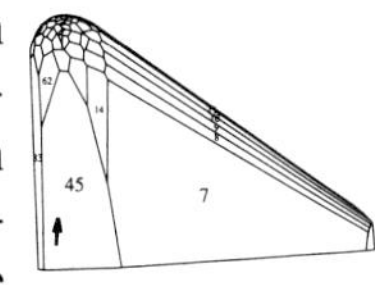

The following images show a series of frames illustrating the path taken by an interior point algorithm in solving a three-variable linear programming problem [6]. The arrow indicates the current descent direction of the algorithm. The polytope changes shape as part of the scaling operation that occurs as part of the interior point algorithm. The sequence continues in Figure 14.13 and is also presented as a flip chart starting at the beginning of this chapter.

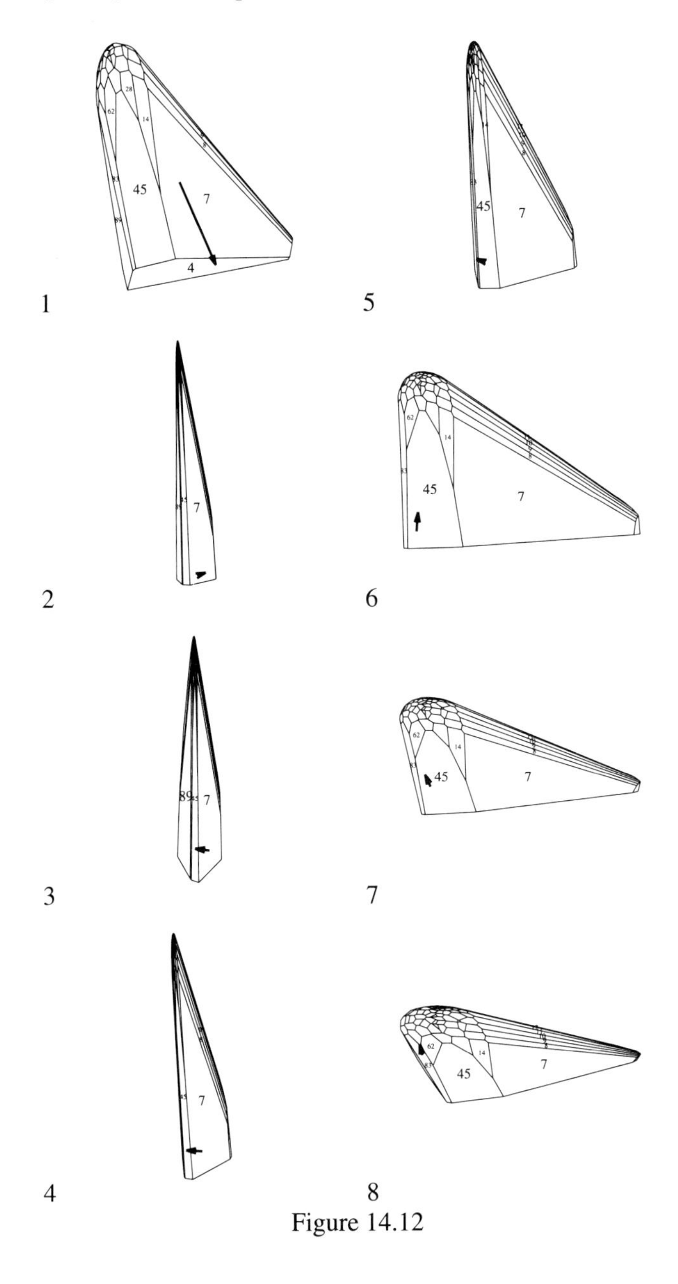

Figure 14.12

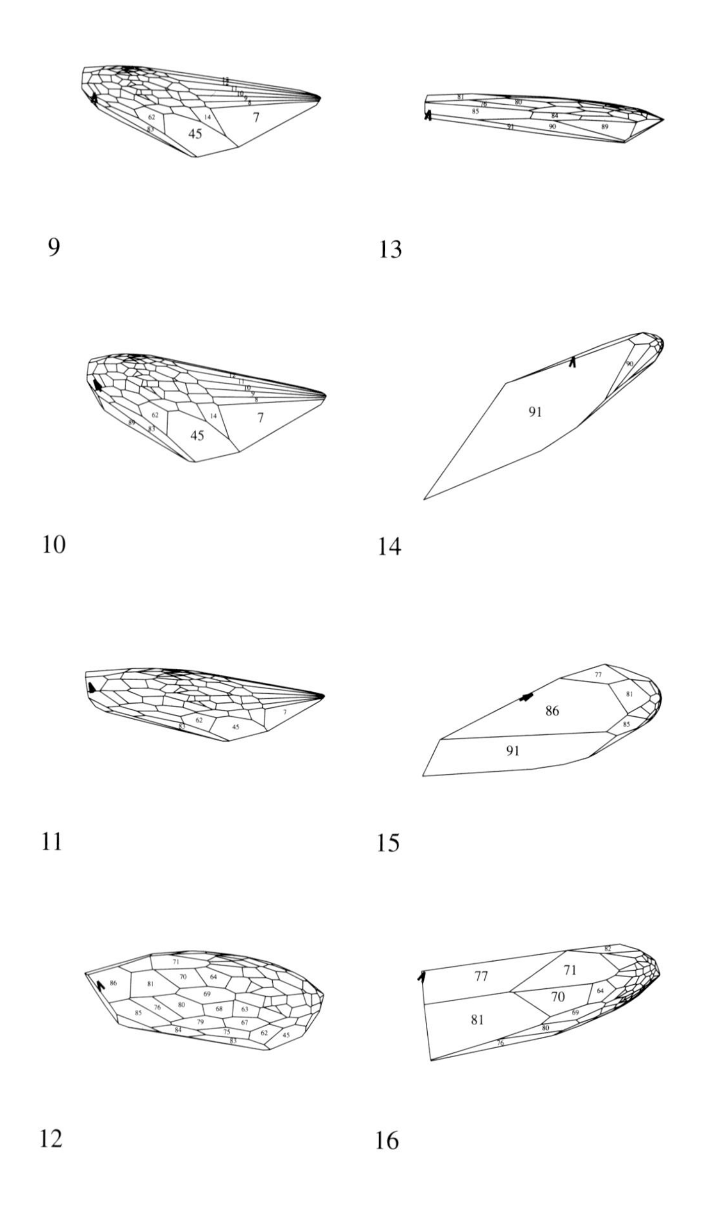

Figure 14.13: The next series of frames continuing the images from Figure 14.12. The series continues in Figure 14.14. Images courtesy D. Gay [6].

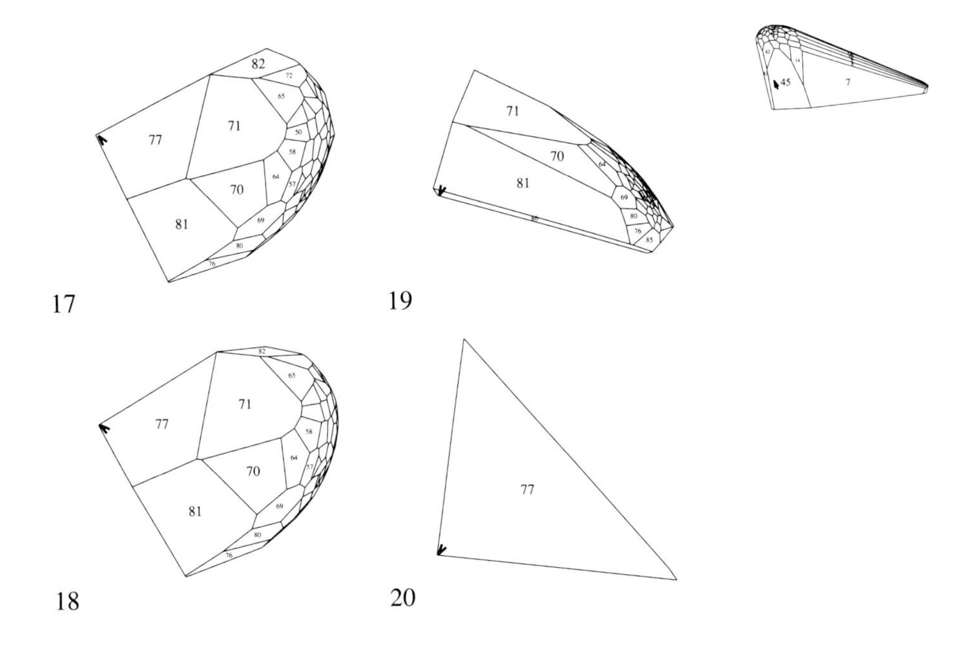

Figure 14.14: The next series of frames continuing the images from Figure 14.13. Images courtesy D. Gay [6].

14.1.3 Summary

Since many polyhedra of interest in optimization do not display the tremendous symmetries found in hypercubes, it seems doubtful that a graph-embedding approach will produce useful representations for higher dimensional polyhedra. Carpendale, Cowperthwaite and Fracchia [8], [9] have explored using distorted views of graphs to visualize n-dimensional polyhedra as discussed in Section 13.3.8. Although their research is preliminary, the technique seems promising.

14.2 Statistics

Statistics has developed a long tradition of research into visualizing higher dimensional data [10], [11], [12]. In what follows, we assume the data consists of a set of m observations of n attributes. More precisely, the data is arranged into an $m \times n$ matrix, A, wherein each row of A represents an observation, and each column an attribute of the observation. Column i will be denoted A^i.

14.2.1 Scatterplots

Many of the representations are based on projections of n-dimensional point clouds or scatterplots into two or three dimensions [12], [13], [14], [15], [16]. One could even plot all combinations of A^i vs A^j, $i = 1, \ldots, n$, $j = i, \ldots n$, $i \neq j$ in separate windows. Call the plot of A^i vs A^j, P_{ij}. A *scatterplot matrix* takes this full set of plots and arranges the P_{ij} into a $n \times n$ matrix. Below is an example of a seven-dimensional scatterplot matrix.

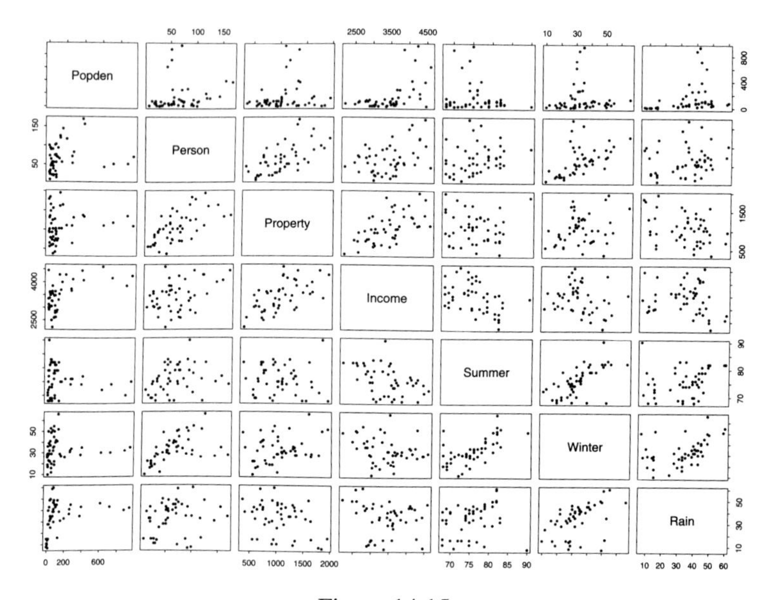

Figure 14.15

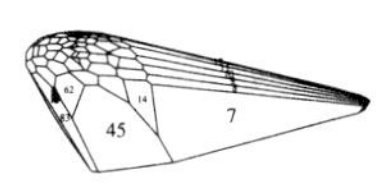

By adding interactivity, one can enhance users' ability to visualize the n-dimensional data. A technique called *brushing* [13], [14] allows users to select points in one two-dimensional plot and simultaneously see their positions in all other views. Below for example is a scatterplot matrix wherein the user has chosen several points that are highlighted in all views. As the user chooses points in one view they are simultaneously highlighted in all views. We shall return to brushing in Chapter 15 when we discuss animated sensitivity analysis and dynamic queries. Brushing highlights the importance of interactivity to visualization.

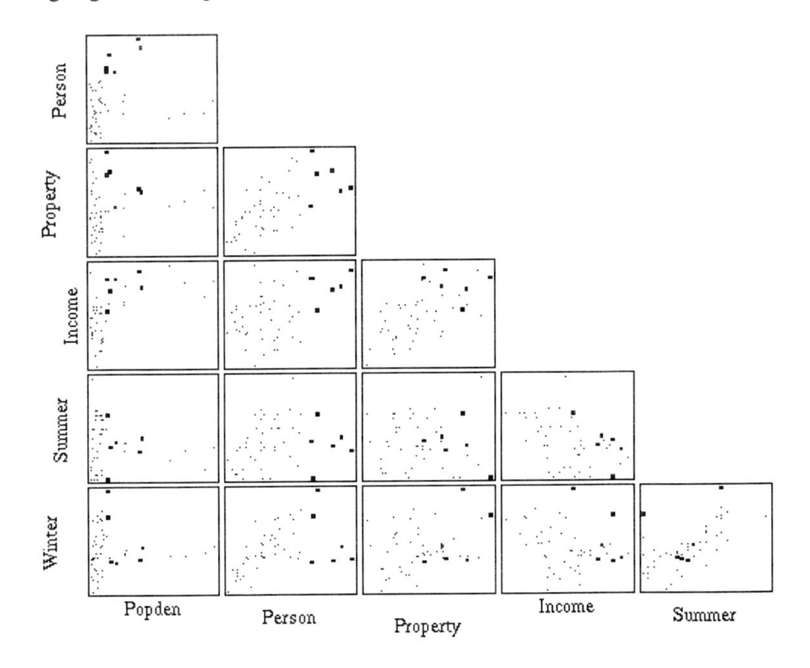

Figure 14.16

14.2.2 Chernoff Faces

Chernoff [17], and Flury and Riedwyl [18] proposed a novel n-dimensional representation wherein each data point is encoded as a face, each coordinate of the point encoding a different characteristic. A total of 15 facial characteristics can be controlled, including the area of the face, the length of the nose, location, curve and width of the mouth, and the shape and location of the eyes. Although arguably too cute, these representations attempt to exploit our well-developed ability to read emotion human faces. Since only 15 characteristics can be represented, this approach is limited to at most 15-dimensional data. Flury and Riedwyl [18] proposed an extension to Chernoff faces that uses asymmetric faces, thereby almost doubling the number of dimensions that can be represented.

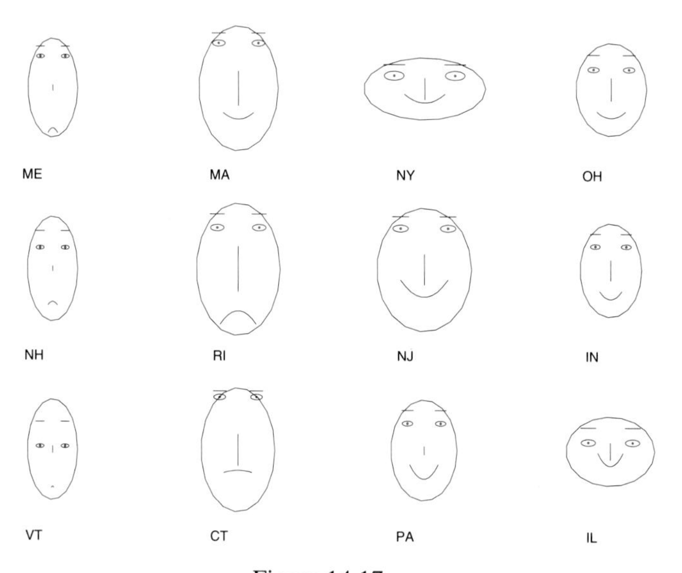

Figure 14.17

Humans also are quite adept at inferring a personality based on the motion of an object. Disney animators have exploited this fact in their marvelous animations of a flying carpet in *Aladdin* or a candlestick in *Beauty and the Beast*. One could imagine using such techniques to be able to display even higher dimensional data than Chernoff faces; for example, the face could change from happy to sad representing a particular trend in the data.

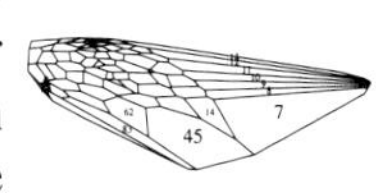

14.2.3 Summary

Statistical visualization techniques are designed to help people identify interesting patterns in data. At the least, since optimization models are huge consumers of data, the techniques are useful for analyzing data input to and output from optimization models.

The techniques from statistics do not seem to scale up to huge numbers of dimensions, however. For example, Chernoff faces allow at most 15 dimensions of information to be displayed. For scatterplot matrices, as the number of dimensions becomes large, the individual matrices in a scatterplot matrix become illegible.

Moreover, it is unclear how well these data visualization techniques would transfer to the problem of visualizing n-dimensional polyhedra or other n-dimensional objects of interest to optimization.

14.3 Parallel Coordinates

In 1981, Inselberg [19] proposed a new representation technique for n-dimensional information. Instead of drawing the n coordinate axes perpendicular to one another as in Cartesian coordinates, the axes are drawn as parallel vertical lines. A great deal of research on Parallel coordinates has been conducted [20], [21], [22] and they have been usefully applied to a variety of areas especially statistics [23]. They have also been applied to linear programming [24]. In this section, we develop some of the basic ideas behind Parallel coordinates and illustrate their application to linear programming.

Since the fundamental building block in Cartesian coordinates is the point, we begin our discussion of Parallel coordinates there. In the discussion that follows, a major source of confusion exists. As will be seen, a point in Cartesian coordinates is represented by a line in Parallel coordinates and a line in Cartesian coordinates is represented by a point in Parallel coordinates. When we speak of a "line" or a "point," therefore, we are careful to distinguish the type of coordinate system to which we refer. This source of confusion enhanced by the fact that we must use Cartesian coordinates to describe Parallel coordinate representations. Since Cartesian coordinates are the norm, we rely on them to explain Parallel coordinates. We hope that we have overestimated the potential for confusion.

14.3.1 Points

To be precise, in Parallel coordinates, given n dimensions numbered $i = 1, \ldots, n$, the Parallel coordinate axis representing the ith dimension is drawn as the vertical line $x = i - 1$ (in Cartesian coordinates).

A data point $(x_1, \ldots, x_n)$ is plotted in Parallel coordinates by locating coordinate x_i vertically along axis i, that is, at point $(i-1, x_i)$. Adjacent points are then connected by a straight line (called a *polyline*). For example, the six-dimensional point $(-5, 3, 4, -2, 0, 1)$ at right in Parallel coordinates is a polyline.

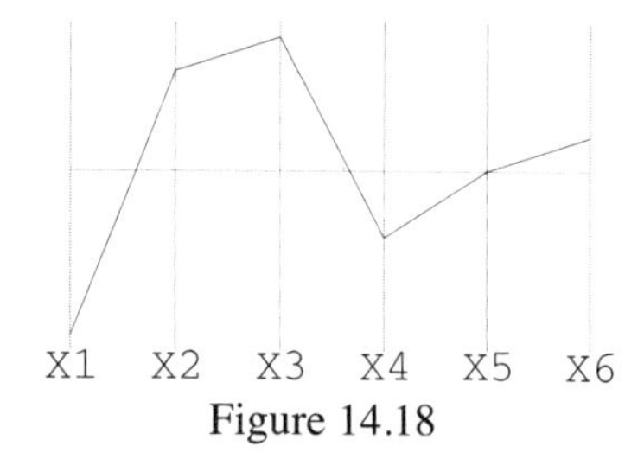

Figure 14.18

Merely by representing n-dimensional points, Parallel coordinates are useful. Consider, for example, the data below which was collected from New York financial markets between 1985 and 1993. The first three axes (from left to right) represent the week, the month and the year. The remaining axes (from left to right) give the price in US dollars of British Pound Sterling (BPS), Deutsch Marks (GDM), Japanese Yen (YEN), 3 month US Treasury bills (TB3M), 30 year US Treasury bills (TB30Y), Gold and the Standard and Poors 500 index. The figure below was generated using IBM's Parallel Visual Explorer [25] (figures courtesy IBM).

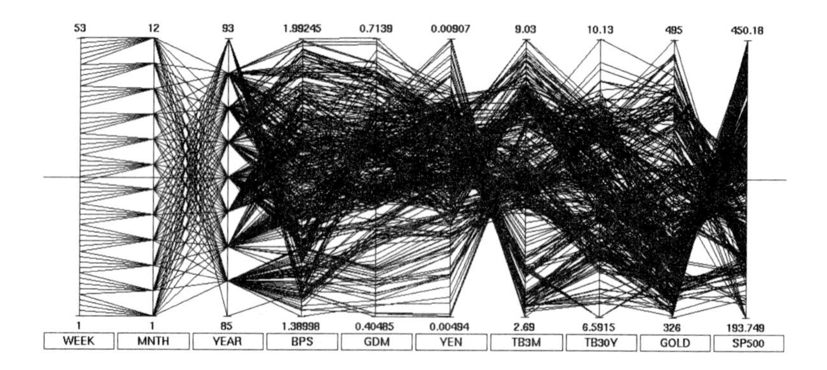

Figure 14.19

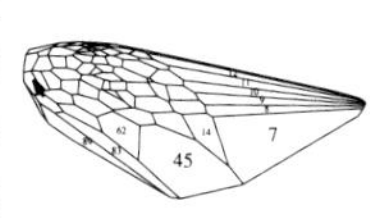

By including axes for years and weeks, one can select ranges of weeks and years. For example, consider the points below for 1986 (black) and 1992 (grey). In 1986, the prices of Yen, Deutsch Marks, and British Pounds were relatively low whereas the price of US Treasury Bills was high. In 1992, the opposite was true. Moreover, in 1986, 30 year Treasury bills were the most volatile, and there was a big jump in the price of gold. In 1992, in contrast, British Pound Sterling was the most volatile, whereas gold was very stable.

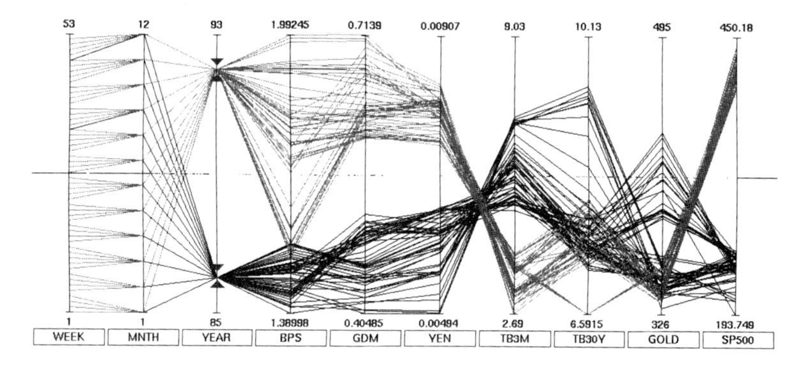

Figure 14.20

If one selects only those data points associated with 1986 and then selects the data from the first few weeks of the year (black lines), one can see that the price of gold at the beginning of the year was relatively low.

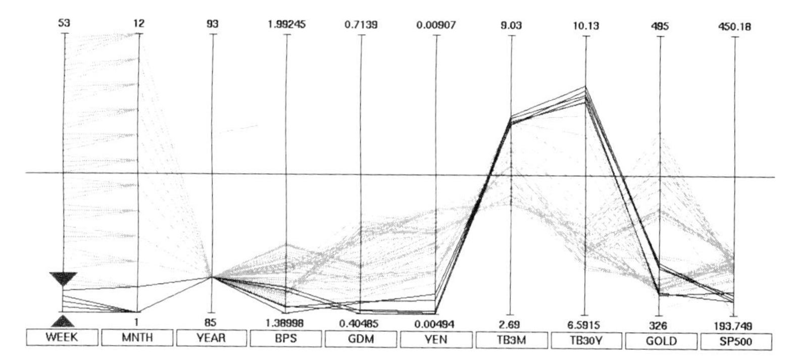

Figure 14.21

As one scans through succeeding weeks, one can detect that the price of gold jumps significantly during the last three weeks of July and the first two weeks of August. On the Gold axis below, note the highest solid black line.

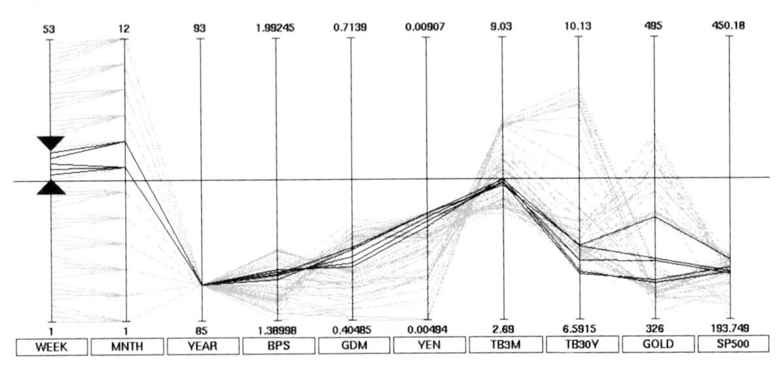

Figure 14.22

14.3.2 Two-Dimensional Lines

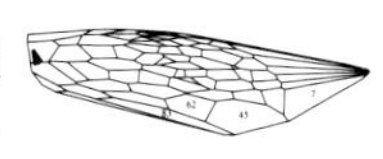

Given this discussion of Parallel coordinate representations of points and an application to data visualization, we now consider how Parallel coordinates can represent lines, planes and hyperplanes.

Parallel coordinates and traditional Cartesian coordinates exhibit some elegant duality. As shown before, points in Cartesian coordinates appear as (poly) lines in Parallel coordinates. Similarly, lines in Cartesian coordinates can be seen to correspond to one or more points in Parallel coordinates. In the two-dimensional case, if one plots a series of points satisfying $x_2 = -3x_1 + 20$, in Parallel coordinates a series of lines arise. The lines intersect at a common point. For the two-dimensional case, the linear equation $x_2 = mx_1 + b$ is represented in Parallel coordinates by the point $(1/1 - m, b/1 - m)$ assuming $m \neq 1$.

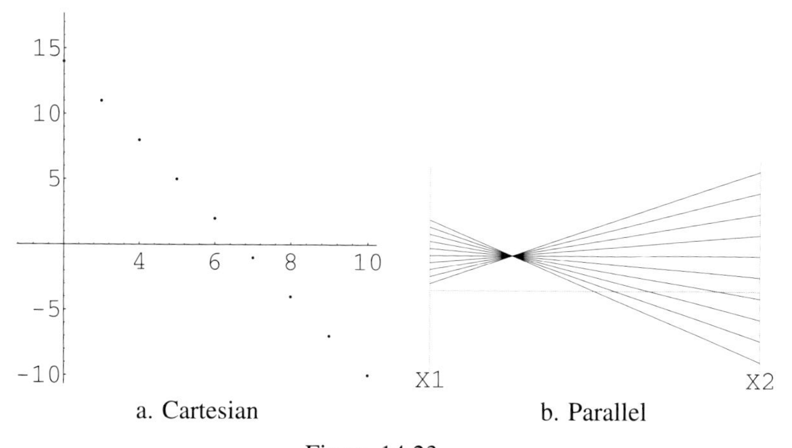

a. Cartesian b. Parallel

Figure 14.23

14.3.3 Lines with Slope 1

For the case $m = 1$, consider a plot of a series of points on the line $y = x + 2$ in Cartesian (a) and Parallel (b) coordinates below. The linear equation with $m = 1$ is represented as a series of parallel lines in Parallel coordinates. The slope of these lines is equal to the y-intercept of the line (in the example, the slope is 2).

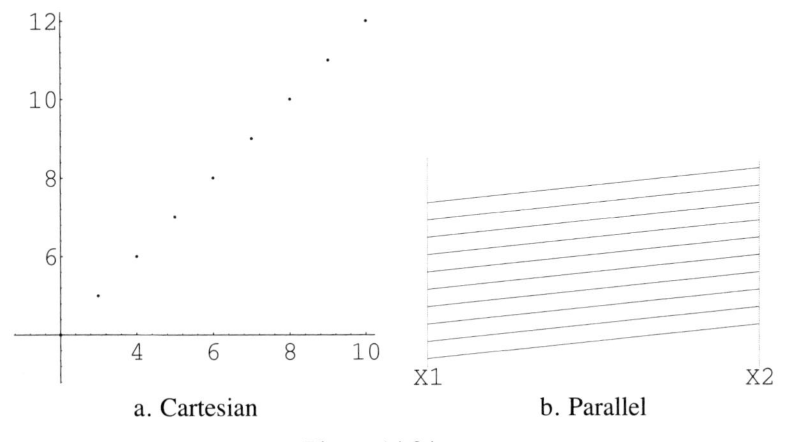

a. Cartesian b. Parallel

Figure 14.24

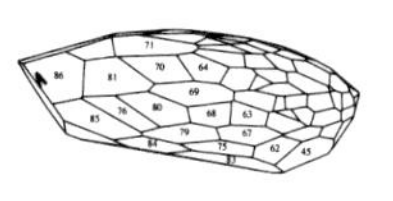

Visually, this seems to be about the best that one can do to represent a line with $m = 1$ in parallel coordinates. Mathematically, however, the apparent singularity of $1/m - 1$ when $m = 1$ can be easily handled by using a *homogeneous* coordinate system. Homogeneous coordinates add an additional coordinate to every point. More precisely, point $p = (x_1, \ldots, x_n)$, can be represented by point $p' = (x_1, \ldots, x_n, 1)$ in homogeneous coordinates. Point $p'' = (y_1, \ldots, y_n, d)$, $d \neq 0$ in homogeneous coordinates represents point $p''' = (y_1/d, \ldots, y_n/d)$ in non-homogeneous coordinates. One can define traditional mathematical operations for homogeneous coordinates (e.g., addition of two points, matrix multiplication) that produce the same answer as for non-homogeneous coordinates. In essence, homogeneous coordinates defer division operations. Parallel coordinates calculations, by using homogeneous coordinates avoids the singularity of division by 0.

One of most practical applications of homogeneous coordinates is three-dimensional computer graphics systems. Three-dimensional computer graphics systems must rotate, scale and translate and project thousands of three-dimensional points in two dimensions as fast as possible. These operations are performed using matrix multiplication. The projection into two dimensions require division operations. By using homogeneous coordinates, the division operations can be deferred. As discussed in Section 4.3.2, the denominator is accumulated into the last coordinate. A single division for each point can then be performed at the last possible moment, thereby accelerating the display of three-dimensional images.

14.3.4 n-Dimensional Lines

Mathematically, an n-dimensional line can be represented as a set of $n - 1$ equations of two variables, i.e.,

$$l_i : x_i = m_i x_{i-1} + b_i \quad i = 2, \ldots, n \tag{14.1}$$

Since each of the individual equations defines a two-dimensional line, a n-dimensional line be represented in Parallel coordinates as $n - 1$ points:

$$\left(i - 1 + \frac{1}{m_i - 1}, \frac{b_i}{m_i - 1}\right) \quad m_i \neq 1 \quad i = 2..n \tag{14.2}$$

For example, consider the 4-dimensional line defined by the following equations:

$$\begin{aligned} x_2 &= -3x_1 + 12 \\ x_3 &= -4x_2 + 48 \\ x_4 &= -2x_3 - 54 \end{aligned} \tag{14.3}$$

If we plot a set of points which lie on the line defined in (14.3) in Parallel coordinates, one can see the three intersection points that define the line.

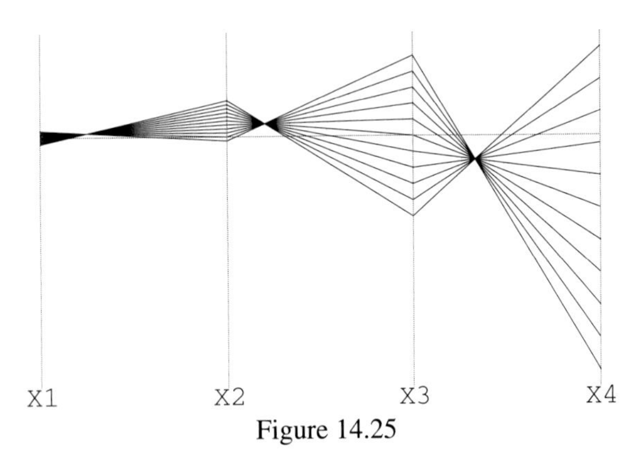

Figure 14.25

14.3.5 Parallel Lines in 2 Dimensions

Representing parallel lines in Parallel coordinates is a stepping stone to representing hyperplanes and half-spaces in Parallel coordinates. Representing hyperplanes and half-spaces will then allow us to represent n-dimensional linear programs in Parallel coordinates. What follows is based on the work of Chatterjee, Das and Bhattacharya [24], who first used Parallel coordinates to represent linear programs.

In 2 dimensions, the line $x_2 = m_2x_1 + b_2$, $(m_2 \neq 1)$ is represented in Parallel coordinates as the point $(1/(1-m_2), b_2/(1-m_2))$. The set of lines parallel to $x_2 = m_2x_1 + b_2$ is defined by the set of points $(1/(1-m_2), b_2/(1-m_2))$, $-\infty \leq b \leq \infty$. In other words, the set of lines in Cartesian coordinates having slope m is represented as a vertical line through $(1/(1-m_2), b_2/(1-m_2))$ in Parallel coordinates. To the right, for example, is a Parallel coordinate plot of the set of lines $x_2 = -3x_1 + b$, $-\infty \leq b \leq \infty$.

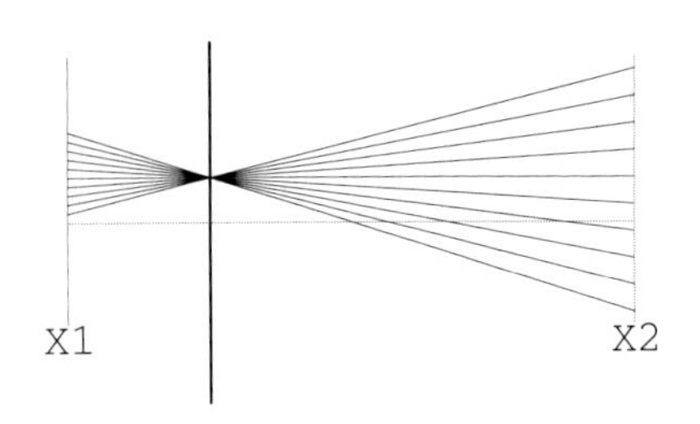

Figure 14.26

14.3.6 Parallel Lines in n-Dimensions

The set of lines parallel to an n-dimensional line (14.1) is represented in Parallel coordinates by a set of $(n-1)$ vertical lines passing through the points (14.2). For example, the set of lines parallel to (14.3) consists of a set of 3 vertical lines passing through the points $(1/4, 3), (6/5, 9.8), (7/3, -18)$ as shown at right.

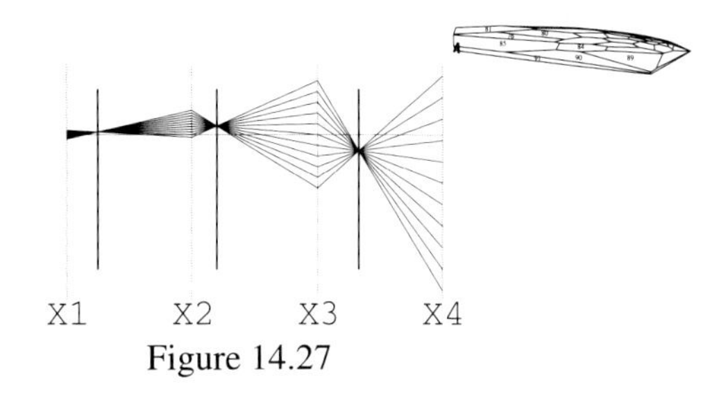

Figure 14.27

14.3.7 Two-Dimensional Half-Spaces

In two dimensions, a half-space H can be defined as the set of points (x_1, x_2) such that $x_2 \leq mx_1 + b$ (without loss of generality, $|m| < \infty$). Equivalently, a half-space H can be defined as the union of parallel lines $x_2 = mx_1 + b'$ where $b' \leq b$. A half-space in Parallel coordinates is therefore represented as a vertical ray. For example, the half-space $x_2 \leq -3x_1 + 20$ is represented by the vertical ray $x = 1/4$ for $y \leq 5$.

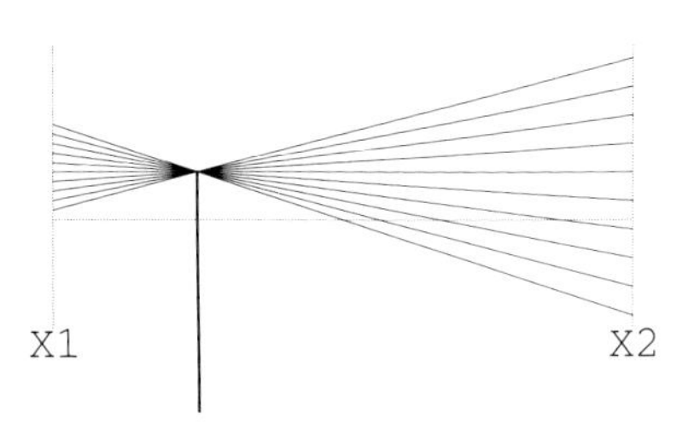

Figure 14.28

Now consider $x_1 = 1$, $x_2 = 0$, which is not in the halfspace $x_2 \leq -3x_1 + 2$ as well as $x_1 = -1, x_2 = 0$, which is. In Parallel coordinates, the line representing $(1,0)$ does not intersect the ray representing the halfspace, whereas the line representing $(-1,0)$ does intersect the ray. Any line intersecting the ray defines a point in the halfspace. Unfortunately, this simple construction for identifying feasible and infeasible points does not extend to more than two dimensions. A more complicated construction will be necessary. This is the topic of the rest of this section.

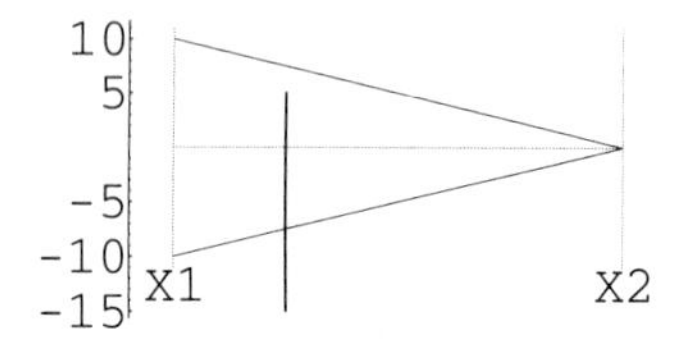

Figure 14.29

14.3.8 n-Dimensional Half-Spaces

We seek a representation of an n-dimensional half-space in Parallel coordinates that will allow us to detect whether or not a given n dimensional point is contained in the halfspace. This is one of the fundamental operations required for linear programming. A more general discussion of this problem can be found in [26].

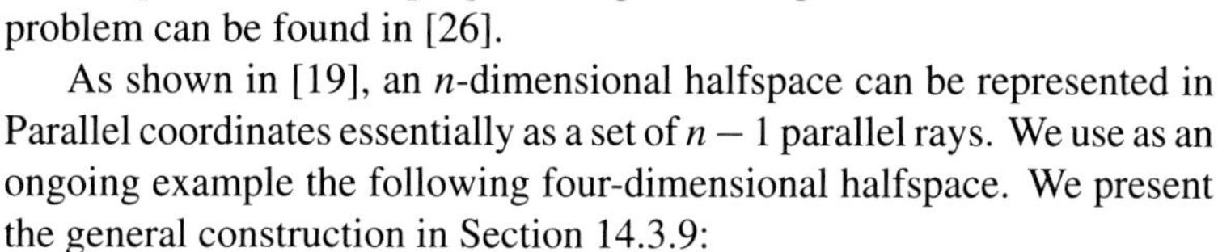

As shown in [19], an n-dimensional halfspace can be represented in Parallel coordinates essentially as a set of $n-1$ parallel rays. We use as an ongoing example the following four-dimensional halfspace. We present the general construction in Section 14.3.9:

$$x_1 + 2x_2 + 3x_3 + 4x_4 \leq 20 \tag{14.4}$$

Given (14.4) consider the following three half-spaces:

$$\begin{aligned} x_1 + 2x_2 &\leq 20 \\ 2x_2 + 3x_3 &\leq 20 \\ 3x_3 + 4x_4 &\leq 20 \end{aligned} \tag{14.5}$$

The half-spaces were constructed by setting all but two of the variables in (14.4) to 0 , leaving only adjacent pairs of variables.

Plot each one of these half-spaces in Parallel coordinates. This produces three vertical rays labeled y_1, y_2 and y_3, respectively.

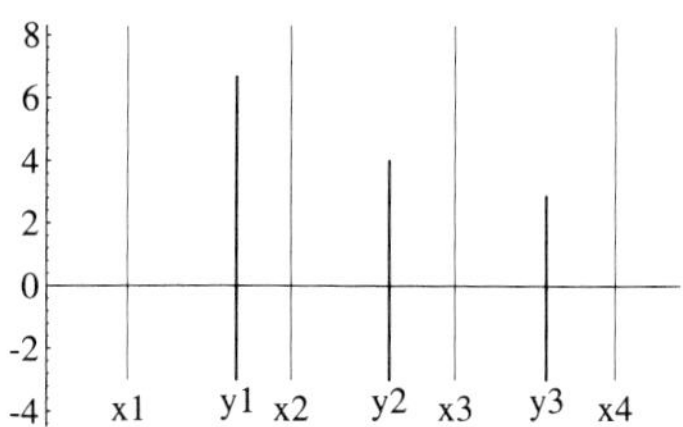

Figure 14.30

At this point, one might hope that any point inside the half-space, when plotted in Parallel coordinates, would intersect each of the vertical rays of Figure 14.30, but this is not the case. Consider, for example, the points $a = (4, 3, 2, 1)$, $b = (4, 3, 2, 3)$ and $c = (4, 3, 2, -1)$. Point a lies on the boundary of the half-space, point b is outside, and point c is strictly inside. Yet all the points, when plotted in Parallel coordinates, intersect each vertical ray. Unfortunately, a more complicated construction is required to determine whether or not a point is inside a given half-space.

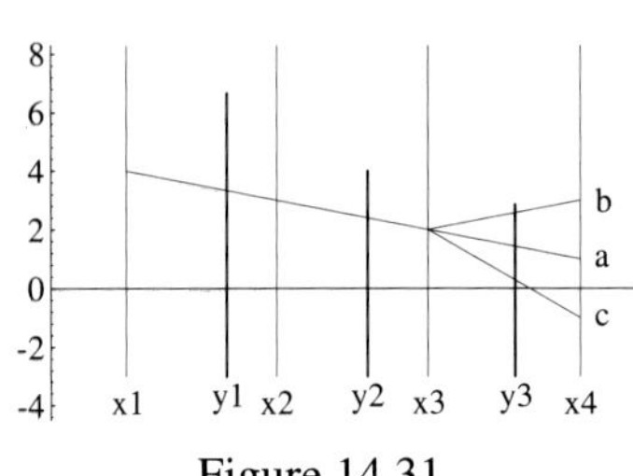

Figure 14.31

The construction consists of drawing a series of lines on the Parallel coordinates. We first present the construction, and then explain why it works. First, draw a line from $x_1 = 4$ (the first coordinate of a labeled A in the figure at right) through the origin of ray y_1 until it intersects the x_2 axis (at point B). Lines intersecting the origin of ray y_1 represent points on the line $x_1 + 2x_2 = 20$. Setting $x_1 = 4$ and solving for x_2 yields $x_2 = 8$.

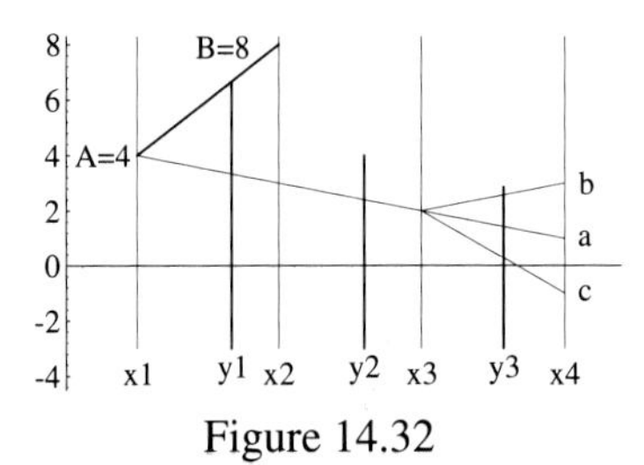

Figure 14.32

Next, draw a line from B to the point $x_3 = 0$ as shown at right. Call the intersection of the line and y_2 point C. Ray y_2 identifies lines of the form $2x_2 + 3x_3 = q$ where $q \leq 20$. For $x_2 = 8$ and $x_3 = 0$, $q = 16$.

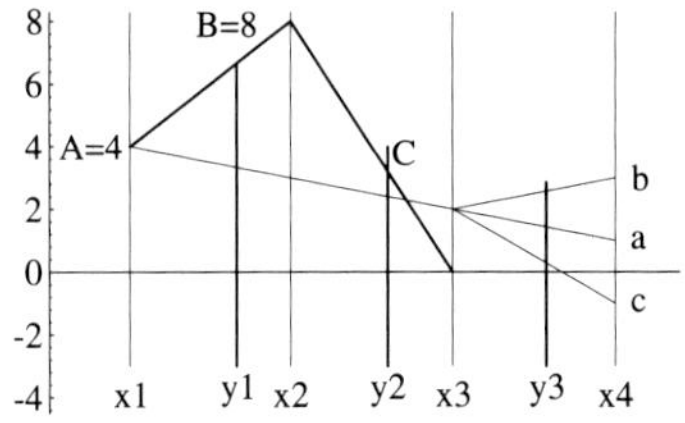

Figure 14.33

Now draw a line from point D where $x_2 = 3$, through point C to the x_3 axis at point E. Since line DE intersects C, it corresponds to a point on the line $2x_2 + 3x_3 = 16$. Setting $x_2 = 3$ and solving for x_3 yields $x_3 = 10/3$.

Next draw a line from point E to the point on the x_4 axis where $x_4 = 0$. This line intersects ray y_3 at point F. Draw a line from $x_3 = 2$ through F to the x_4 axis. Call the intersection of the line and the x_4 axis G. Since point b intersects x_4 higher than G, it is infeasible; since point c intersects x_4 lower than G it is (strictly) feasible; finally, since point a intersects x_4 at G, point a lies on the boundary of the half-space.

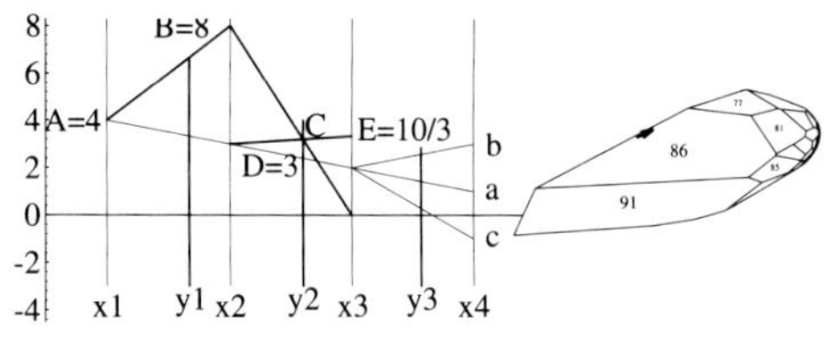

Figure 14.34

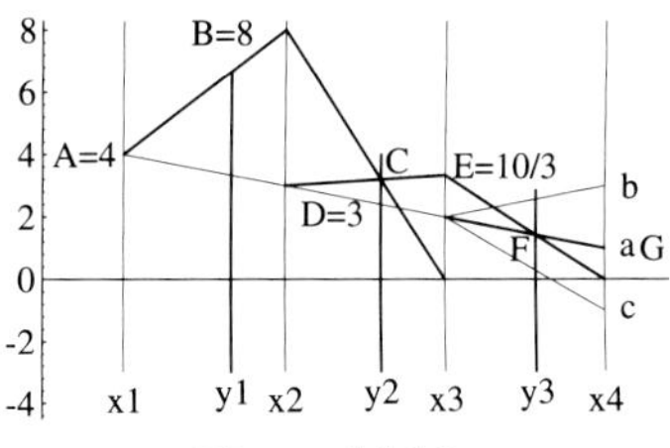

Figure 14.35

14.3.9 The Construction in Detail

Why does the construction illustrated in the previous section work? Consider a hyperplane of the form:

$$\sum_{i=1}^{n} a_i x_i \leq b \tag{14.6}$$

We assume that $a_i \neq 0$, $i = 1, \ldots, n$. If $a_i = 0$, then we simply delete the x_i axis from the plot and renumber the remaining axes sequentially. We first construct a series of $n - 1$ vertical rays from the following inequalities:

$$y_i : a_i x_i + a_{i+1} x_{i+1} \leq b \quad i = 1, \ldots, n-1 \tag{14.7}$$

Given a point $x = (\alpha_1, \ldots, \alpha_n)$, we wish to determine if x satisfies (14.6). Using simple arithmetic, one could just calculate $z = \sum_{i=1}^{n} a_i \alpha_i$ and then compare z with b. The geometric construction calculates z and compares it with b graphically rather than arithmetically.

Having drawn the rays representing the inequalities (14.7), the construction starts by drawing a line from the point $x_1 = \alpha_1$ through the origin of ray y_1 to axis x_2. Label the intersection of this line with axis x_2, z_2. Note that

$$z_2 = \frac{b - a_1 \alpha_1}{a_2}.$$

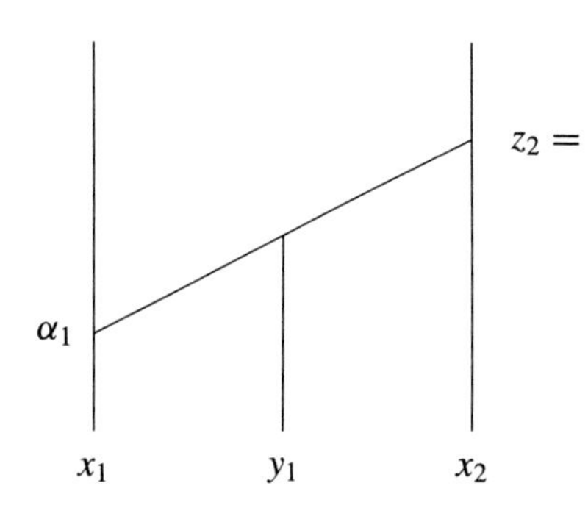

Figure 14.36

This is true since the line connecting α_1 and z_2 intersects the origin of ray y_1 and the origin of ray y_1 is defined by the equation $a_1 x_1 + a_2 x_2 = b$. In other words, z_1 was determined by setting x_1 to α_1 in $a_1 x_1 + a_2 x_2 = b$ and solving for x_2.

For axes $i = 2, \ldots, n-1$, draw a line from z_i on axis i to 0 on axis $i+1$. Call the point of intersection of this line with y_i (or its extension), W_i. Also draw a line from α_i on axis x_i through W_i to axis $i+1$, yielding point z_{i+1}.

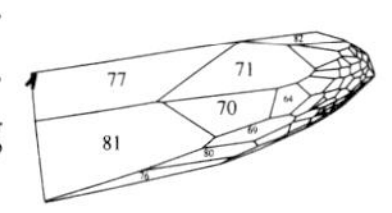

W_i corresponds to the linear equation $a_i x_i + a_{i+1} x_{i+1} = a_i z_i$. This is true because every point on ray y_i defines a line of the form

$$a_i x_i + a_{i+1} x_{i+1} = q$$

For $x_i = z_i$, $x_{i+1} = 0$, $q = a_i z_i$. If we assume for now that

$$z_i = \frac{b - \sum_{j=1}^{i-1} a_j \alpha_j}{a_i}$$

(it is certainly true for $i = 2$), then

$$q = b - \sum_{j=1}^{i-1} a_j \alpha_j.$$

Since $x_i = \alpha_i$ and $x_{i+1} = z_{i+1}$ lie on the line $a_i x_i + a_{i+1} x_{i+1} = b - \sum_{j=1}^{i-1} a_j \alpha_j$,

$$z_{i+1} = \frac{b - \sum_{j=1}^{i} a_j \alpha_j}{a_{i+1}},$$

which corresponds to the assumed form for z_i.

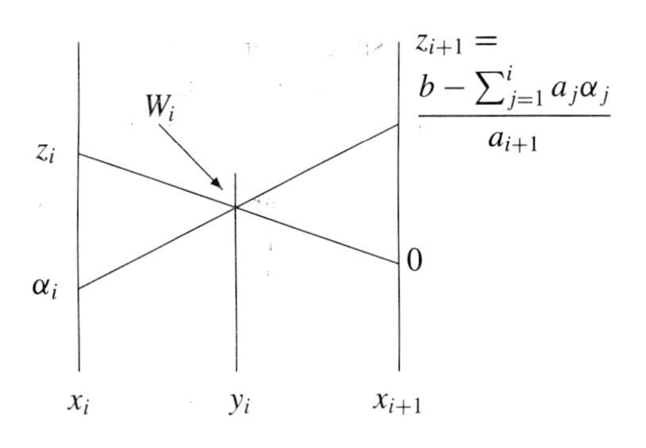

Figure 14.37

By using this construction, we eventually calculate graphically

$$z_n = \frac{b - \sum_{j=1}^{n-1} a_j \alpha_j}{a_n}. \tag{14.8}$$

By comparing α_n to z_n, which can be done visually, one can determine if the original point x is inside or outside the halfspace. Without loss of generality, assume that $a_n > 0$. Then x is feasible if and only if $\alpha_n \leq z_n$. This is true since if $\alpha_n \leq z_n$ implies that $\alpha_n \leq \frac{b - \sum_{j=1}^{n-1} a_j \alpha_j}{a_n}$, which through simple algebra can be rewritten as $\sum_{i=1}^{n} a_i \alpha_i \leq b$, the desired result. So if α_n is at the same height or below z_n, then the point is feasible, otherwise it is not.

It would of course be preferable if one could detect whether or not a point is feasible through simple inspection of the Parallel coordinate plot. The construction described in the previous section takes some time, and if the lines are drawn by hand, small errors can be amplified during the construction. On the other hand, the construction is relatively straightforward, and at least Parallel coordinates provides a graphical mechanism for determining feasibility. No equivalently simple, graphical method is known for projections of n-dimensional hyperplanes into two-dimensional Cartesian coordinates. Perhaps a different representation of a hyperplane in Parallel coordinates can be developed that would make more obvious the task of graphically determining feasibility.

14.3.10 Linear Programming

Given Parallel coordinate representations of half-spaces it is possible to represent linear programming problems in Parallel coordinates. Chatterjee, Das and Bhattacharya [24] explored this idea.

Consider the following two-dimensional linear programming problem:

$$\begin{array}{llcl} \max & z = x_1 + x_2 & & \\ \text{subject to:} & & & \\ c_1: & -2x_1 + 5x_2 & \leq & 20 \\ c_2: & x_1 + 4x_2 & \leq & 45 \\ c_3: & 2x_1 + x_2 & \leq & 27 \\ c_4: & 3x_1 - x_2 & \leq & 24 \\ & x_1, x_2 & \leq & 0 \end{array} \tag{14.9}$$

In traditional Cartesian coordinates, the feasible region and optimal solution appears at right. A Simplex algorithm could trace the series of points (0,0), (8,0), (51/5,33/5) to the optimal solution (115/12,47/6), with binding constraints c_2 and c_3 and objective function value 209/12.

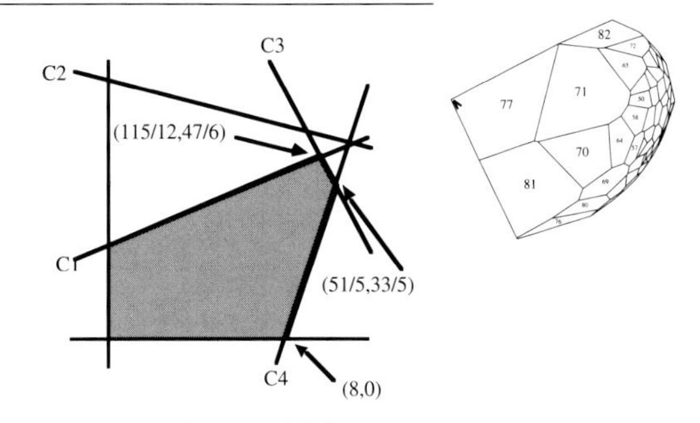

Figure 14.38

In Parallel coordinates, each constraint is replaced by a vertical ray. The objective function (Z) appears as a vertical line.

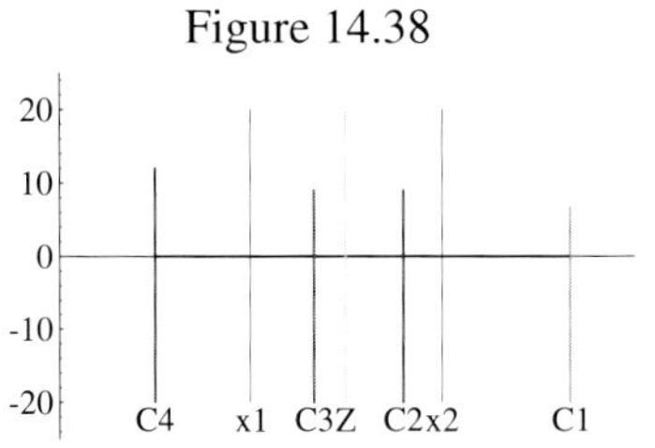

Figure 14.39

Below is a series of Parallel Coordinate images showing a set of lines representing the extreme points found by the Simplex algorithm. The objective function seeks to move the line representing the solution point as high as possible. The line representing the solution must intersect all the vertical rays that represent constraints.

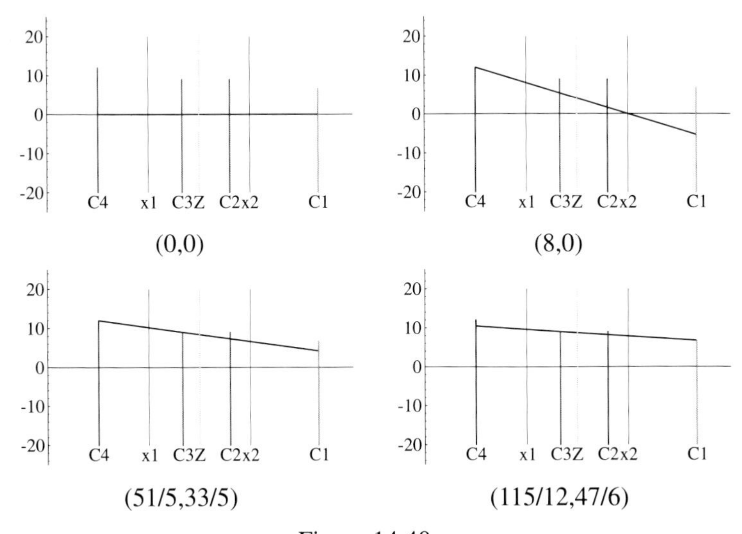

Figure 14.40

Since this is a two-dimensional problem, all feasible points intersect each vertical ray. For higher dimensional problems, one must use the construction described in the previous section. However, if each constraint has at most 3 non-zeros (aside from the right-hand side), the simple intersection test will work even for higher dimensions. In Chatterjee, *et al.*'s paper [24], they exploit this fact when illustrating Parallel coordinates applied to a 5-dimensional linear programming problem.

14.3.11 Hypercubes and Hyperspheres

Since we began the discussion of n-dimensional representations with some (purposely confusing) illustrations of n-dimensional hypercubes, it seems fair to end the discussion of Parallel coordinates with an exploration of how well hypercubes can be represented in Parallel coordinates. Consider the figure at right which shows the extreme points of a square.

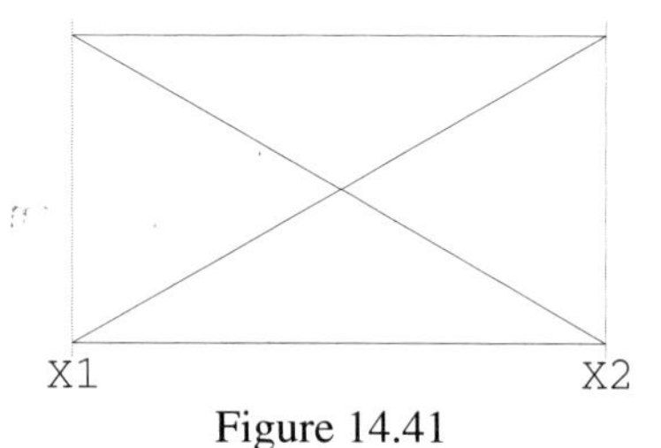

Figure 14.41

The extreme points of a cube are shown at right.

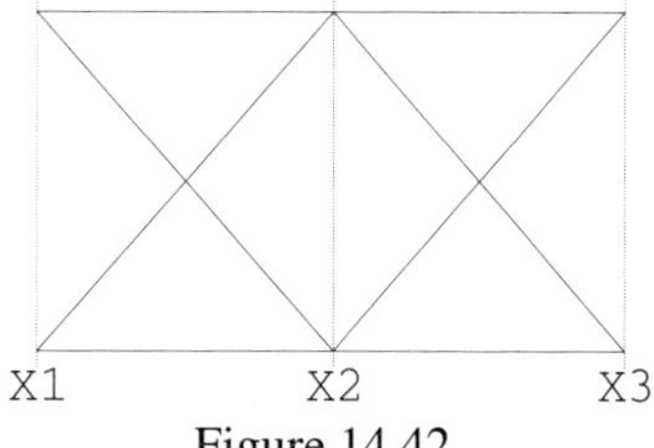

Figure 14.42

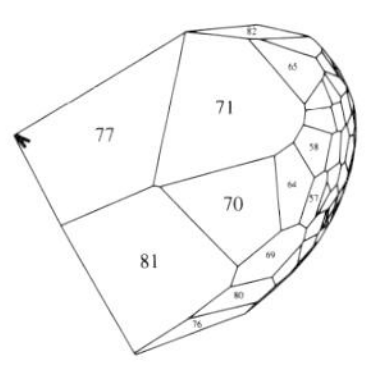

This pattern continues for larger dimensions. Below is an eight-dimensional hypercube plotted in Parallel coordinates. In general, one creates higher dimensional hypercubes merely by repeating the pattern established by a two-dimensional cube.

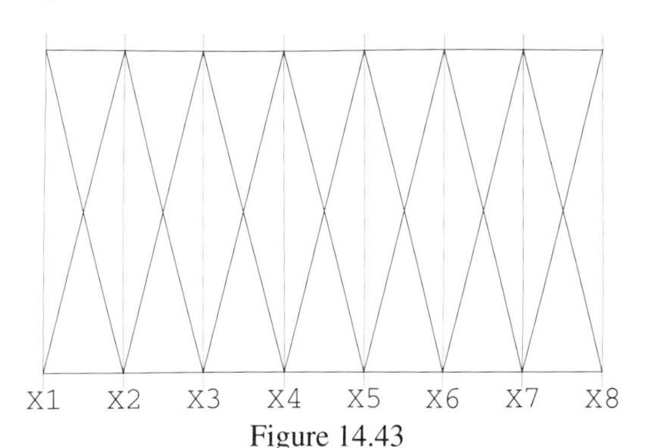

Figure 14.43

This pattern also generalizes to other shapes. Below are Cartesian and Parallel coordinate plots of 36 points equally spaced around a circle.

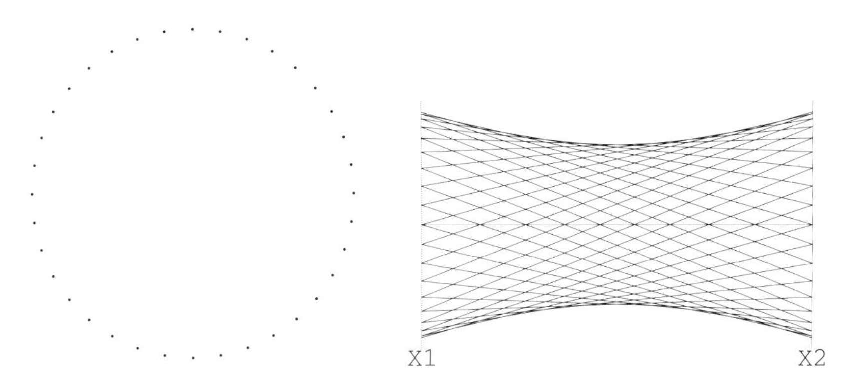

Figure 14.44

A Parallel coordinate plot of points on a sphere is shown below.

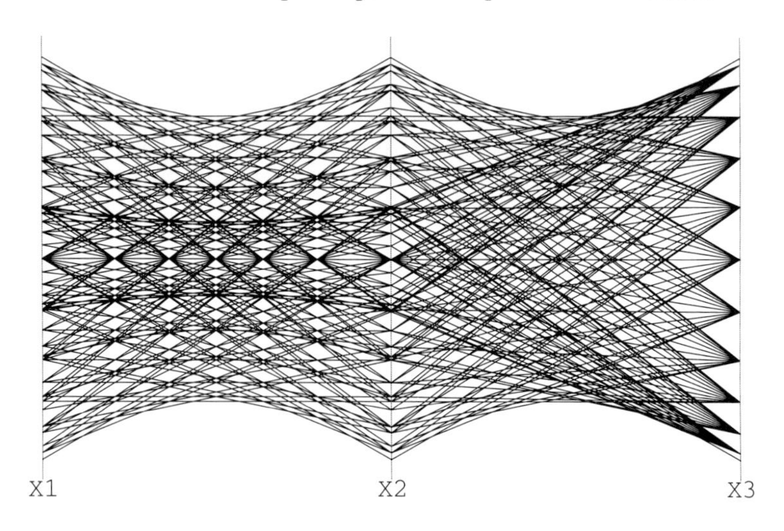

Figure 14.45

Parallel coordinates provides perhaps the cleanest representation of hypercubes (and hyperspheres).

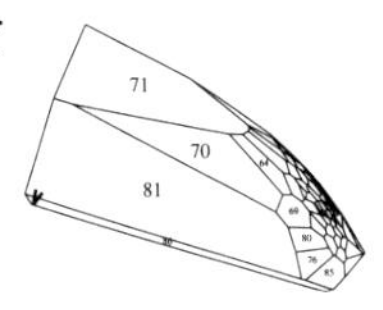

14.3.12 Summary of Parallel Coordinates

Parallel coordinates are not a panacea. Lines with slope equal to 1 (common in many mathematical programming models) are difficult, if not impossible to represent graphically. In addition, the display of several n-dimensional objects can easily cause information overload. For example, an n-dimensional line is represented by a *set* of points in Parallel coordinates. If one were to plot several n-dimensional lines in Parallel coordinates, some labeling, shading or coloring scheme would have to be used to group together the set of points associated with a particular n-dimensional line. As discussed in Section 4.1.1, as the number of lines increases, our ability to distinguish such groups of points diminishes significantly. Similarly, the order in which the axes are arranged can have a significant effect. Different permutations of axes will produce quite different plots.Although Parallel coordinates allow one to represent n-dimensional objects, human cognition still has its limits.

Finally, the Parallel coordinate representation of half-spaces, a fundamental component of linear programming problems, is not as elegant as one might like. An intricate construction is required merely to demonstrate the feasibility of a point.

Tan and Burton [27] proposed an extension to Parallel coordinates that they called *Parallel Planes* coordinates. To plot n-dimensional information, one constructs $n(n-1)/2$ parallel planes where each plane represents one pair of two-dimensional axes $i < j$, $1 \leq i, j \leq n$. To plot a point $(\alpha_1, \ldots, \alpha_n)$, one connects the points (α_i, α_j) on each of the parallel planes. The actual usefulness of this proposal, at this point, is unclear.

Despite their shortcomings, Parallel coordinates have been successfully used for a variety of applications, particularly for exploratory data analysis. Whereas Cartesian coordinates have been used for several hundred years, Parallel coordinates were proposed only 15 years ago. Even in that short time, however, Parallel coordinates have proven to be a useful alternative to Cartesian coordinates for visualizing information with more than 3 dimensions.

14.4 Worlds within Worlds

Feiner and Beshers [28], [29] proposed the use of nested three-dimensional coordinate systems for visualizing functions of the form $f : \Re^n \mapsto \Re$. If $n = 2$, then f is traditionally viewed as a two-dimensional surface. If

$n > 2$, they propose using a set of nested axes. For example, consider the case of calculating a monthly mortgage payment as a function of five values ($n = 5$); Price, Tax, Insurance, Term (length of the mortgage) and the Interest Rate. That is,

$$Payment : \{\text{Price, Tax, Insurance, Term, Interest}\} \mapsto \Re.$$

In the figure below (from [30]), an outer set of axes is drawn for coordinates Insurance, Price and Tax. The user then chose a point where Price=\$87,000, Tax=260 and Insurance=36. At that point, three orthogonal axes are drawn; the two horizontal axes represent Term and Interest, and the vertical axis represents the Payment. The surface plotted gives Payment(87000,260,36,Term,Interest). The user can interactively change the origin of the inner axes, and the surface plot of the function will adjust simultaneously. Image copyright 1993, Stephen Feiner and Clifford Beshers.

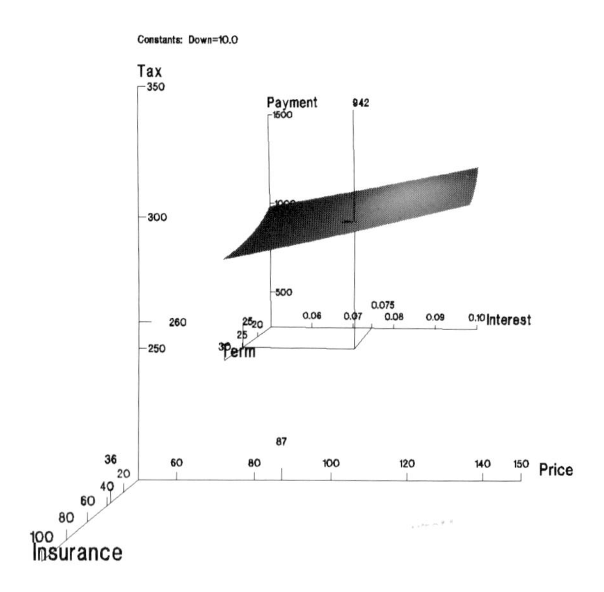

Figure 14.46

For higher dimensional functions, one can nest coordinate axes inside other coordinate axes to whatever level necessary. Note that one can display several inner worlds simultaneously. Furthermore, if the number of dimensions is not a multiple of three, one can use one or two-dimensional worlds as necessary. From this nesting arises Feiner and Beshers's name for the technique, *Worlds within Worlds*.

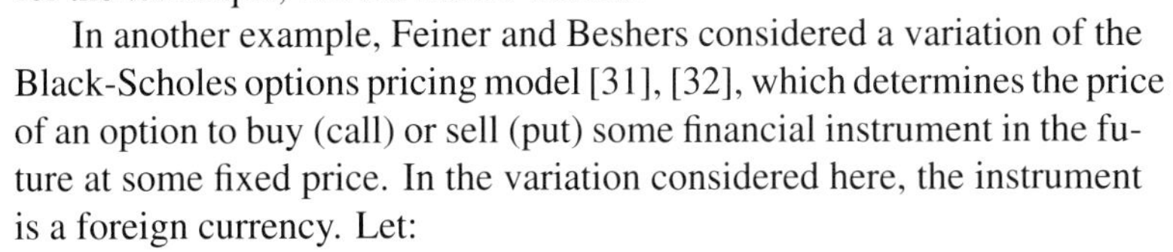

In another example, Feiner and Beshers considered a variation of the Black-Scholes options pricing model [31], [32], which determines the price of an option to buy (call) or sell (put) some financial instrument in the future at some fixed price. In the variation considered here, the instrument is a foreign currency. Let:

- $S =$ the current price of the currency. S is called the *spot* price.
- $X =$ the price of the currency when the option can be exercised. X is called the *strike* price.
- $r_f =$ the foreign interest rate.
- $r =$ the domestic interest rate.
- $\sigma =$ the volatility in the market.
- $t =$ the current time.
- $T =$ the time when the option matures.

The price of a call option, c, is defined by the following formula [32]:

$$c = f(S, X, r_f, r, \sigma, T) = Se^{-r_f(T-t)}\Phi(d_1) - Xe^{-r(T-t)}\Phi(d_2) \quad (14.10)$$

where:

$$d_1 = \frac{\ln(S/X) + (r - r_f + \sigma^2/2)(T-t)}{\sigma\sqrt{T-t}} \quad (14.11)$$

$$d_2 = \frac{\ln(S/X) + (r + r_f - \sigma^2/2)(T-t)}{\sigma\sqrt{T-t}} \quad (14.12)$$

$$\Phi(x) = \frac{1}{\sqrt{2\pi}}\int_{-\infty}^{x} e^{t^2}\,dt \quad (14.13)$$

that is, $\Phi(x)$ is the cumulative distribution function for a standard Normal probability distribution.

Clearly, equation (14.10) is complicated. How does the value of the option depend on each of its parameters?

The figure below illustrates the Black-Scholes model using several worlds nested inside one world for $T = 1$. The outer axes represent r, r_f and X. Each set of inner axes represent T, S and σ for particular values of r, r_f and X. A point cloud is drawn in each inner world, the brighter the color, the higher the value of the option. The figure below is probably too busy to be immediately comprehensible; it was mainly included to illustrate the concept of nested three-dimensional worlds. A typical user would probably create fewer inner worlds at one time, and would interactively explore the behavior of the option price as a function of its input parameters. Image copyright 1993, Stephen Feiner and Clifford Beshers.

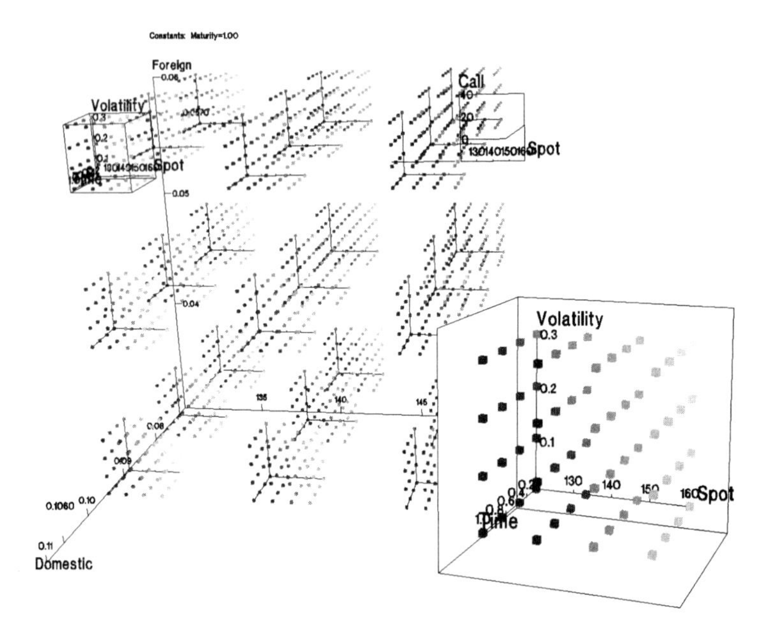

Figure 14.47

14.4.1 Summary of Worlds within Worlds

Theoretically, the Worlds within Worlds technique allows an arbitrary number of dimensions to be visualized. It is unclear how well the technique will work as the number of dimensions increases. No formal experimentation has been reported in the literature. The technique also illustrates the importance of interaction. Users need to probe an information space interactively rather than just seeing a single static image. This is similar in spirit to dynamic queries (Section 15.1.1) or animated sensitivity analysis discussed in Section 15.1. Perhaps the Worlds within Worlds technique will have use in optimization; to our knowledge, no one has yet explored this possibility.

14.5 Summary

Although Parallel coordinates, Worlds within Worlds and other techniques may help visualize larger dimensions, drawing polyhedra with the hundreds, thousands or millions of dimensions for optimization problems now routinely solved still seems almost hopeless. One might conclude, therefore, that visualizing complex, n-dimensional problems is doomed. Displaying n-dimensional polyhedra, although a challenging goal, may be a red herring. A n-dimensional polyhedron, if it could be shown in a comprehensible manner, would display all possible solutions to a particular linear programming problem. A single solution point, though, itself consists of multiple dimensions. Displaying a single solution point in a comprehensible manner poses a challenge by itself. Graphical representation formats for single solutions are usually highly dependent on the particular application. We discussed a variety of application dependent representations for single solution points in Chapter 10.

In a sense, optimization itself tries to untangle n-dimensional complexity. After all, an optimization algorithm navigates through a treacherous n-dimensional space in order to find the best possible solution. But the solution itself is n-dimensional, so understanding it is also a problem. Aside from application specific solutions such as Gantt charts and various types of maps, more generic solutions have been proposed to display n-dimensional information. Statistics has made contributions such as Chernoff faces and scatterplot matrices. Interactive computer graphics has enabled the development of techniques such as Worlds within Worlds. Parallel coordinates represents another attempt. Even so, most of the attempts to display higher dimensional objects have not even attempted to represent objects having the hundreds, thousands, or even millions of dimensions found in actual mathematical programming problems. Helping people to understand complexity remains an immense challenge.

Bibliography

[1] Banchoff TF. Visualizing two-dimensional phenomena in four-dimensional space: A computer graphics approach. In Wegman J, DePriest DJ, editors, *Statistical Image Processing and Graphics*, pages 187–202. New York:Marcel Dekker, 1986.

[2] Banchoff TF. *Beyond the Third Dimension: Geometry, Computer Graphics and Higher Dimensions*. New York:W. H. Freeman, 1990.

[3] Rucker R. *The Fourth Dimension and How to Get There*. New York:Viking Penguin, 1985.

[4] Dao M, Habib M, Richard JP, Tallot D. Cabri, an interactive system for graph manipulation. In Tinhofer G, Schmidt G, editors, *Graph-Theoretic Concepts in Computer Science*, pages 58–67. Berlin:Springer-Verlag, 1986.

[5] Skiena S. *Implementing Discrete Mathematics: Combinatorics and Graph Theory with Mathematica*. Redwood City (CA):Addison-Wesley, 1990.

[6] Gay DM. Pictures of Karmarkar's linear programming algorithm. Technical Report 136, Murray Hill (NJ):AT&T Bell Laboratories, Computer Science, 1987.

[7] Lustig I. Applications of interactive computer graphics to linear programming. In Sharda R, Golden BL, Wasil E, Balci O, Stewart W, editors, *Proceedings of the Conference on Impact of Recent Computer Advances in Operations Research*, pages 183–189, New York:1989. North-Holland.

[8] Carpendale MST, Cowperthwaite DJ, Fracchia FD. 3-dimensional pliable surfaces: For the effective presentation of visual information. Technical report, Burnaby (BC), V5A 1S6 CANADA:School of Computing Science, Simon Fraser University, 1995.

[9] Carpendale MST, Cowperthwaite DJ, Fracchia FD. Graph folding: Extending detail and context viewing into a tool for subgraph comparisons. Technical report, Burnaby (BC), V5A 1S6 CANADA:School of Computing Science, Simon Fraser University, 1995.

[10] Cleveland WS. Research in statistical graphics. *J. of the American Statistical Association*, 1987;82(398):419–423.

[11] Funkhouser HG. Historical development of the graphical representation of statistical data. *Osiris*, 1937;3:269–404.

[12] Tukey J. *Exploratory Data Analysis*. Reading (MA):Addison-Wesley, 1977.

[13] Becker RA, Cleveland WS. Brushing scatterplots. *Technometrics*, 1987;29(2):127–142.

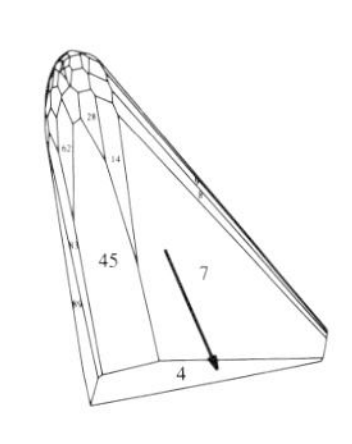

[14] Becker RA, Cleveland WS, Wilks AR. Dynamic graphics for data analysis. *Statistical Science*, 1987;2(4):355–395.

[15] Carr DB, Littlefield RJ, Nicholson WL, Littlefield JS. Scatterplot matrix techniques for large n. *J. of the American Statistical Association*, 1987;82(398):424–436.

[16] Huber PJ. Experiences with three-dimensional scatterplots. *Journal of the American Statistical Association*, 1987;82(398):448–453.

[17] Chernoff H. The use of faces to represent points in k-dimensional space graphically. *J. of the American Statistical Association*, 1971;68:361–368.

[18] Flury B, Riedwyl H. Graphical representation of multivariate data by means of asymmetrical faces. *Journal of the American Statistical Association*, 1981;76(376):757–765.

[19] Inselberg A. n-dimensional graphics, part I–lines and hyperplanes. Technical Report G320-2711, IBM Scientific Center, 9045 Lincoln Boulevard, Los Angeles (CA), 900435:IBM Los Angeles Scientific Center, 1981.

[20] Inselberg A. The plane with parallel coordinates. *The Visual Computer*, 1985;1:69–91.

[21] Inselberg A, Dimsdale B. Multidimensional lines 1: Representation. *SIAM Journal on Applied Mathematics*, 1994;54(2):559–577.

[22] Inselberg A, Dimsdale B. *Parallel Coordinates: A Tool for Visualizing Multivariate Relations*, chapter 9, pages 199–233. New York:Plenum Publishing Corporation, 1991.

[23] Wegman E. Hyperdimensional data analysis using parallel coordinates. *Journal of the American Statistical Association*, 1990;411(85):664–675.

[24] Chatterjee A, Das PP, Bhattacharya S. Visualization in linear programming using parallel coordinates. *Pattern Recognition*, 1993;26(11):1725–1736.

[25] IBM Corporation. *Parallel Visual Explorer for AIX Release 1.0 User's Guide*. IBM Corporation Software and Visualization Solutions, E70A, Yorktown Heights (NY) 10598:IBM Corporation, 1995.
URL: *http://www.ibm.com/News/950203/pve-01.html*

[26] Eickemeyer JS, Inselberg A, Dimsdale B. Visualizing p-flats in n-space using parallel coordinates. Technical Report G320-3581, Palo Alto (CA):IBM Palo Alto Scientific Center, 1992.

[27] Tan X, Burton RP. Parallel planes coordinates for hyperdimensional computer graphics. *The Journal of Imaging Science and Technology*, 1992;36(6):591–595.

[28] Feiner S, Beshers C. Worlds within worlds: Metaphors for exploring *n*-dimensional virtual worlds. In *Proceedings of the ACM Symposium on User Interface Software and Technology*, pages 76–83, Snowbird (UT):October 1990. Association for Computing Machinery.
URL: *http://www.cs.columbia.edu/~beshers/nvision.html*

[29] Feiner S, Beshers C. Visualizing *n*-dimensional virtual worlds with *n*-vision. *Computer Graphics*, 1990;24(2):37–38.

[30] Beshers C, Feiner S. Autovisual: Rule-based design of interactive multivariate visualizations. *IEEE Computer Graphics and Applications*, 1993;13(4):41–49.

[31] Black F, Scholes M. The pricing of options and corporate liabilities. *Journal of Political Economy*, 1973;81:637–654.

[32] Hull J. *Options, Futures and Other Derivative Securities*. Englewood Cliffs (NJ):Prentice-Hall, 1993.

Chapter 15

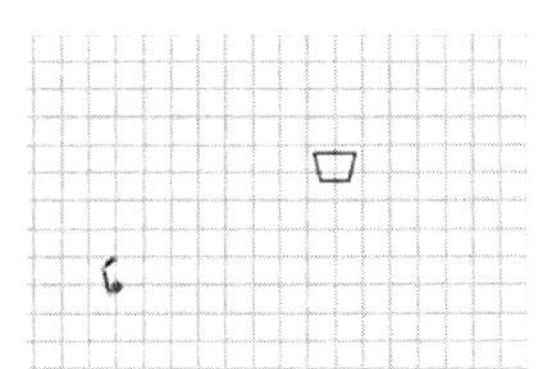

Animation

When images are updated quickly enough (ideally over 30 times per second [1]), *animation* occurs. With the development of standards for digital movies such as [2] and QuickTime [3] animation is become as easy to include in applications as text and graphics. As discussed in detail Section 9, algorithm animation can help algorithm designers and users understand the behavior of algorithms. Algorithm animation illustrates the behavior of an algorithm as it tries to find a good solution. It helps algorithm designers understand how the algorithm works. In contrast, the first part of this chapter (Section 15.1) discusses a different application of animation to optimization. This style of animation, animated sensitivity analysis, is intended mainly for non-technical people interested in the behavior of the solutions produced by algorithms, rather than the intricacies of the algorithms themselves. The second part of the chapter turns the tables and discusses how optimization techniques have been used to help create animations (Section 15.2). Optimization has proved useful in creating natural looking animation.

15.1 Animation for Optimization

Jones [4] proposed a different style of animation intended more for non-technical people. Instead of animating the internals of the algorithm, this style of animation, known as *animated sensitivity analysis*, animates how the solution changes as input parameters are changed continuously. Jones demonstrated applications of this idea to teaching the graphical solution technique for two-variable linear programming problems and understanding solution heuristics for traveling salesman problems and production scheduling problems.

For example, in a two-variable linear programming application, users can change a right-hand side value by dragging the line representing the constraint with a mouse (see Figure 15.1). As the line is dragged, the optimal solution, feasible region, and dual values are recalculated and redisplayed simultaneously.

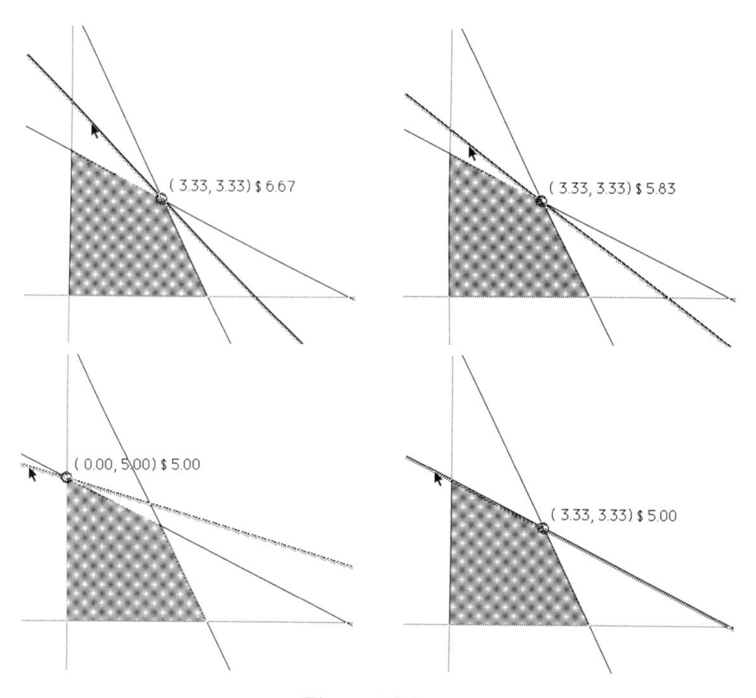

Figure 15.1

In the Animated Interactive Modeling System (AIMS) developed by Buchanan and McKinnon [6] (below), model inputs (control variables) and outputs are displayed as vertical bars. Users change the values of control variables by dragging their corresponding bars with the mouse. The values of the output variables computed by the model are then simultaneously updated and displayed. Although AIMS allows a user to minimize or maximize a given output variable, it does not allow the user to manipulate a given input and simultaneously see the optimium values of output variables.

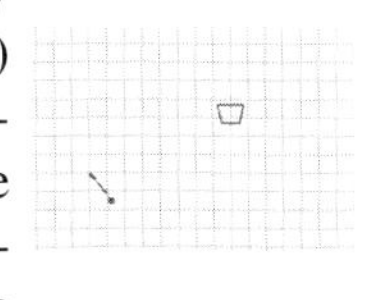

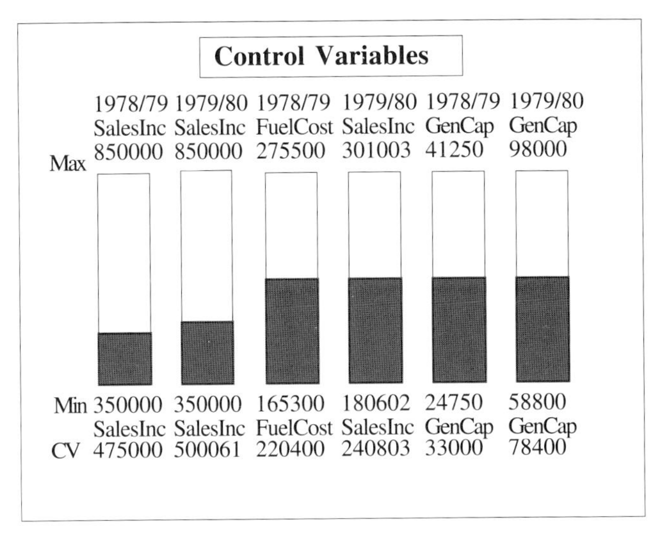

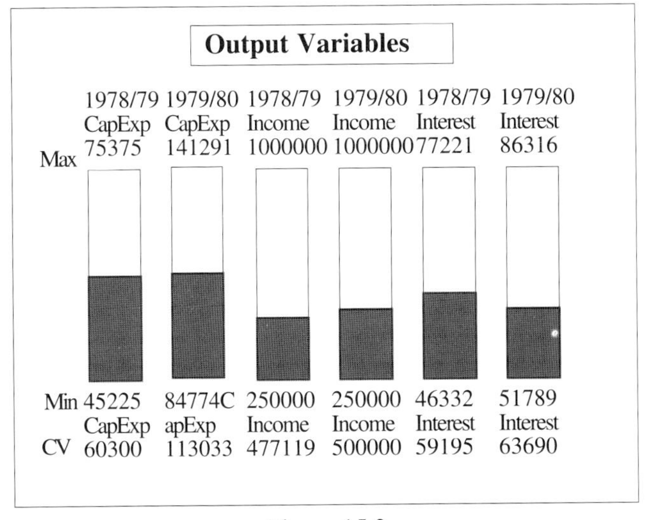

Figure 15.2

The system developed by Belton, Elder and Meldrum[7], Visual Interactive Linear Programming (VILP), is similar to AIMS, but computes the new optimum solution simultaneously. Targetted specifically to linear programming, in VILP, the values of the right-hand sides, objective coefficients, decision variables, dual prices and objective are represented as bars. For example, the linear program,

$$\begin{array}{lrcrcl} \text{min} & 80\text{CAB} & + & 50\text{BAR} & & \\ \text{subject to:} & & & & & \\ & & & \text{BAR} & \leq & 80 \\ & \text{CAB} & + & \text{BAR} & \leq & 110 \\ & 8\text{CAB} & + & 4\text{BAR} & \leq & 720 \\ & \text{CAB} & , & \text{BAR} & \geq & 0 \end{array} \tag{15.1}$$

has optimal solution CAB $=70$ and BAR $=40$, with objective function value 7600.

In VILP, this linear program would be represented as shown below. The objective function appears as the horizontal bar at the top. The values of the decision variables are represented as the vertical bars in the upper left. The objective function coefficients are shown by the vertical bars in the lower left. The right-hand side values are shown in the lower right. Finally, the values of the slack variables are shown by the vertical bars in the upper right.

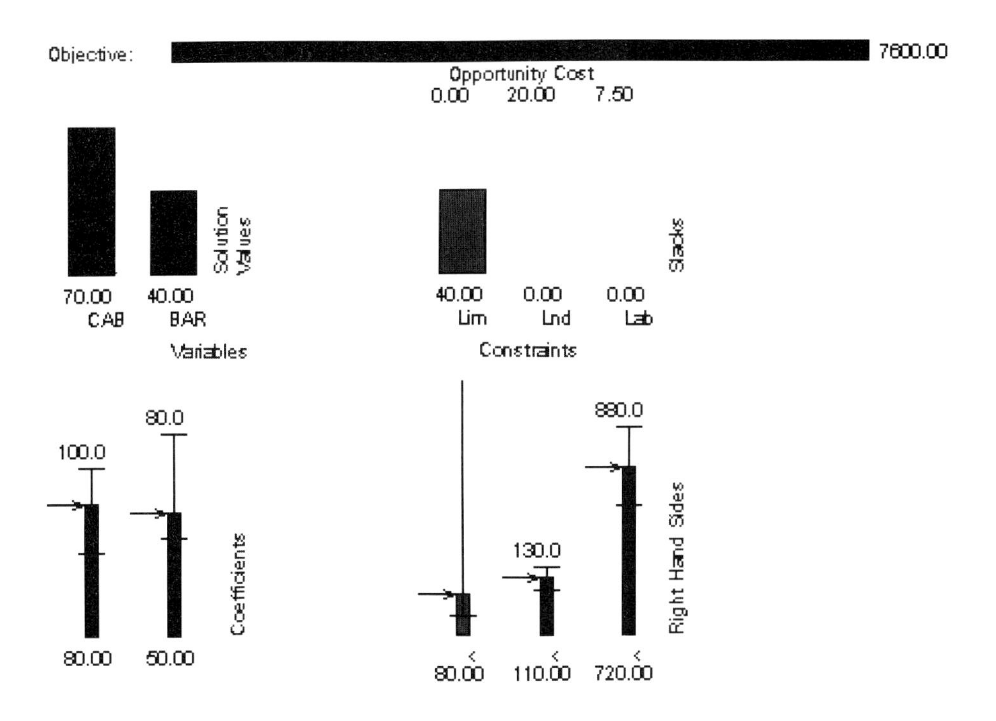

Figure 15.3

For the objective function coefficients and right-hand side values, an arrow points to its current value. The horizontal lines associated with each bar represent the minimum and maximum allowed value, that is, the standard range information for linear programming problems.

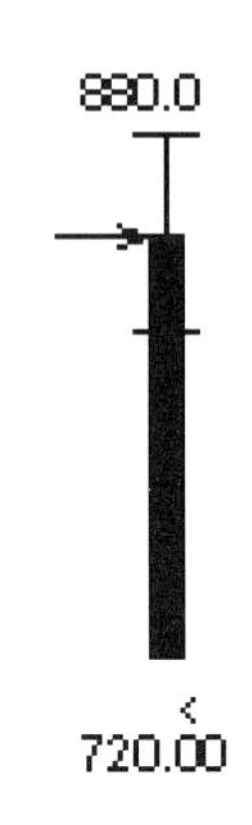

Figure 15.4

Using a mouse, users can change the value of right-hand side values and objective coefficients simply be dragging on the corresponding bar. As the change is made, the solution is recalculated and the sizes of all other relevant bars are updated.

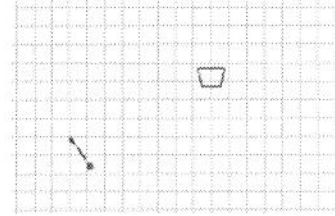

For example, as the user decreases right-hand side, b, from a value of 720 to 603, the screen changes as shown below.

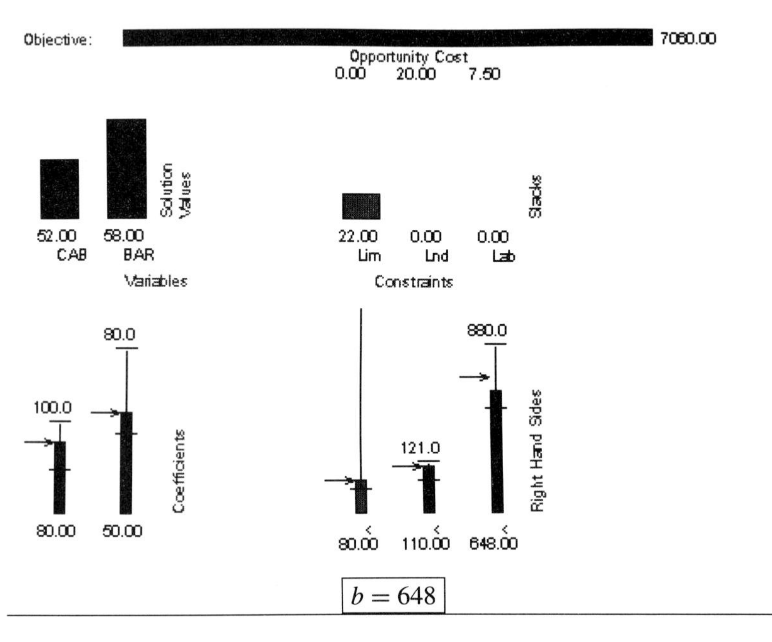

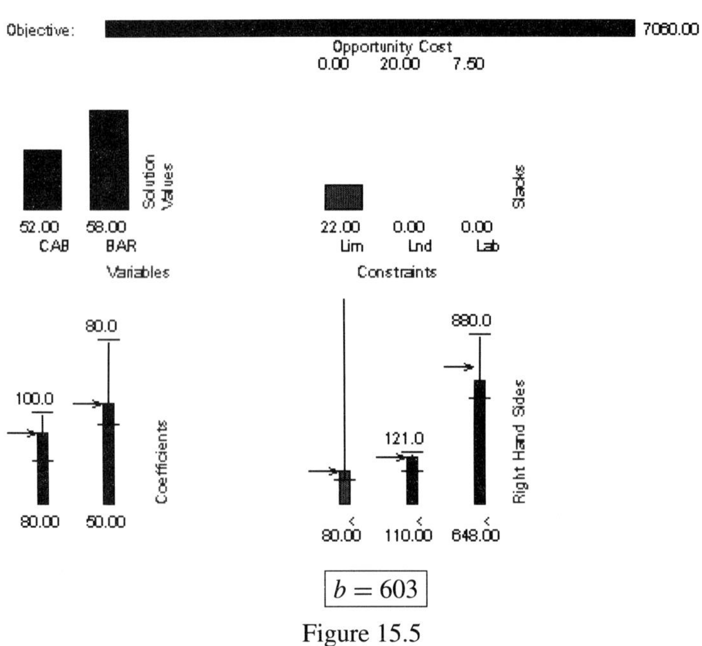

Figure 15.5

Unlike the work of [4], VILP is not limited to problems of two variables, but can handle problems of arbitrary size.

15.1.1 Dynamic Queries

Animated sensitivity analysis is closely related to *dynamic queries*, a representational style proposed by Ahlberg, Williamson and Shneiderman [8]. Dynamic queries allows a user to query a database by changing one or more parameters continuously. As the parameters are changed, the results of the query are simultaneously displayed, thereby producing animation. Dynamic queries is similar to animated sensitivity analysis, but dynamic queries replaces an optimization algorithm with a database query engine.

Ahlberg, WIlliamson and Shneiderman [8] illustrated the use of dynamic queries for a real-estate application. A map (below) shows the locations of properties for sale. Each dot represents a house, townhouse or condominium for sale in the Washington, DC area. Washington DC itself is represented by the square region in the center. By manipulating sliders representing the selling price, number of bedrooms and other characteristics of the desired house, houses appear and disappear according to the values of the sliders.

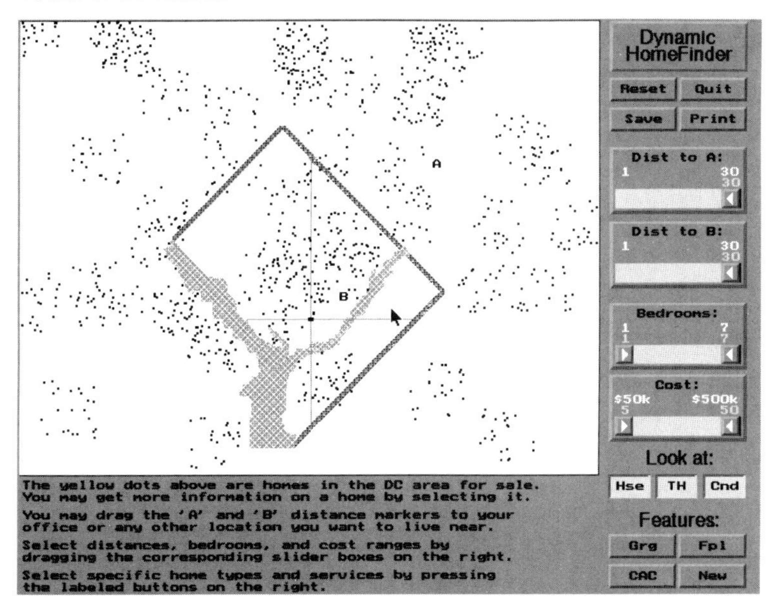

Figure 15.6

The figures below shows a sequence of images from the system as the user increases the minimum selling price for the home from \$60,000 to \$300,000. This allows the user to see quickly where the inexpensive and expensive real estate is located. For example, the less expensive homes cluster in Washington, DC, or far out in Virginia (shown in the upper left image below) and the more expensive homes cluster in Virginia (shown in the lower right image below).

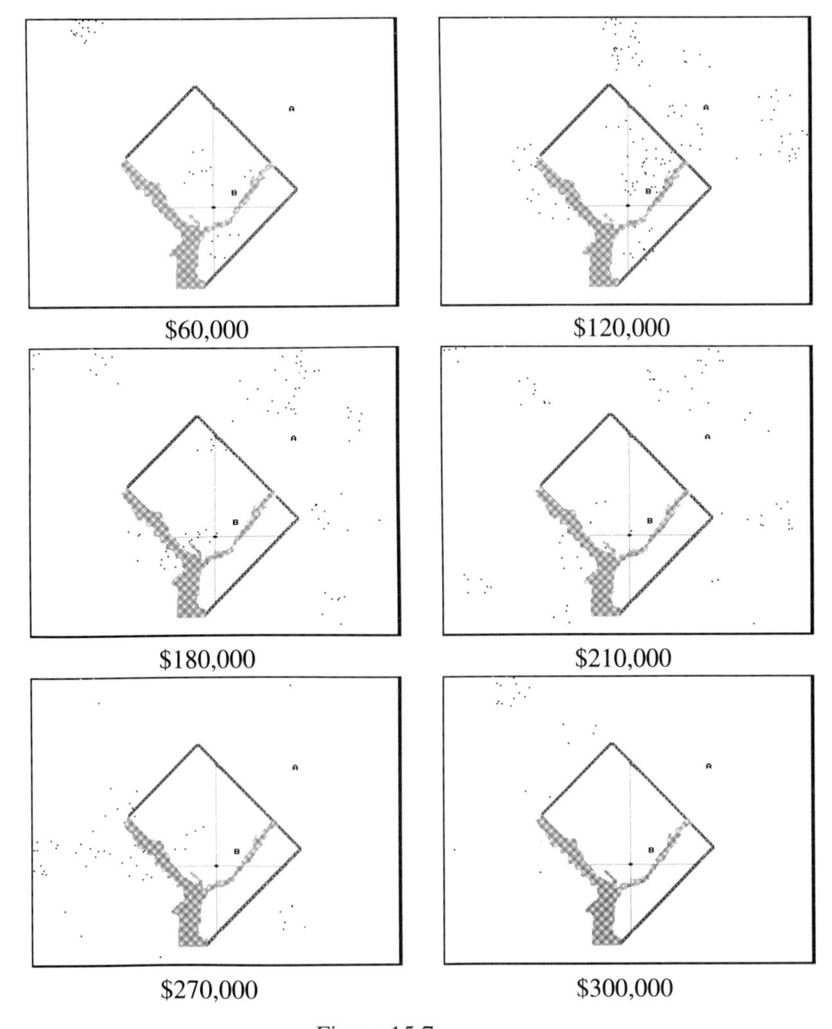

Figure 15.7

Dynamic queries are quite closely related to brushing, promoted in the statistics community (Section 14.2.1). Recall that brushing allows a user to pick out points in one view of the data and simultaneously those points are highlighted in all other views. Brushing is really just dynamic queries applied to a scatterplot matrix.

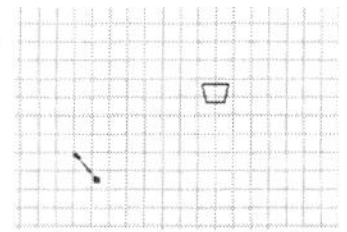

15.1.2 Animated Sensitivity Analysis and the Traveling Salesman Problem

In both algorithm animation and animated sensitivity analysis, it often helps to produce a static representation of the history of the animation. For example, in the visualization of the planar traveling salesman problem [4] discussed in Chapter 9, users can move a city and simultaneously see the new tour. As a city moves, the sequence of cities on the tour changes only at discrete points. The regions in which the tour does not change are drawn in "batch" mode.

Animation is now used in a different way, however. In the original version, only a single city moves, say city 1, with the other city locations fixed. What happens if the other cities are allowed to move as well? For each incremental set of city positions, the regions are drawn, assuming city 1 is allowed to move with respect to the other cities. This set of images can then be played back, producing an animation. A flip chart of the images both for farthest insertion and an optimal algorithm begins Part III of this book. From viewing the animation, very preliminary observations suggest that optimal algorithms still produce more stable solutions than most classic approximation algorithms. Snapshots from the animation appear below. Observe how as city 5 moves from inside the convex hull of cities to lie on the boundary of the convex hull: the regions become visible much less complex. If all cities lie on the boundary of the convex hull (as in the bottom row), the optimal tour is equal to the convex hull. Perhaps this fact explains why the regions are simpler in the bottom row.

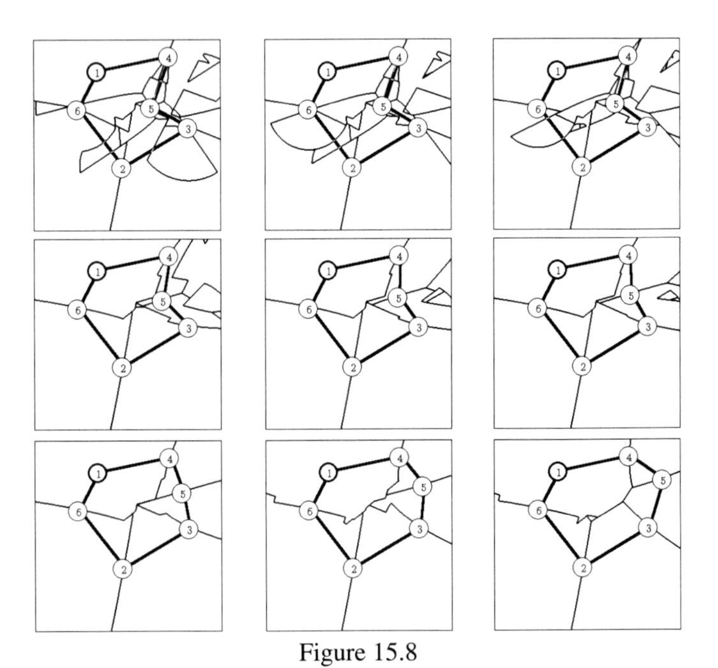

Figure 15.8

15.1.3 An Application of Animated Sensitivity Analysis to Chemistry

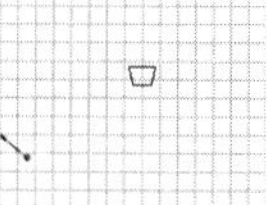

Surles *et al.* [9], [10] used non-linear optimization and animated sensitivity analysis (though he did not use that term) to explore molecular dynamics. In particular, chemists are interested in how molecules behave when subject to forces. In the visualization system that resulted from Surles's work, SCULPT, chemists "pull" on an atom in a molecule. As the atom is pulled, the shape of the molecule distorts based on forces among the atoms in the molecule and other constraints. The distortion is calculated by a non-linear optimization model. The non-linear program seeks to minimize the molecule's potential energy subject to a variety of constraints, for example, the minimum angular separation between two atoms.

Below is a series of snapshots of SCULPT in action. Without going into a great deal of chemistry, the images illustrate a "two-stranded, antiparallel β-ribbon" at the top that is being manipulated into a "six-stranded, Greek-key β-barrel" at bottom.

Figure 15.9

Surles's work is an example of animated sensitivity analysis because the chemist interactively changes an input parameter, e.g., the force on an atom, an optimization algorithm is executed, and the results displayed sufficiently quickly to produce a smoothly changing representation (animation) of the molecule. Using parallel processing, and exploiting a special structure for the non-linear optimization model, Surles was able to obtain update rates of 0.62 seconds per frame for models having over 9,000 decision variables and 8,000 constraints 15.9. Surles's work indicates that animated sensitivity analysis can be practical for realistic problems.

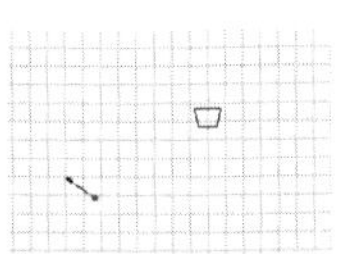

15.2 Optimization for Animation

The computer graphics community has used optimization techniques to generate animations. In using computer graphics to produce films, a key challenge involves planning the motion of the characters (for example, simulated people, dinosaurs, or aliens) in the scene. Most realistic characters possess many interconnected movable parts such as arms, legs, hands and fingers. Traditionally, an animator would use an interactive system to manually trace out paths in three-dimensions for each part of the character to follow over time. From these paths, the motion can be replayed, with the animator making adjustments as necessary. The task of constructing these paths is called *motion planning*.

Unfortunately, this manual approach (albeit with some computer assistance) can be tedious. For example, to have a human character throw a basketball into the basketball hoop in a natural manner requires carefully coordinating the motion of the arms, hands, fingers of the character as well as the basketball.

Starting with the work of Witkin and Kass [11] and expanded upon in, [12] and [13], optimization has been used for motion planning. The objective function classically minimizes the amount of energy that a character needs to expend in order to accomplish a particular motion, for example, throwing the basketball through the hoop. Constraints are added to the problem based on the geometry of the character. For example, the maximum angle of the human elbow is about 180°. Constraints on the position, speed and acceleration of objects in the scene are also added. For example, the basketball should be traveling down through the hoop 3 seconds from the start of the scene. These problems are typically non-linear, which implies that local optimal solutions may be produced.

Witkin and Kass give a simple example of such a formulation. Consider a single particle having mass m. Attached to the particle is a jet engine. The particle needs to move from position $\mathbf{a}$ at time t_0 to position $\mathbf{b}$ at time t_1. Gravitational pull is also a factor. From Newton's laws of motion, we know that

$$m\frac{\partial^2 \mathbf{x}(t)}{\partial t^2} - \mathbf{f}(t) - m\mathbf{g} = 0 \tag{15.2}$$

where $\mathbf{x}(t)$ is the position of the particle and $\mathbf{f}(t)$ is the force produced by the jet engine at time t, $\mathbf{x}(t_0) = \mathbf{a}$, and $\mathbf{x}(t_1) = \mathbf{b}$.

The objective is to minimize the total amount of fuel consumed, R, required for the jet engine, or, for simplicity,

$$R = \int_{t_0}^{t_1} |f(t)|^2 \, dt \tag{15.3}$$

Witkin and Kass attacked the problem using non-linear optimization. First, they discretize both $\mathbf{x}(t)$ and $\mathbf{f}(t)$. In other words, they model $\mathbf{x}(t)$ and $\mathbf{f}(t)$ as a sequence of discrete points $\mathbf{x}_0, \ldots \mathbf{x}_n$ and $\mathbf{f}_0, \ldots \mathbf{f}_n$, respectively with a time interval of h between samples. In the equation of motion (15.2), $\partial^2\mathbf{x}(t)/\partial t^2$ is approximated as:

$$\frac{\partial^2 \mathbf{x}(hi)}{\partial t^2} \approx \frac{\mathbf{x}_{i+1} - 2\mathbf{x}_i + \mathbf{x}_{i-1}}{h^2} \qquad i = 1, \ldots, n-1 \tag{15.4}$$

Next, substituting approximation (15.4) in (15.2) produces constraints:

$$\mathbf{p}_i = m\frac{\mathbf{x}_{i+1} - 2\mathbf{x}_i + \mathbf{x}_{i-1}}{h^2} - \mathbf{f}_i - m\mathbf{g} = 0 \quad i = 1, \ldots, n-1 \tag{15.5}$$

The objective function becomes:

$$R = \sum_{i=0}^{n} |\mathbf{f}_i|^2 \tag{15.6}$$

In short the full mathematical program becomes:

$$\begin{array}{ll} \text{Choose} & \mathbf{x}_i, f_i, i = 0, \ldots n \text{ to} \\ \text{minimize} & R = \sum_{i=0}^{n} |\mathbf{f}_i|^2 \end{array}$$

subject to:

$$\begin{aligned} m\frac{\mathbf{x}_{i+1} - 2\mathbf{x}_i + \mathbf{x}_{i-1}}{h^2} - \mathbf{f}_i &= m\mathbf{g} \quad i = 1, \ldots, n-1 \\ \mathbf{x}_0 &= \mathbf{a} \\ \mathbf{x}_n &= \mathbf{b} \end{aligned} \tag{15.7}$$

The example contains a quadratic objective function and linear constraints. More complicated problems, however, contain non-linear constraints. Witkin and Kass used a sequential quadratic programming algorithm to find an optimal solution.

Several extensions to Witkin and Kass's basic idea have been proposed. A problem with Witkin and Kass's approach involves the choice of h, which controls the level of discretization. Too small value of h, although increasing accuracy, can lead to an intractable problem, whereas too large a value of h may lead to an inaccurate model. Liu, Gortler and Cohen [13] proposed techniques to dynamically vary the level of discretization depending on the needs of the model. Cohen [12] also developed techniques to integrate the optimization techniques with traditional manual motion control systems.

Below is a sequence from a film by Liu, Gortler and Cohen [13] of a single arm throwing a ball through a basketball hoop. The sequence of images shows the solution from the first iteration of the non-linear optimization algorithm. The ball is thrown directly at the hoop and much too hard. Images © copyright 1994 Association for Computing Machinery, Inc.

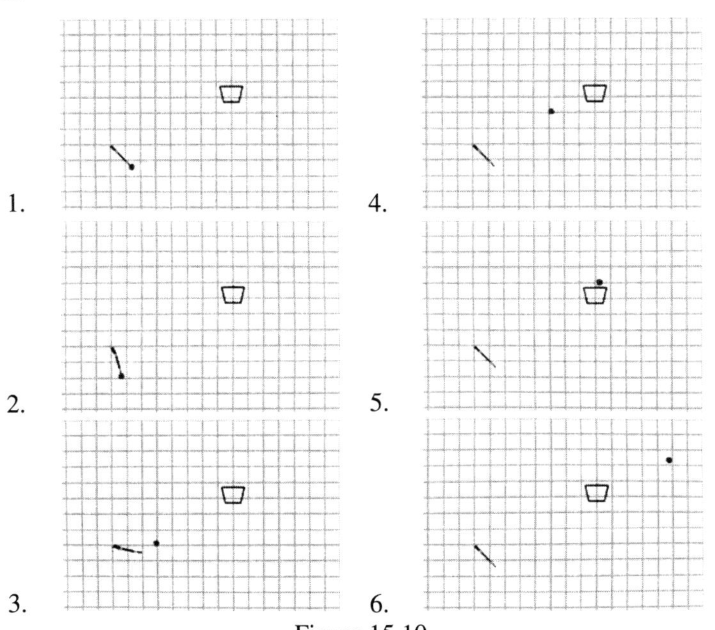

Figure 15.10

The figure below shows a sequence of images for the solution from the second iteration of the non-linear optimization algorithm. The ball is thrown towards above the hoop, but again much too hard. Images © copyright 1994 Association for Computing Machinery, Inc.

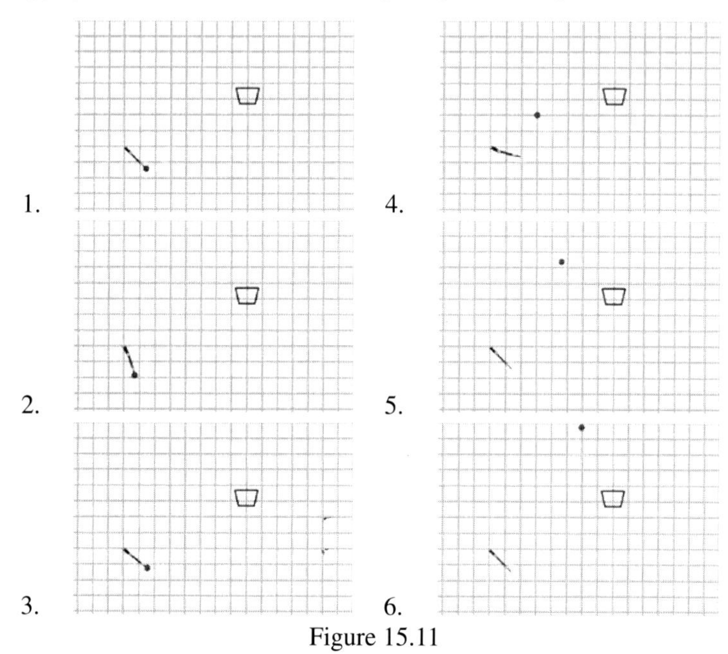

Figure 15.11

The figure below shows the solution from the eighth iteration of the non-linear optimization algorithm. The ball now is thrown too softly and it falls short of the hoop. Images © copyright 1994 Association for Computing Machinery, Inc.

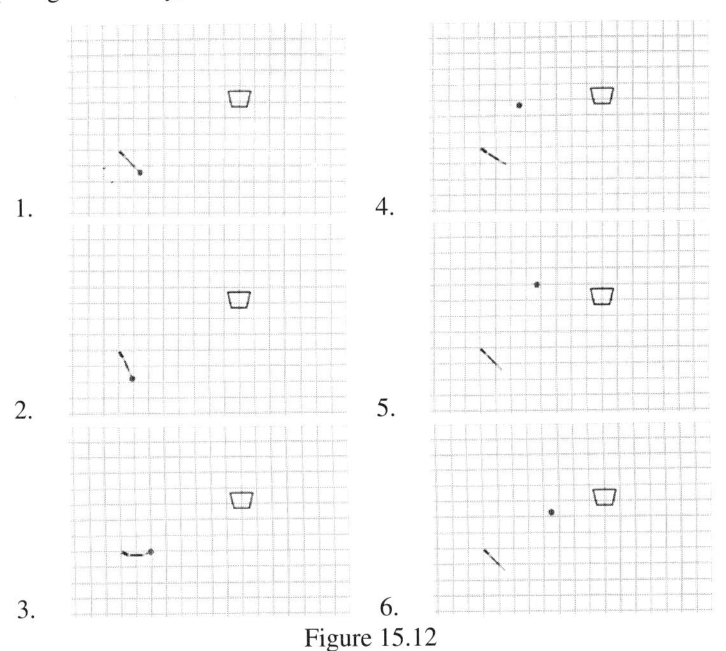

Figure 15.12

The figure below shows the solution from the sixteenth iteration of the non-linear optimization algorithm. Score! Note the "windup" on the arm as it prepares to throw the ball. This images are also shown as the flip chart that begins this chapter. Images © copyright 1994 Association for Computing Machinery, Inc.

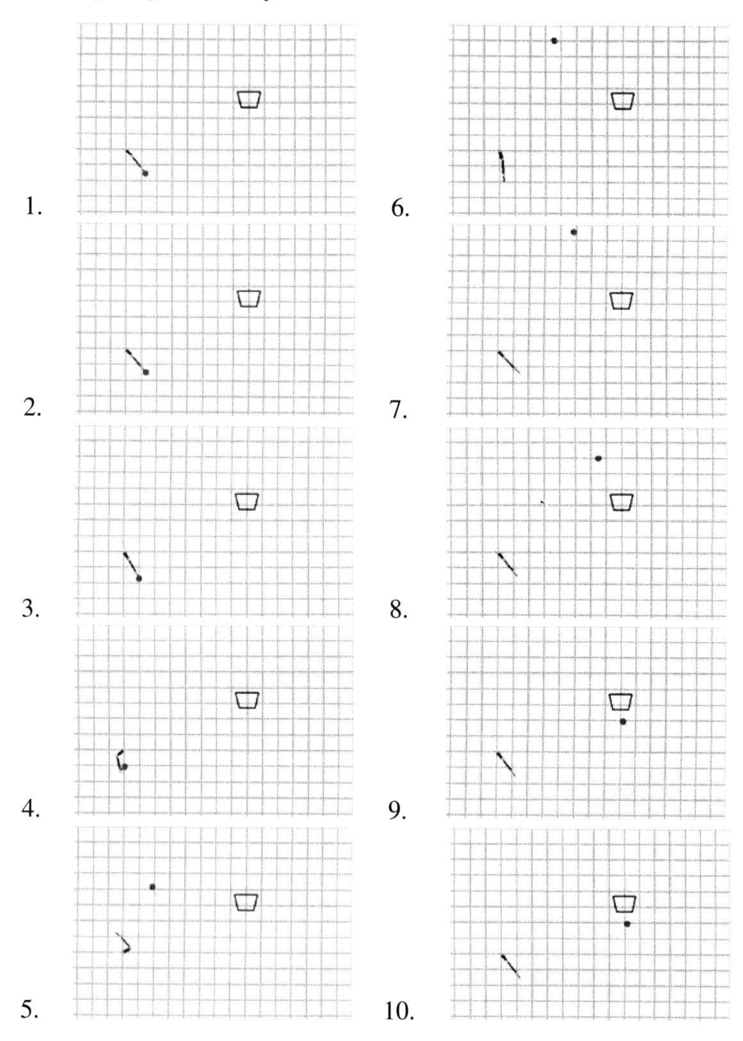

Figure 15.13

15.3 Summary

With the advent of multimedia standards such as MPEG [2] and QuickTime [3], animation is becoming as widespread as graphics. For example, one can now include animations in spreadsheets, word processors, and even other programs such as Mathematica.

Guidelines for proper use of animation are rare, as discussed in Chapter 4. The most well-developed guidelines were developed by animators from Walt Disney, with uncertain applicability to to visualization and optimization. The algorithm animation community [14], [15], [16], [17] has provided anecdotal recommendations for that technique. The experimental evidence for dynamic queries supports the possibility of the usefulness of animated sensitivity analysis, but formal experimentation has not been conducted. Animation is an extremely appealing medium that will become increasingly exploited for optimization modeling. Much more research must be conducted, however, to identify exactly how it should be used effectively.

Bibliography

[1] Card S, Moran TP, Newell A. *The Psychology of Human-Computer Interaction*. Hillsdale (NJ):Lawrence Erlbaum Associates, 1983.

[2] LeGall D. MPEG: A video compression standard for multimedia applications. *Communications of the ACM*, 1991;34(4):46–58.

[3] Apple Computer. *QuickTime Developer's Kit, Version 1.0.* Cupertino (CA):Apple Computer, 1991.

[4] Jones CV. An integrated modeling environment based on attributed graphs and graph-grammars. *Decision Support Systems*, 1993;10:255–275.

[5] Buchanan I, McKinnon KIM. An animated interactive modeling system for decision support. In *Operational Research '87*, pages 111–118. Amsterdam, The Netherlands:Elsevier Science Publishers, 1987.

[6] Buchanan I, McKinnon K. An animated interactive modelling system for decision support. *European Journal of Operational Research*, 1991;54:306–317.

[7] Belton V, Elder M, Meldrum D. Using VILP. Technical report, Glasgow, United Kingdom:Strathclyde University, 1993.

[8] Ahlberg C, Williamson C, Shneiderman B. Dynamic queries for information exploration: An implementation and evaluation. In Bauersfeld P, Bennett J, Lynch G, editors, *Human Factors in Computer Systems CHI '92*

Conference Proceedings, pages 619–626, Reading (MA):1992. Addison-Wesley.

[9] Surles MC. An algorithm with linear complexity for interactive, physically-based modeling of large proteins. *Computer Graphics*, 1992;26(2):221–230.

[10] Surles MC, Richardson JS, Richardson DC, Brooks FP. Sculpting proteins interactively: Continual energy minimization embedded in a graphical modeling system. *Protein Science*, 1994;3(2):198–210.

[11] Witkin A, Kass M. Spacetime constraints. *Computer Graphics*, 1988;22(4):159–168.

[12] Cohen MF. Interactive spacetime control for animation. *Computer Graphics*, 1992;26(2):293–302.

[13] Liu Z, Gorleter SJ, Cohen MF. Hierarchical spacetime control. In *Computer Graphics, Proceedings of SIGGRAPH '94*, Orlando (FL):July 1994. ACM SIGGRAPH, Association for Computing Machinery.

[14] Brown MH. *Algorithm Animation.* Cambridge (MA):MIT Press, 1988.

[15] Brown MH. Zeus: A system for algorithm animation and multi-view editing. Technical Report 75, Palo Alto (CA):Digital Systems Research Center, Digital Equipment Corporation, 1992.
URL: *http://www.research.digital.com/SRC/zeus/home.html*

[16] Brown MH, Hershberger J. Color and sound in algorithm animation. Technical Report 76a, Palo Alto (CA):Digital Systems Research Center, Digital Equipment Corporation, 1991.

[17] Stasko J. A practical animation language for software development. In *Proceedings of the IEEE International Conference on Computer Languages*, pages 1–10, Los Alamitos (CA):1990. IEEE Computer Society Press.
URL: *http://www.cc.gatech.edu/gvu/softviz/algoanim/algoanim.html*

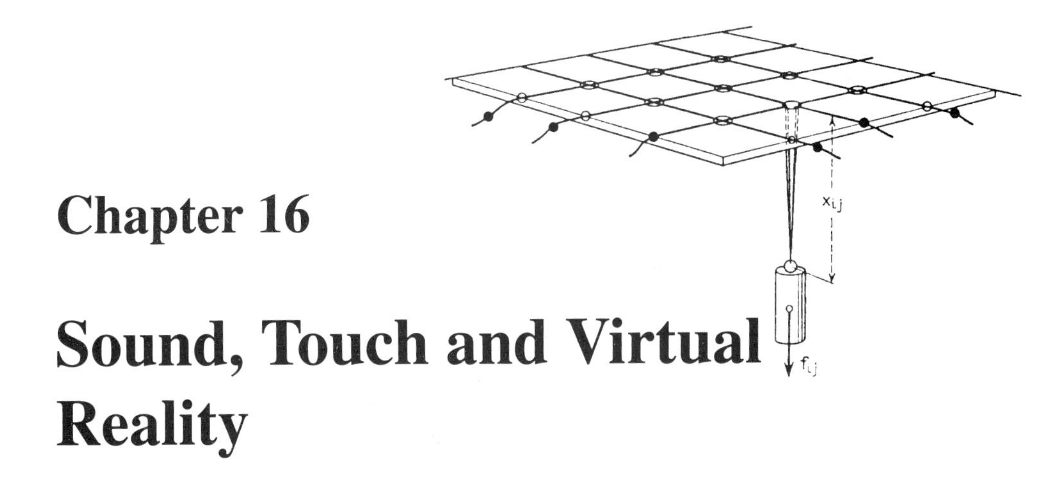

Chapter 16

Sound, Touch and Virtual Reality

This chapter considers the use of two of the non-visual senses–hearing and touch–for representing complex information in the context of optimization. Although much research has been conducted on these topics applied often to entertainment and sometimes to scientific activities, very few applications of sound and touch have been made that are relevant to optimization. We briefly discuss some of the relevant work. The final part of the chapter discusses how optimization is making contributions to the field of virtual reality.

16.1 Sound

More and more computers provide high-quality sound input and output capabilities. Several spreadsheets [1], [2] now allow users to annotate the spreadsheets with recorded sounds. Mathematica [3] allows voice annotation as well, but also allows one-dimensional mathematical functions to be played back as sounds. The user can also plot the sound as a line chart.

Although the use of sound in user interfaces is still emerging, some idea of its potential can be described [4], [5], [6], [7], [8]. Perhaps most importantly, sound provides another means of communication between user and computer. That is, sound provides a communications channel that is independent of the visual channel. Police cars and ambulances exploit this phenomenon: one need only hear the ambulance to know it is nearby. Sound is frequently used in computer programs to signal unusual conditions, if only as a beep when the user commits an error.

Brown and Hershberger [9] explored the use of sound in algorithm animation. They have used sound, for example, in an animation of a sorting algorithm. The magnitude of each item is encoded as the pitch of a sound. Whenever the item is moved, its corresponding note is played. Different sorting algorithms using the encoding scheme produce audibly different patterns.

16.1.1 Summary

Although sound has been used in computer simulation [10], it has rarely been used for mathematical programming. It has potential because it offers another communication channel between human and computer.

We have informally experimented with the use of sound in our work on animated sensitivity analysis for the planar traveling salesman problem (Section 15.1). As a city is moved, the tour can change. An alert tone can be sounded whenever the tour changes. Also, a tone can be played whose pitch is proportional to the objective function value. Whether these sounds are actually useful or merely produce an annoying cacophony is unclear.

16.2 Touch

Brooks [11] and Brooks, *et al.* [12] have used a force-feedback robot arm for "molecular docking." Molecular biologists need to understand how molecules interact and how they bond together or "dock." Interactive computer graphics systems that allow users to see and manipulate molecules in three dimensions are widely used. In Brooks's system, the molecular biologists not only see the molecules, but feel the forces exerted on the molecules.

Brooks showed that the addition of force-feedback cuts the time to determine the docking configuration by a factor of two. The molecular docking problem can also be modeled as a nonlinear mathematical program, but at the time of Brooks's article, the best mathematical programming approach required over a day of CPU time, whereas molecular biologists could determine the proper configuration in a few minutes [13].

For optimization, the use of force-feedback, touch sensitive (*haptic*) displays is at best speculative. They await a reduction in price and a clearer understanding of their applicability. Surprisingly, in the 1950's Sinden [14] described several mechanical devices for solving mathematical programming problems. Consisting of weights intricately connected by strings and pulleys, they could theoretically solve any linear programming problem.

For example, the figure below shows a schematic for a device proposed by [14] to solve the Hitchcock-Koopmans transportation problem. Strings are arranged at right angles over a horizontal board with perforations at every intersection. One set of parallel strings represents sources; the orthogonal strings represent sinks. The length of each source string i is set equal to the supply, s_i, at source i. Similarly, the length of each sink string j corresponds to the demand, d_j, at sink j. Weights f_{ij} are attached to the strings at every perforation, each weight attached to both the corresponding horizontal and vertical string. The size of the weight corresponds to the cost to transport a unit of material from source i to sink j. The laws of physics can be used to show that, at equilibrium, the displacement of each weight f_{ij} beneath the board gives the value of the decision variable x_{ij} indicating the amount to ship from source i to sink j. Moreover, the tension in each string corresponds to the dual variables.

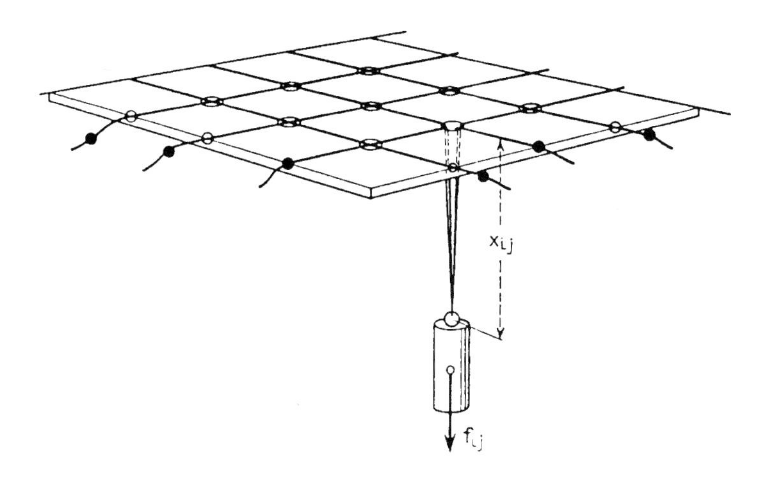

Figure 16.1

Models of the proposed mechanical devices were actually built, but reported "difficulties with friction." The use of haptic displays for optimization may be just as impractical as the mechanical devices described by Sinden.

16.2.1 Summary

At this point, applications of haptic displays to optimization are only speculative. However, force-feedback might be used to illustrate hard and soft constraints physically. As a soft constraint is violated more and more, the force-feedback would increase proportionately.

16.3 Virtual Reality

Virtual Reality applied to optimization modeling is essentially non-existent. One can certainly speculate about creating virtual worlds to navigate through representations of multi-dimensional polyhedra. It is unclear whether the great effort required, even today, to create such a virtual world would yield significant benefits. Some of the work that in virtual worlds that is relevant to optimization has been discussed in other sections, for example, Worlds within Worlds (Section 14.4). In contrast, this section discusses how optimization algorithms been used to help generate virtual environments.

Virtual reality systems classically allow users to move through complex artificial worlds consisting of a variety of three-dimensional objects. A key challenge, therefore, for virtual reality is updating the display of the artificial world at sufficient speed to produce smooth motion. If the virtual world contains hundreds of thousands or even millions of polygons (the most commonly used geometric object for computer graphics), the fastest available computer graphics hardware alone cannot update the scene sufficiently quickly to produce smooth motion. That is, each new image should be rendered within a certain amount of time, $t_{\max}$. A great deal of research has been devoted to the problem of removing objects from a scene that are obscured by other objects (this is the classic *hidden surface* problem). For complex scenes, however, even if all obscured objects are removed, many thousand objects can still be visible to the observer. The problem of updating the display quickly remains.

A common approach to allow more complicated worlds to be rendered involves adjusting the quality of the rendering of the visible objects. For example, a sphere could be represented using just a few polygons or with a great many. The rendering using more polygons would look more realistic, but would require more processing time to produce. Similarly, a variety of different algorithms have been developed that increase the realism of the rendering, but that also increase the time to create the rendering.

A simple heuristic for choosing the level of quality for a particular object considers the size of the object as seen by the observer. Small objects or objects in the distance would be rendered at low quality, whereas large objects or objects near the observer would be rendered at high quality. If objects are sufficiently small to the observer, they need not be rendered at all—the lowest quality of all possible renderings.

However, what is the maximum size of a "small" object? This threshold will determine the smoothness of the update algorithm. If the threshold is fixed, some scenes may contain too many visible objects to allow smooth motion, whereas scenes with few visible objects will be rendered at a lower quality than they could have been.

An improvement to this approach varies the threshold dynamically based on the time it took to render the last image; if the time was greater than a required minimum, then the threshold is increased. Although this approach improves on the previous approach, oftentimes the complexity of a scene can change rapidly, for example, if one opens a door into a new room. The size threshold adjustment, being retrospective, does not respond quickly, and the first few frames of the new scene may take more than the maximum desired time to render.

Funkhouser and Séquin [15] used integer programming techniques to provide smoother animation of virtual worlds. Each object to be rendered, $o \in O$, can be rendered at a different level of detail $l \in L$ using a different rendering algorithm $r \in R$. For each triple (o, l, r) they develop an estimate of the time c_{olr} to render object o at level of detail l using algorithm r. The time estimate is based on a linear regression model that considers the number of polygons, vertices and pixels in object o at level of detail l. They also develop an estimate of the benefit b_{olr} of rendering object o at level of detail l using rendering algorithm r. Decision variables x_{olr} are defined to be equal to 1 if object o should be rendered at level of detail l using algorithm r and to be equal to 0 otherwise. The integer linear programming formulation appears below:

$$\begin{aligned}
\max \quad & \sum_{o\in O}\sum_{l\in L}\sum_{r\in R} b_{olr}x_{olr} \\
\text{subject to:} \quad & \\
& \sum_{o\in O}\sum_{l\in L}\sum_{r\in R} c_{olr}x_{olr} \leq t_{\max} \\
& \sum_{l\in L}\sum_{r\in R} x_{olr} = 1 \quad \forall o \in O \\
& x_{olr} \in \{0, 1\}
\end{aligned} \tag{16.1}$$

This is just a multiple-choice knapsack problem [16]. Of course the multiple-choice knapsack problem is *NP*-hard, which does not seem to bode well for its use in a virtual reality system requiring fast response. Funkhouser and Sèquin, however, do not solve the problem to optimality, but rather employ a greedy heuristic to find an approximate answer. The heuristic sorts the ratios $v_{olr} = b_{olr}/c_{olr}$ in descending order. It then chooses the level of detail and rendering algorithm for each object o that has the highest ratio. Since there is usually a great deal of similarity between adjacent frames, rather than sort the ratios for each frame, they rely on an incremental algorithm to adjust the sorted ratios. Using this technique, they were able to maintain a stable frame rate even for scenes containing over 80,000 visible polygons on what was then (1993) state-of-the-art computer graphics processing engine.

16.4 Summary

Although they show great promise, the technologies of virtual reality—sound, haptic displays, three-dimensional images–have yet to be widely applied to optimization, although optimization has been usefully applied to virtual reality. One can expect that the hardware and software to support these capabilities will continue to decline in price and improve in quality. Interactive three-dimensional computer graphics is now becoming readily available on personal computers, as are sound generation capabilities. Perhaps virtual reality will provide better representations of multi-dimensional information than are currently available.

Bibliography

[1] Lotus Development Corporation. *Improv for Windows Release 2.0 Reference Manual.* Cambridge (MA):Lotus Development Corporation, 1993.

[2] Microsoft Corporation. *Microsoft Excel User's Guide Version 5.0.* Redmond (WA):Microsoft Corporation, 1994.

[3] Wolfram S. *Mathematica: A System for Doing Mathematics by Computer.* Redwood City (CA):Addison-Wesley, 2nd edition, 1991.

[4] Baecker RM, Buxton WAS. *Readings in Human-Computer Interaction.* Los Altos (CA):Morgan Kaufmann, 1987.

[5] Blattner MM, Sumikawa DA, Greenberg RM. Earcons and icons: Their structure and common design principles. *Human-Computer Interaction,* 1989;4(1):11–44.

[6] Buxton W, Bly SA, Frysinger SP, Lunney D. Communications with sound. In Borman L, Curtis B, editors, *Proceedings of the ACM SIGCHI Human Factors in Computing Systems Conference*, pages 115–119, New York:1985. ACM Press.

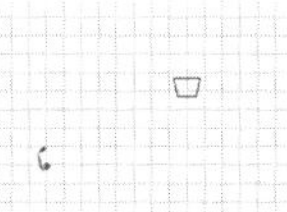

[7] Gaver WW. The SonicFinder: An interface that uses auditory icons. *Human-Computer Interaction*, 1989;4(1):67–94.

[8] Gaver WW, Smith RB, O'Shea T. Effective sounds in complex systems: The Arkola simulation. In Robertson SP, Olson GM, Olson JS, editors, *Reaching Through Technology: Human Factors in Computing Systems; ACM SIGCHI '91 Conference Proceedings*, pages 85–90, New York:1991. ACM Press.

[9] Brown MH, Hershberger J. Color and sound in algorithm animation. Technical Report 76a, Palo Alto (CA):Digital Systems Research Center, Digital Equipment Corporation, 1991.

[10] Conway RW, Maxwell WL, Worona SL. *User's Guide to XCELL Factory Modeling System.* Palo Alto (CA):The Scientific Press, 1986.

[11] Brooks FP. Grasping reality through illusion: Interactive graphics serving science. In Soloway E, Frye D, Sheppard SB, editors, *Proceedings of the ACM SIGCHI Human Factors in Computing Systems Conference*, pages 1–1, New York:1988. ACM Press.

[12] Brooks FP, Ouh-Young M, Batter JJ, Kilpatrick PJ. Project GROPE—haptic displays for scientific visualization. *Computer Graphics*, 1990;24(4):177–185.

[13] Ming O, Beard DV, Brooks FP. Force display performs better than visual display in a simple 6-D docking task. In *Proceedings of the IEEE Robotics and Automation Conference*, volume 3, 1989.

[14] Sinden FW. Mechanisms for linear programs. *Operations Research*, 1959;7:728–739.

[15] Funkhouser TA, Séquin CH. Adaptive display algorithm for interactive frame rates during visualization of complex virtual environments. In *Computer Graphics Proceedings*, pages 247–254, 1993.

[16] Martello S, Toth P. *Knapsack Problems: Algorithms and Computer Implementations*. Chichester, UK:Wiley, 1990.

Chapter 17

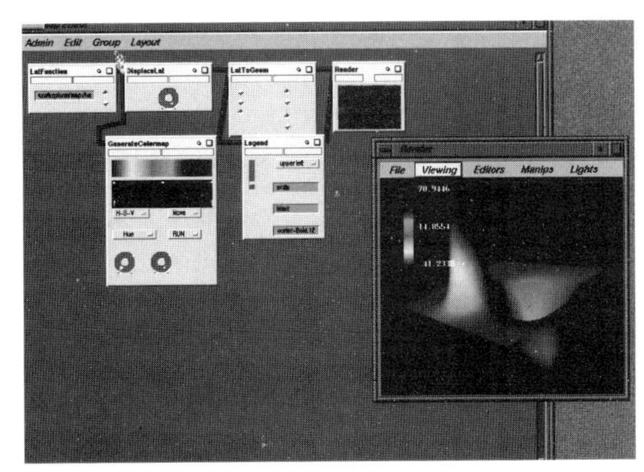

Visualization Tools

With the advent of scientific visualization, a variety of software tools and environments have been developed to facilitate creating visualizations. This chapter discusses some of their basic characteristics.

There are three broad classes of visualization systems. Some rely on textual languages; others rely on a visual (usually graph-based) language; still others use a spreadsheet-like (tabular) metaphor; many visualization tools are highly specialized for particular classes of representation. The careful reader will note that this classification scheme is similar to the classification scheme used in Section 8 when discussing languages for optimization. Unlike for optimization, however, visual languages for visualization have enjoyed much commercial success. Tabular languages, however, have only recently been proposed.

17.1 Visual Languages

Visual languages for visualization are types of dataflow programming languages [1]. Traditional programming languages concentrated on the sequence of operations that occur in a program—the flow of control. Dataflow programming languages, (e.g., Prograph [2]), in contrast, concentrate on how the flow of the data during execution of the program. Commonly represented as graphs and networks, a dataflow program has data flowing into one part of a network, each node transforming the data, with a the result appearing at the end of the network.

Since Haeberli's [3] paper, a variety of visualization languages have been developed that rely on dataflow representations. Nodes represent transformations of various sorts, such as reading data, coloring data, and then finally displaying the data. Edges connect transformations representing the flow of data and other information between transformations. Such visualization environments are sometimes also called *modular visualization environments* [4].

17.1.1 An Example

The image below, for example, illustrates Silicon Graphics's Iris Explorer [5], [6]. It shows a visualization of a mathematical function in the window at right. The node in the upper left ("LatFunction") samples datapoints from the two-dimensional function. The data consists of a "lattice" representation a two-dimensional surface—a standard data type understood by the system. The lattice is then fed into two other nodes, "DisplaceLat" and "GenerateColorMap" immediately to the right. The upper node, "DisplaceLat," displaces the data vertically. The lower node, "GenerateColorMap," generates a color scale based on the data. The color scale will be used to color code the two-dimensional surface that finally results. The color scale is fed to two nodes to the right, "LatToGeom" and "Legend." The lower node, "Legend," generates a legend for the color scale. The upper node also uses as input the scaled lattice from node "DisplaceLat" and generates a geometric representation that can be drawn using the node at right, "Render." Node "Render" produces the final visualization shown. Note that the user can double-click on any node and change any of the parameters associated with the node. When the change is made, the visualization is then regenerated interactively.

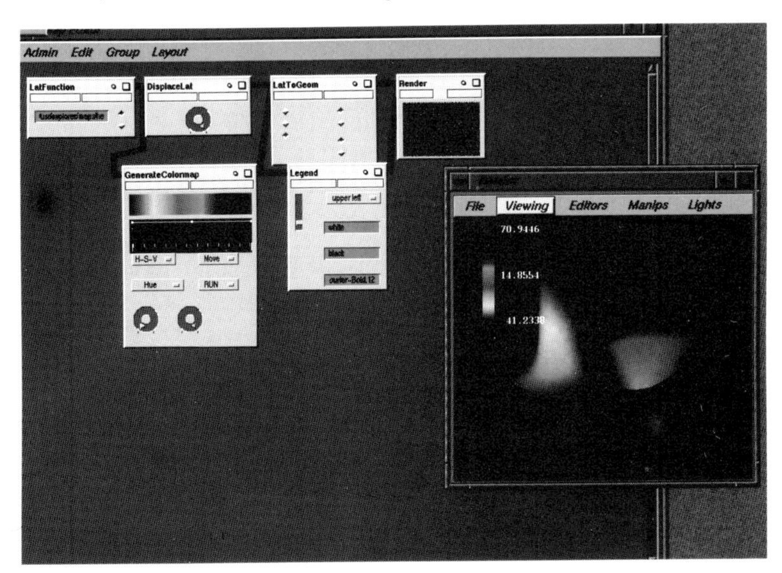

Figure 17.1

These visual languages have enjoyed substantial commercial success. Other examples of such languages include IBM's Information Visualizer [7], [8], the Application Visualization System [9], [10] and Khoros [11], [12]. There are differences among the products. For example, at least one of the systems allows polymorphic data. That is one module can accept as input a variety of data types, performing the appropriate operation based on the type of data input. Other systems require a different node for each different type of data [8]. Although there are differences, their basic metaphor is the same, a graph-based language based on a dataflow metaphor.

17.1.2 Problems with Visual Visualization Languages

Although visual languages have been commercially successful, challenges remain. First, they were originally too slow to work with very large data sets (over 100Mb) [10]. Solutions proposed to accommodate larger data sets include increased use of parallelism [13], optimizing compilers for the dataflow languages, and more intelligent storage and partitioning of the data during execution [8].

17.2 Textual Languages

The visual programming languages for visualization are not well integrated with algebraic modeling languages. Those comfortable with algebraic languages must learn the visual programming language in order to make use of the visualization tools. As [13] wrote:

> To be blunt: what is the better use of human resources (a) training ocean physicists to be numerical analysts and computer specialists while computer scientists design compilers that recognize Runga-Kutte implementations; or (b) letting engineers and scientists experiment with their flux theory (or whatever) in a canonical language and building the combined expertise of numerical analysts and computational scientists into an end-user system? Science would certainly be better served by the latter.

One solution to this disparity would incorporate algebraic models as nodes in the visual programming language [13]. Another solution would simply do the visualization using the algebraic modeling language itself. Symbolic mathematics systems such as Mathematica and Maple (discussed in Section 11.3), provide extensive capabilities for creating visualizations. Like any text-based language, the facilities provided are powerful, but they require careful attention to the language's syntax in order to be used properly. Many of the images in this book were generated using Mathematica, for example.

For another example, consider the following visualization generated using MatLab of the path taken by a steepest descent algorithm in finding the minimum of the two dimensional "banana" function $f(x, y) = 100(y - x^2)^2 + (1 - x)^2$ [14], [15]. This "banana demo" was developed using the MatLab Optimization Toolbox. Graphic provided courtesy of The MathWorks, Inc.

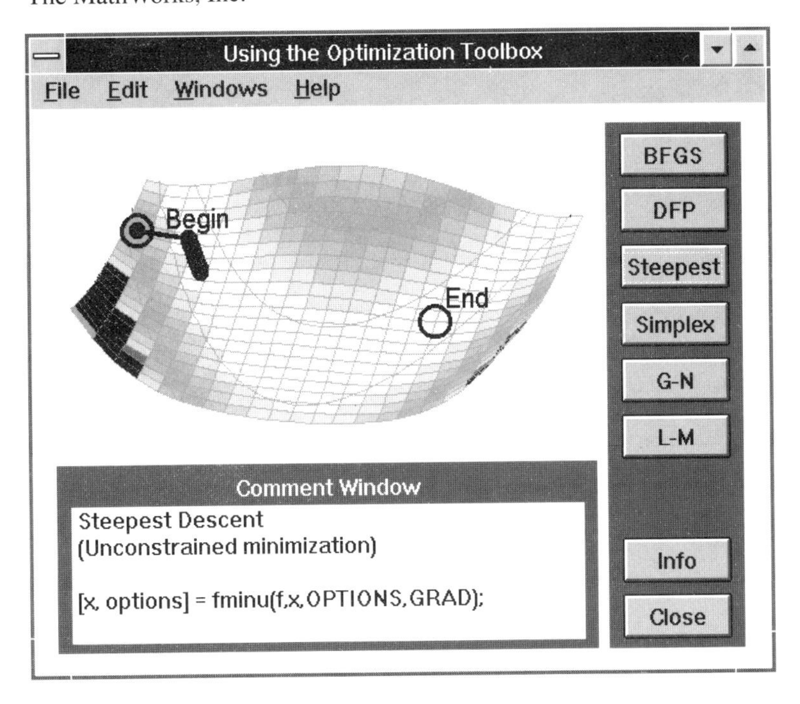

Figure 17.2

The actual minimum lies at the point $(x, y) = (1, 1)$; the steepest descent algorithm, even after many iterations remains far from the true minimum. Compare Figure 17.2 to the image below showing the path taken by a quasi-Newton method (Broyden-Fletcher-Goldfarb-Shanno).

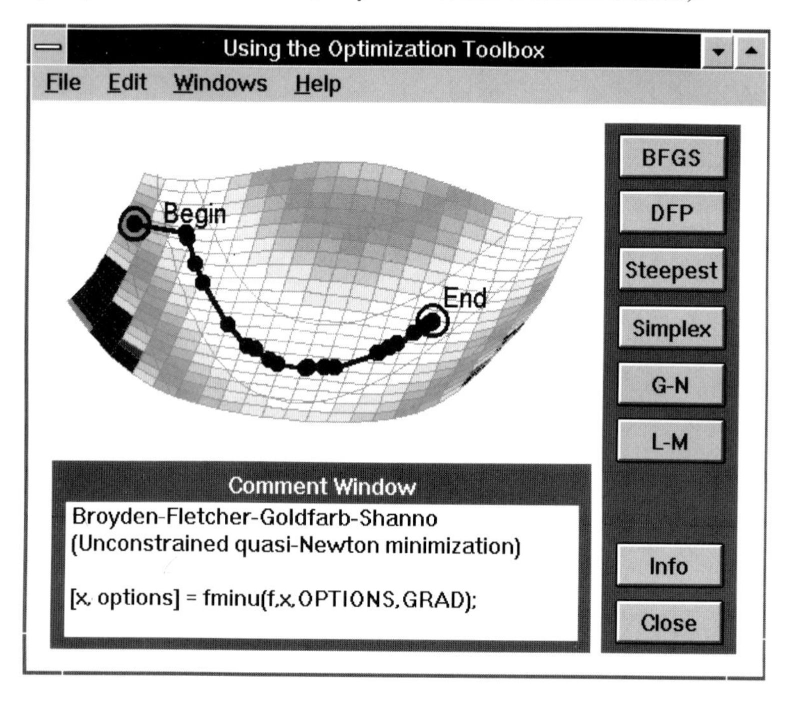

Figure 17.3

Statistical software provides a variety of visualization techniques including scatterplot matrices and Chernoff faces, in addition to simple two-dimensional point clouds, as discussed in Section 14.2.

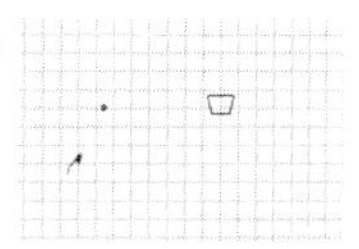

17.3 Tabular Languages

Levoy [16] proposed the use of a spreadsheet metaphor for visualization. The spreadsheet metaphor is extended to allow cells to display images. Formulas underlying cells can manipulate images. Moreover, the tabular metaphor allows for single operators to be applied simultaneously to multiple images, or conversely, multiple operators to be applied simultaneously to a single image. For example, the image below shows a spreadsheet used to analyze a volumetric medical dataset. The data is loaded into cell A2. Cell B1 contains a slider that is used to select a particular slice, which is displayed in cell B2. Cell C1 contains a filter that is used to classify the pixels in the image, in this case, highlighting the brighter parts (e.g., bone), and darkening the other parts of the image. The result of the classification is shown in cell C3. By clicking on the button in cell D1, the user can apply the filter to every slice in the image. The brightest slices are displayed in row 4. Image © copyright 1994 Association for Computing Machinery, Inc.

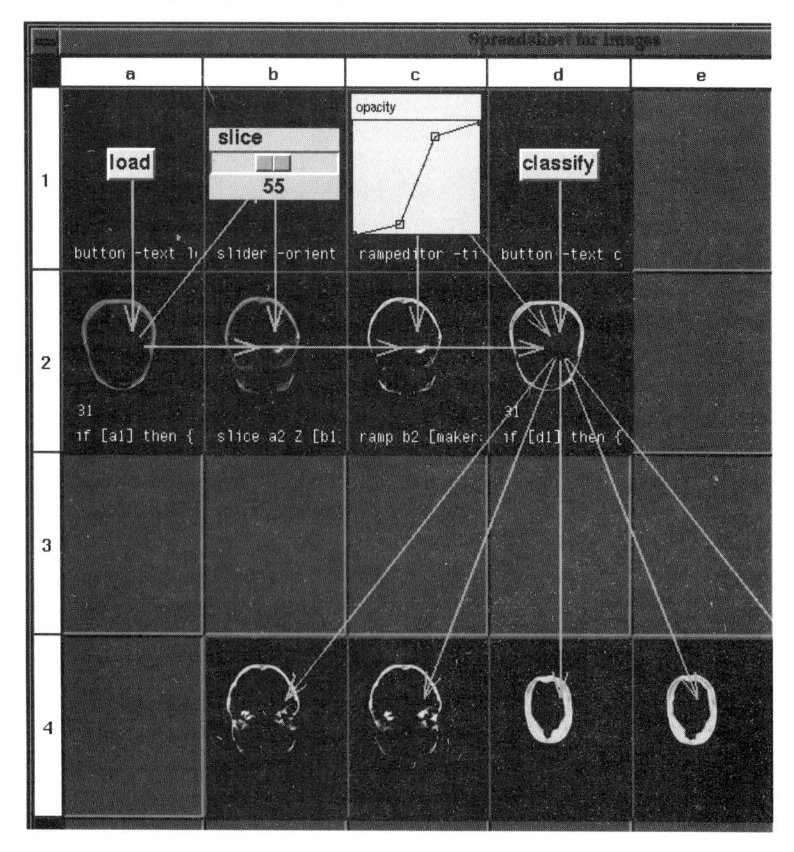

Figure 17.4

This proposal has not yet seen commercial application. Clearly, if multiple images are simultaneously being transformed and displayed in a spreadsheet, the recalculation time could be quite large. Yet, the spreadsheet metaphor is quite commonly used, and, as Levoy notes, "spends its screen space on operands rather than operators, which are usually more interesting to the user."

17.4 Summary

Creating visualizations has never been easier. A variety of useful commercial systems now exist that run on personal computers. They either provide a algebraic programming environment, like MatLab, Mathematica or Maple or a visual data-flow-like language (Iris Explorer, IBM Information Visualizer, Khoros). There are of course specialized visualization tools used in particular application areas such as statistics and the entertainment industry. We discuss some of the visualizations developed by statistics in Chapter 14.

Bibliography

[1] Sharp JA, editor. *Data-Flow Computing: Theory and Practice*. Norwood (NJ):Ablex Publishing Corporation, 1992.

[2] Pictorius Incorporated . *Prograph CPX Reference*. 2745 Dutch Village Road, Suite 200 Halifax, Nova Scotia B3L 4G7 Canada:1995.
URL: *http://192.219.29.95/home.html*

[3] Haeberli P. ConMan: A visual programming language for interactive graphics. *Computer Graphics*, 1988;22(4).

[4] Cameron G. Modular visualization environments: Past, present, and future. *Computer Graphics*, May 1995;29(2):3–4.

[5] Silicon Graphics, Inc. *IRIS Explorer User's Guide*:Silicon Graphics, Inc., 1993.

[6] Foulser D. IRIS Explorer: A framework for investigation. *Computer Graphics*, May 1995;29(2):13–16.

[7] Lucas B, Abram GD, Collins NS, Epstein DA, Gresh DL, McAuliffe KP. An architecture for a scientific visualization system. In *Proceedings IEEE Visualization '92*, pages 243–249, October 1992.

[8] Abram G, Treinish L. An extended data-flow architecture for data analysis and visualization. *Computer Graphics*, May 1995;29(2):17–21.

[9] Upson C, Faulhaber T, Kamins D, Laidlaw D. The application visualization system: A computational environment for scientific visualization. *IEEE Computer Graphics and Applications*, 1989;9(4):30–42.

[10] Lord HD. Improving the application development process with modular visualization environments. *Computer Graphics*, May 1995;29(2):10–12.

[11] Rasure R, Kubica S. The Khoros application development environment. In Christensen H, Crowley J, editors, *Experimental Environments for Computer Vision and Image Processing*:World Scientific, 1994.
URL: *http://www.khoros.unm.edu/*

[12] Young M, Argiro D, Kubica S. Cantata: Visual programming environment for the khoros system. *Computer Graphics*, May 1995;29(2):22–23.

[13] Frost R. High-performance visual programming environments: Goals and considerations. *Computer Graphics*, May 1995;29(2):29–32.

[14] The MathWorks, Inc. MathWorks home page. Technical report, Natick (MA):The MathWorks, Inc., 1995.
URL: *http://www.mathworks.com*

[15] The MathWorks, Inc. *The MatLab Expo*. Natick (MA):The MathWorks, Inc., 1993.

[16] Levoy M. Spreadsheets for images. In *Computer Graphics, Proceedings of SIGGRAPH '94*, pages 139–146, Orlando (FL):July 1994. ACM SIGGRAPH, Association for Computing Machinery.

Chapter 18

Integration

In previous chapters, we discussed the use of visualization from the perspective of the different tasks involved, and the different representation formats possible. We now attempt to tie the tasks and representations together. In Section 18.1, we first discuss experimental evidence supporting the use of multiple representations, and then discuss how optimization systems can support multiple representations (Section 18.2). Finally, we discuss how these representations can be built and maintained by multiple users perhaps working in multiple locations at different points in time (Section 18.3).

18.1 Experimental Evidence

Vessey's cognitive fit hypothesis [1], [2] states that decision making is enhanced when the representations used to present the information about a problem match the problem itself and the cognitive style of the decision maker. With respect to optimization, Orlikowski and Dhar [3] studied expert and novice mathematical programmers and discovered that different groups used different cognitive processes to formulate linear programming models. This further supports the notation that different people may benefit from different representations.

Experimental evidence that examines the effectiveness of different graphical representations has largely been equivocal, however. Field research suggests that problem solvers like graphical representations [4], but results from formal experimentation have not shown conclusively that graphical representations consistently produce measurably better performance. Most of the studies have explored the effectiveness of different types of presentation graphics such as bar charts versus pie charts versus tabular data. Although most authors now agree that the effectiveness of a particular format depends not only on the format, but also on the users and the tasks at hand, little experimentally-based guidance has been developed. In a detailed survey in 1985, DeSanctis [5] stated that "...relatively little is known about the actual utility of graphics as decision aids." Since that survey, several other authors have weighed in with studies that claim to have developed more consistent, applicable guidelines [6], [7]. Mackinlay [8] even attempted to build a system that would select an appropriate representation format automatically.

Guidelines for emerging representation formats such as sound [9], [10], [11] and animation [12] are much less well developed than those for static graphics. Formal experimentation that explores the relative effectiveness of these representation formats are even less developed, although some is emerging [13]. Given the difficulty in producing consistent experimental results even for simple graphics, one might expect that it will be some time before consistent, experimentally-verified guidelines are available for the newer representation formats.

To support multiple, simultaneous representations, windowed user interfaces were developed [14] and are now common (Apple Macintosh, Microsoft Windows, and X Window [15]). Any modern interface for mathematical programming will involve a *multiple window user interface*. Sadly, commercial optimization tools were latecomers to such user interfaces, even though it has been widely available since the mid 1980s.

In short, no one representation is useful for all tasks and all users [16], [17], [3]. Therefore, we need multiple representations for multiple tasks and users.

18.2 Multiple Representations

If one is to provide multiple representations, two approaches are possible. One approach assumes a base of a single underlying representation, with alternative representations built on top of the base. A classic example of this approach is Structured Modeling [18]. The other approach attempts to link together heterogeneous representations that were originally developed separately, perhaps with vastly different assumptions. We call the former *homogeneous* and the latter *heterogeneous*.

The homogeneous approach provides great economies of scale. Any tool developed, perhaps originally for a specific application, is applicable automatically to all other types of applications. This idea has been exploited in widely different settings. Unix, for example, uses text as its single underlying representation. Unix provides extensive facilities to search for patterns of text in files, allowing the text output by one command to be fed (or "piped," in Unix terms) into another command. Relational database systems store not only the data as relations, but also as models of the data, with security information controlling who has access to the data, and the location of the data. Spreadsheets provide another example, since everything represented in a spreadsheet—data, formulas, macros—is stored in the same format.

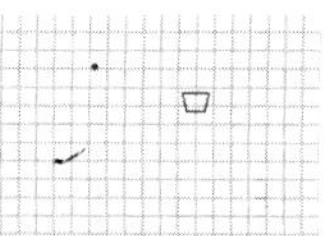

For optimization (and more general OR models), the homogeneous approach has been advocated by several authors including, [19], [20], [18], [21], [22], and [23]. Several implemented, commercial systems [24], [25], [26] also rely on a homogeneous approach. Much academic research has been conducted as well. For example, Bhargava and Kimbrough [20] based their hypertext modeling environment on formal logic. In Bhargava and Kimbrough's framework, models, data, and information about models are represented in the same way. When a new feature is provided, for example, hypertext navigation, it can be used for the data, the models, and the information about the models. Jones [27], [28], [29] based his approach on attributed graphs and graph-grammars, which is a topic of formal language theory.

Given the proliferation of different modeling tools including spreadsheets, symbolic mathematics environments, and word processors, as well as environments specifically targeted to mathematical programming, it seems unlikely that a single environment will completely dominate. Although the homogeneous approach has produced notable successes, even the homogeneous proposals must coexist with other systems. Ultimately, heterogeneous representations must be supported [30], [31].

Academic research exploring the heterogeneous approach is just emerging. Derrick and Balci have proposed a heterogeneous environment for simulation [32]. Greenberg and Murphy discuss the different representations required for mathematical programming [22].

If no single system dominates, then what takes responsibility for linking together heterogeneous representations? The answer is: the computer operating system. All the different representations must coexist with some operating system anyway. If the operating system manages system resources such as memory, storage, and processor time, it can also provide facilities to promote linking together different representations from different systems.

Features to link together different representations are now being added to operating systems at a rapid pace. For example, exchanging certain types of data such as text, spreadsheet data, and graphics is relatively easy. One can now readily copy text, pictures, movies and sounds from application and paste them into another application.

Until recently, however, the copy was separate from the original. If the original was changed, the copy was not updated. Similarly, if the copy need to be changed, it would be necessary to first move the copy back to the creating application, make the changes, and then copy the revised version back into the destination application. The development of technologies such as Object Linking and Embedding (OLE) [33] and OpenDoc [34], [35] seek to overcome these problems. For example, OLE allows a spreadsheet to be incorporated directly as part of another document, say a presentation. When the user double clicks on the spreadsheet, without leaving the presentation, the user can alter the spreadsheet. Imagine when the boss questions the validity of a model. The presenter need only double click on the spreadsheet in the presentation tool to make the spreadsheet "live" enabling the boss to explore the spreadsheet. Or, the numbers in a report could be linked to an underlying model. When the model is changed, the numbers in the report are automatically updated. These capabilities are become more and more tightly integrated with the operating system as evidenced by emerging (or existing) object-oriented operating systems such as Taligent [36] and Cairo from Microsoft [37] as well as NeXTStep from NeXT [38].

Implementing multiple, heterogeneous representations in multiple windows poses several major challenges. If several representations merely represent different views of the same problem, then changes made to one representation should be reflected in the other representations. Current spreadsheets provide such capabilities.

Incorporating such a capability in a multiple window mathematical programming environment remains in its infancy. For example, if an algebraic representation of the problem is displayed in one window, and the instance version is displayed in another window, then when the user deletes a decision variable in the algebraic formulation, the instance window will need to be changed, perhaps radically. An alternative approach would not change the old instance automatically, but would force the user either to ask explicitly for the dependent representations to be updated, or to ask explicitly for new versions of the representations to be created. Kendrick [39] developed a system that linked an algebraic and graphical representation of the transportation problem. Some commercial systems such as AIMMS [40], MIMI/G [41], and MPL [42] are beginning to provide such capabilities.

Linking together multiple representations in this fashion is closely related to *constraint-based* programming. Originally proposed by Borning in [43], (see [44] and [45] for more recent research) one specifies a set of mathematical relationships or *constraints*, usually mathematical equations. As input values are changed, the set of constraints is solved again. Surprisingly, techniques from mathematical programming have rarely been used to solve the set of constraints. Often, the variables in these equations are tied to properties of graphical objects such as the location or size of an icon. If the value of one or more variables is changed, perhaps through mouse input, the set of mathematical relationships is solved, with the graphics updated as quickly as possible.

This simple idea has been used to help provide sensitivity analysis [46], and to translate a network into a bar chart [47], and is now seeing application in commercial mathematical programming systems [41]. It could also be used to provide animated sensitivity analysis for representations. By defining constraints across multiple representations and multiple formats, one can provide automatic updating capabilities for linked representations.

The visualization environments discussed in Chapter 17 provide perhaps the most advanced commercial implementation of this idea. Essentially they extend the notion of Unix pipes, which allow for the transmission of text from one Unix command to another, to visualization. Users draw a network of pipes linking data nodes to filters for processing data, and finally to visualization nodes which actually present the information.

For optimization, however, the only common interchange format is the MPS format, which has been showing its age for far too long. MPS can only represent problem instances, not generic structure and its variable names are limited to eight characters. Many mathematical programming systems provide incredibly rich capabilities for squeezing a variety of information into those eight characters. The row and column names in Figures 10.2-10.3, are restricted to eight characters. The column names begin with the letter 'T,' because those variables represent transportation activities. The next two characters represent the sources, and the final two characters represent the sinks. For large problems having hundreds of sources and sinks, cryptic code names must be used. Furthermore, MPS can only represent linear programs. Conn, Gould, and Toint [48] have proposed an extended version of the MPS standard that is upwardly compatible with the existing MPS standard. It also allows nonlinear constraints to be specified. It retains for compatibility's sake, however, much of the conventions of MPS, including variable names limited to a maximum of eight characters.

It seems imperative that members of the mathematical programming community join together to develop a more modern standard. The new standard should allow for linear, integer, and nonlinear programming models to be specified. Identifiers should not be limited to eight characters. It should allow the interchange not only of individual problems (instance level), but also problems at the generic level.

More specifically, the new format should contain information about the sets and tables that were used to generate the problem. For example, in MPS an objective function coefficient for a variable might appear as T1L3L4P8. It might actually represent the transport of product 8 from location 3 to location 4 in time period 1. In other words, this model apparently contains variables that are indexed by time period, product, and location. This type of information should be represented in the interchange standard.

An appropriate interchange format is particularly important for interfacing modeling languages with solvers. Currently each solver vendor provides its own application programming library. The author of a modeling environment is then faced with writing interface code in order to use any desired solver. A key problem involves non-linear models, wherein the solver needs to be able to calculate gradients. The gradients are best calculated by the modeling environment, since they have the symbolic representation.

With the proliferation of algebraic languages, it is an appropriate time to begin defining a general interchange standard for these languages. At least one modeling language vendor (AMPL) has published specifications including source code for linking to solvers [49] but there may be little commercial incentive to pursue a new standard. After all, a standard interchange format would make it easier for users to migrate from one language to another, which is not necessarily in the best interest of the commercial vendors. The success of de facto standards such as Unix, DOS, Postscript, and Microsoft Windows point to the demand for standards. Upward compatibility with MPS should not be a difficult problem, since a simple conversion program from the old format to the new format should suffice. Without a better standard, the vision of multiple, linked representations from multiple vendors will eventually be realized but it will take longer than necessary.

18.3 Multiple Users

Most problem-solving projects are not conducted by one person, but by groups of people. Clearly, groups of people have been using computers for quite some time. A field study of spreadsheet development [50], for example, found that most spreadsheets were developed in a collaborative process involving several individuals.

The field of *computer-supported cooperative work* (CSCW) [51], [52] attempts to support groups of people in their activities. Many different flavors of CSCW are currently emerging. People might work together simultaneously or *synchronously* on a project; for example, they might edit the same text at the same time [53]. In contrast, people may work on the project at different times, that is, *asynchronously*. They could be in the same room or dispersed geographically. In teleconferencing, meetings are conducted using video to link people from far away (see [54] for an extreme use of teleconferencing).

An increasingly common form of CSCW has been dubbed a *Group Decision Support System* (GDSS) [55], [56], [57], [58]. At the University of Arizona [58], meeting participants work together in a room using a network of computers. Guided by a facilitator, the participants use a variety of tools provided by the GDSS to attack a problem. Tools for brainstorming ideas, organizing ideas into categories, and voting to settle points are some of the many techniques provided.

18.4 Summary

Ideas from CSCW have not found their way into the mathematical programming community. This seems somewhat surprising since mathematical programming projects often involve coordinating not only large models and large databases, but also large numbers of people. One could imagine a system for editing mathematical programming formulations that allows multiple users to perform editing simultaneously.

Spreadsheets, as well as at some mathematical programming systems [59], [25], now provide some capabilities for managing multiple versions of models and data. Holsapple, Park, and Whinston [60] proposed an architecture for decision support that could possibly support multiple, simultaneous users. We could also imagine an asynchronous system for mathematical programming formulations that tracks the changes made by each user, much like source code control systems that support computer program development.

Bibliography

[1] Vessey I. Cognitive fit: A theory-based analysis of the graphs versus tables literature. *Decision Sciences*, 1991;22:219–241.

[2] Vessey I, Galletta D. Cognitive fit: An empirical study of information acquisition. *Information Systems Research*, 1991;2(1):63–86.

[3] Orlikowski W, Dhar V. Imposing structure on linear programming problems: an empirical analysis of expert and novice modelers. In *Proceedings of the National Conference on Artificial Intelligence*, volume 1, 1986.

[4] Kirkpatrick P, Bell PC. Visual interactive modeling in industry: Results from a survey of visual interactive model builders. *Interfaces*, 1989;19(5):71–79.

[5] DeSanctis G. Computer graphics as decision aids: Directions for research. *Decision Sciences*, 1984;15:463–487.

[6] Benbasat I, Dexter AS, Todd P. An experimental program investigating color-enhanced and graphical information presentation: An integration of the findings. *Communications of the ACM*, 1986;29(11):1094–1105.

[7] Jarvenpaa SL, Dickson GW. Graphics and managerial decision making: Research based guidelines. *Communications of the Association for Computing Machinery*, 1988;31(6):764–774.

[8] Mackinlay J. Automating the design of graphical presentations of relational information. *ACM Transactions on Graphics*, 1986;5(2):110–141.

[9] Baecker RM, Buxton WAS. *Readings in Human-Computer Interaction.* Los Altos (CA):Morgan Kaufmann, 1987.

[10] Blattner MM, Sumikawa DA, Greenberg RM. Earcons and icons: Their structure and common design principles. *Human-Computer Interaction*, 1989;4(1):11–44.

[11] Buxton W, Bly SA, Frysinger SP, Lunney D. Communications with sound. In Borman L, Curtis B, editors, *Proceedings of the ACM SIGCHI Human Factors in Computing Systems Conference*, pages 115–119, New York:1985. ACM Press.

[12] Thomas F, Johnston O. *Disney Animation: The Illusion of Life.* New York:Abbeville Press, 1984.

[13] Stasko J, Badre A, Lewis C. Do algorithm animations assist learning? an empirical study and analysis. In *Human Factors in Computing Systems, INTERCHI'93 Conference Proceedings, Conference on Human Factors in Computing Systems, INTERACT '93 and CHI '93*, pages 61–66, Amsterdam, The Netherlands:April 1993.

[14] Smith DC, Irby C, Kimball R, Verplank B, Harslem E. Designing the star user interface. In Degano P, Sandwell E, editors, *Integrated Interactive Computing Systems*, pages 297–313. Amsterdam, The Netherlands:North-Holland, 1983.

[15] Scheifler RW, Gettys J. The X window system. *ACM Transactions on Graphics*, 1987;5(2):79–109.

[16] Bell PC, Chau PYK. An empirical assessment of three types of simulation models used in developing decision support systems. Technical report, London, Ontario:Western Business School, University of Western Ontario, 1992.

[17] Ghani J. *The Effects of Information Representation and Modification on Decision Performance.* [dissertation], Philadelphia (PA):The Wharton School, The University of Pennsylvania, 1981.

[18] Geoffrion AM. An introduction to structured modeling. *Management Science*, 1987;33:547–588.

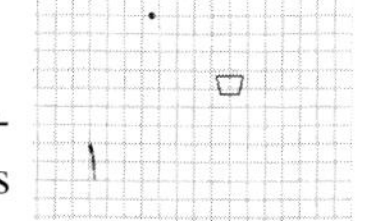

[19] Baldwin D. *Principles of Design for a Multiple Viewpoint Problem Formulation Support System.* [dissertation], Lubboch (TX):School of Business Administration, Texas Tech University, 1989.

[20] Bhargava HK, Kimbrough SO. On embedded languages for model management. In Nunamaker JF, editor, *Proceedings of the Twenty-Third Annual Hawaii International Conference on System Sciences*, pages 443–452, Los Alamitos (CA):1990. IEEE Computer Society Press.

[21] Geoffrion AM. The formal aspects of structured modeling. *Operations Research*, 1989;37(1):30–51.

[22] Greenberg HJ, Murphy FH. Views of mathematical programming models and their instances. *Decision Support Systems*, 1993;13(1):3–34.

[23] Kendrick DA. Parallel model representations. *Expert Systems with Applications*, 1990;1(4):383–389.

[24] Chesapeake Decision Sciences. *MIMI: Manager for Interactive Modeling Interfaces: User's Manual.* New Providence (NJ):Chesapeake Decision Sciences, 1993.

[25] Palmer KH, Boudwin NK, Patton HA, Rowland AJ, Sammes JD, Smith DM. *A Model Management Framework for Mathematical Programming.* New York:Wiley, 1984.

[26] SAS Institute. *SAS/OR User's Guide Version 6.* Cary (NC):SAS Institute, 1st edition, 1989.

[27] Jones CV. An introduction to graph-based modeling systems, part I: Overview. *ORSA Journal on Computing*, 1990;2(2):136–151.

[28] Jones CV. An introduction to graph-based modeling systems, part II: Graph-grammars and the implementation. *ORSA Journal on Computing*, 1991;3(3):180–206.

[29] Jones CV. An integrated modeling environment based on attributed graphs and graph-grammars. *Decision Support Systems*, 1993;10:255–275.

[30] Garlan D. *Views for Tools in Integrated Environments.* [dissertation], Pittsburgh (PA):Computer Science Department, Carnegie-Mellon University, 1987.

[31] Geoffrion AM. Integrated modeling systems. *Computer Science in Economics and Management*, 1989;2:3–15.

[32] Derrick EJ, Balci O. DOMINO: A multifaceted conceptual framework for visual simulation modeling. Technical Report TR-92-43, Blacksburg (VA):Department of Computer Science, Virginia Polytechnic Institute and State University, 1992.

[33] Microsoft Corporation. *Object Linking and Embedding Programmer's Reference Version 1*. Redmond (WA):Microsoft Corporation, 1994.

[34] Novell, Inc. OpenDoc technical white paper. Technical report, 1555 N. Technology Way, Orem (UT) 84057:Novell, Inc., 1995.
URL: *http://www.cilabs.org/pub/novell/tech/whitep23.html*

[35] Apple Computer. OpenDoc–one architecture fits all. Technical report, Apple Computer, Cupertino (CA):Apple Computer, 1995.
URL: *http://www.info.apple.com/dev/appledirections/jun95/techopendoc.html*

[36] Taligent, Inc. Taligent, Inc. WWW home page. Technical report:Taligent, Inc., 1995.
URL: *http://www.taligent.com/*

[37] Quinlan T, Scannell E. Industry sees object system as inevitable. *InfoWorld*, 1993;15(4):13.

[38] NeXT Computer, Inc. *NeXTSTEP General Reference*. Reading (MA):NeXT Computer, Inc., 1992.

[39] Kendrick DA. A graphical interface for production and transportation system modeling: PTS. *Computer Science in Economics and Management*, 1991;4(4):229–236.

[40] Bisschop J, Entriken R. *AIMMS The Modeling System*. PO Box 3277, 2001 DG Haarlem The Netherlands:Paragon Decision Technology, 1993.

[41] Chesapeake Decision Sciences. *MIMI/G User's Manual*. New Providence (NJ):Chesapeake Decision Sciences, 1993.

[42] Maximal Software, Inc. *MPL Modeling System*. Arlington (VA):Maximal Software, Inc., 1994.

[43] Borning A. Thinglab - a constraint-oriented simulation laboratory. *ACM Trans. on Programming Languages and Systems*, 1981;6(4):353–387.

[44] Leler W. *Constraint Programming Languages*. Reading (MA):Addison-Wesley, 1988.

[45] J. Maloney AB, Freeman-Benson B. Constraint technology for user-interface construction in ThingLab II. *SIGPLAN Notes*, 1989;24(10):381–388.

[46] Clemons E, Greenfield A. The SAGE system architecture: A system for the rapid development of graphics interfaces for decision support. *IEEE Computer Graphics and Applications*, 1985;5(11):38–50.

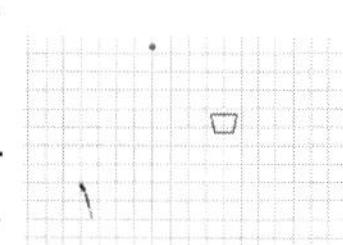

[47] Jones CV. Attributed graphs, graph-grammars, and structured modeling. *Annals of OR*, 1992;38:281–324.

[48] Conn AR, Gould N, Toint PL. *LANCELOT: A FORTRAN Package for Large-Scale NonLinear Optimization*. Berlin:Springer-Verlag, 1993. to appear.

[49] Gay DM. Hooking your solver to AMPL. Technical Report Numerical Analysis Manuscript 93-10, Murray Hill (NJ):AT&T Bell Laboratories, 1994.

[50] Nardi BA, Miller JR. Twinkling lights and nested loops: Distributed problem solving and spreadsheet development. In Greenberg S, editor, *Computer-Supported Cooperative Work and Groupware*, pages 29–54. London:Academic Press, 1991.

[51] Greenberg S, editor. *Computer-supported Cooperative Work and Groupware*. London:Academic Press, 1991.

[52] Greif I. *Computer-supported Cooperative Work: A Book of Readings*. San Mateo (CA):Morgan Kaufmann, 1988.

[53] Hill RD, Brinck T, Patterson JF, Rohall SL, Wilner WT. The Rendezvous language and architecture. *Communications of the Association for Computing Machinery*, 1993;36(1):62–67.

[54] Bly SA, Harrison SR, Irwin S. Mediaspaces: Bringing people together in a video, audio, and computing environment. *Communications of the ACM*, 1993;36(1):28–47.

[55] Dennis AR, George JF, Jessup LM, Nunamaker J, Vogel DR. Information technology to support electronic meetings. *MIS Quarterly*, 1988;12:591–624.

[56] DeSanctis GS, G.Gallupe. A foundation for the study of group decision support systems. *Management Science*, 1987;33:589–609.

[57] Jessup LM, Valacich JS, editors. *Group Support Systems*. New York:MacMillan, 1993.

[58] Nunamaker JF, Dennis AR, Valacich JS, Vogel DR, George JF. Electronic meeting systems to support group work. *Communications of the ACM*, 1991;34(7):40–61.

[59] MathPro, Inc. *MathPro Usage Guide: Introduction and Reference*. Washington (DC):MathPro, Inc., 1989.

[60] Holsapple CW, Park S, Whinston AB. Framework for DSS interface development. Technical report, Lexington (KY):Center for Robotics and Manufacturing Systems, University of Kentucky, 1991.

Chapter 19

Research and Future Directions

The ideas presented in this book represent just a snapshot in time. Given the rapid pace of technological development, the representations presented here are likely to improve significantly in the next few years. New ways of visualizing information, not yet envisioned, will surely be invented. This chapter begins with a discussion of how one can *successfully* conduct research on visualization and optimization (Section 19.1). The chapter ends with fearless predictions of developments that can be reasonably expected in the near future (Section 19.2).

19.1 Research

Whereas optimization prizes insightful theorem that lead to faster algorithms yielding optimal solutions, visualization seeks to help people understand the solution. Those seeking to conduct academic research in visualization and optimization face a significant challenge. Traditional researchers in optimization can often dismiss the work as just pretty pictures lacking any scientific foundation. That criticism, all too often, is justified. It *is* possible to do research in visualization that will earn respect, but it requires dedication, hard work and a careful management of one's research program. This section presents some guidelines to help prospective researchers achieve success. The guidelines do not guarantee success. Clear, cogent writing, careful attention to detail—the same characteristics expected of any scholarly work apply as always. The guidelines should provide some direction to those who find this area as exciting and interesting as the author does.

Theory Merely presenting an interesting representation of complicated information is insufficient. Almost always, there is some theory from some scientific field of study (usually not optimization) that can help explain the importance or usefulness of the representation.

Appropriate theory may come from cognitive psychology, if only to explain some of the color choices, or the layout of a window. Many authors attempt to finesse this issue by including the chestnut "a picture is worth a thousand words" somewhere in their introduction. Although a picture is sometimes worth a thousand words, sometimes that picture *is* a thousand words of text.

Appropriate theory may come from fields such as logic or formal languages. Bhargava and Kimbrough *et al.*'s [1], [2] work on hypertext-based modeling environments relied on formal logic, for example. Jones [3], [4] based his research on formal language theory applied to graph-based representations.

One can of course propose new theories about how one can represent complex information. For example, Geoffrion's Structured Modeling proposed a new theory for representing models in a beautifully written paper [5]. Theories, though, need to be verified or justified in some fashion. Geoffrion's original paper on Structured Modeling first argued persuasively for particular characteristics that a modeling environment should possess, especially a theoretically sound foundation, and only then proposed his theoretical foundation for modeling—Structured Modeling.

Implementation Papers frequently present intriguing ideas, but lack a demonstrable implementation as software. For example, many papers have proposed an interesting graphical modeling language. It can be difficult to assess the expressive power of proposed language. Demonstrating a variety of interesting cases in an actual implementation readily silences critics.

Although a computer implementation often seems necessary to demonstrate the practicality of the ideas proposed, implementations place a great burden on a researcher since they require special skills and a great deal of time. An approach that does not require an implementation provides an extensive library of problems expressed in the proposed language, available by request to the author. Geoffrion [6] followed this strategy in developing Structured Modeling. But by having an implementation that can at least be demonstrated, and better yet, distributed to other researchers, the quality of the ideas can be more thoroughly assessed.

Formal Experimentation If one claims that a visualization technique is useful, why not demonstrate it through experimentation? Does the representation actually help people understand the underlying problem? Do they make better decisions? Do they make faster decisions? Are they more satisfied? Many papers in visualization and optimization have been written that present new representations without any formal (or even informal) experimentation to demonstrate how helpful they actually are.

Developing such experiments, however, is more difficult than it might at first seem. For example, suppose one has developed a new graphical language for modeling and would like to see how well people are able to make use of it. One might hope that the language would allow people to express their models more easily.

This vague statement must be made much more concrete. Which people are relevant? What level of modeling expertise do they possess already? Perhaps one means "expert" modelers, but what is an expert modeler? Perhaps an expert modeler is someone who has developed optimization models for 5 or more years in industry or who has a PhD in the area, yet PhD's and actual practitioners often have many differences in temperament. Ignoring the difficulty in attracting such modelers to a study (students are usually much easier to find), each of the experts would certainly have their own individual tastes as to a modeling language and would probably not be as comfortable or proficient with the new language.

Assuming the problems of defining and attracting a sufficiently homogeneous group of subjects are solved, what does one mean by "express their models more easily." Is a "correct" model created more quickly? Do modelers make fewer mistakes? Are modelers more satisfied? It is possible to develop instruments to measure each of these questions, but they must be carefully crafted.

Even if one attracts a sufficiently homogeneous group of modelers and is able to develop appropriate measures for the experiment, other factors may confound the experiment. Suppose one is attempting to compare two graphical languages for modeling, Language A and Language B. Suppose the experiment shows that Language A allows modelers to express models faster and with fewer errors than Language B. One might conclude, therefore, that Language A should be preferred to Language B. That conclusion may be too hasty, however. Language A may have a superb user interface, whereas Language B may have a lousy user interface. Language B's poor rating may be attributable more to its user interface than to any intrinsic properties of the language.

Some people have argued that if one must conduct a formal experiment to demonstrate the superiority of one proposal over another, then the benefits of the new proposal are not sufficient to merit its consideration anyway. In other words, great ideas do not need formal experimentation to demonstrate that they are great. Spreadsheets, for example, became a standard analytical tool without formal experimentation. Such experimentation was actually done [7], but long after their initial invention.

In order to avoid these difficulties, most formal experimentation in cognitive psychology attempts to answer relatively simple questions. For example, how many colors can a person easily distinguish? How many moving objects can a person easily track? These lower level questions are much easier to student than the question of which modeling languages is better.

Novelty Research should, of course, study the unexplored. With respect to visualization, the development of technology provides enhanced ways of representing information. For example, the development of inexpensive computer graphics equipment has caused an explosion in the use of visualization. Today, virtual reality, hypermedia, sound, and touch are attracting researchers. This typically produces papers with titles of the from "A Virtual Reality Approach for X," where X can be any topic of interest. While the medium remains new, such research can have value, since it is necessary to explore (fool around with) the new medium to see if anything interesting develops. This phase of exploration, however, should remain brief. This allows the new medium to be explored, but eventually theory and experimentation should dominate the research.

Most researchers and practitioners in optimization are unaware of useful results from visualization. Any research hoping to do research at the intersection of both fields should be well versed in both—certainly a difficult challenge.

None of these recommendations is sufficient by itself to guarantee success. A novel idea without theoretical underpinnings or a demonstrable implementation may be oft-cited (e.g., Vanevar Bush's 1945 *Atlantic Monthly* article [8] foreshadowing hypermedia), but may not live up to academic standards. Similarly, an implementation of an idea without sufficient novelty is just "reinventing the wheel." High-quality, well-regarded research in visualization combines two or more of the above characteristics. Even so, it can be difficult to have one's paper accepted into a traditional OR/MS journal. One could submit the work to journals that regularly publish articles on visualization, but tenure committees for OR/MS departments may not feel comfortable relying on the quality of those journals. Luckily, several respected OR/MS journals have published work on visualization and optimization. Success is not impossible, just difficult.

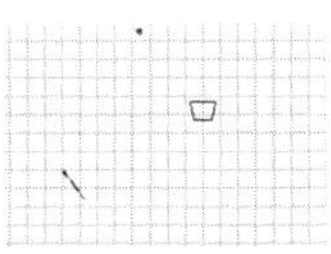

19.2 Future

The goals of visualization and OR are quite similar: to provide insight into complex problems. OR uses mathematical algorithms whereas visualization studies representations. However, the techniques that OR has developed rest on the solid foundation of mathematics and computer science. We can usually prove that a particular solution to a mathematical program is optimal, that an algorithm will converge, or that the solution, when not optimal, is at least close to the optimal solution. Mathematical proofs of the effects of visualization seem less likely.

At best, we can conduct carefully designed experiments to establish the quality of different forms of visualization. That experimental style of research does not fit into the mathematical traditions of OR. Although reference disciplines, including cognitive psychology, linguistics, human-computer interaction, and formal logic can contribute well-developed theories, many of these disciplines have not traditionally been considered part of the mainstream of OR.

As discussed throughout the previous chapters, representation forms a key component of the life cycle of mathematical programming projects. From conceptual models, through formulations, databases, data structures to facilitate solution, outputs of algorithms, debugging information, and final presentation, an OR project is really just a process of transforming one representation into another.

Given the variety of representations needed, the difficulty in constructing them, and most importantly, the need to tie them together, research in this area is essential to the growth of OR. Research on visualization in optimization should explore the following directions.

Empirical Evidence. Several authors have presented articulate, detailed discussions of how to represent complicated datasets and problems. Bertin [9], [10] developed a taxonomy of different graphic representations based on their ability to encode different types of data. Tufte's books [11], [12] developed several recommendations for creating effective, parsimonious graphical representations of data. In Section 18.3 we discussed the lack of experimental evidence concerning the effectiveness of different representations. With new hardware and software capabilities, the constraints on representation capabilities are relaxing—this could lead to better ways of visualizing problems. Although such research in the past has been fraught with inconsistencies, the need to understand which representations work best for which tasks and users remains. Perhaps researchers have considered problems at too high a level when they try to answer a question such as: Do problems represented as bar charts lead to better understanding than the same problems represented as tables? Experimental research that has asked more fundamental questions (such as which encoding schemes are perceived more accurately) seems to have enjoyed greater success. Better experimental evidence, specifically targeted to optimization, is essential.

Formalizations. Several fields, such as formal languages and formal logic, have concentrated on developing theories of representation. These have formed the basis of several homogeneous (as defined in Section 18) representations used for modeling environments. Formal language theory has been used to help develop and describe computer programming languages. Recently, these theories have been applied in the realm of mathematical programming. Syntax-directed editors are one result that have been developed for mathematical programming languages. Kimbrough *et al.* [2], [13], Krishnan [14], Hong, Mannino, and Greenberg [15], and Muhanna and Pick [16] have relied on formal logic for their approaches. Graph-grammars, a formal theory for visual languages, have been applied to describe the variety of graph-based models that are employed in mathematical programming [3], [4], [17], [18], [19], [20], [21]. Baldwin [22] and Greenberg and Murphy [23] have proposed the use of a single, central representation format to support a wide variety of different representations or views of the underlying problem. Geoffrion [5], [24], [25], in developing Structured Modeling, has attempted to provide a single, formal framework of systems supporting multiple representations. In fact, Geoffrion proposes a variety of useful representations based on Structured Modeling. Formalization brings legitimacy to the field, as well as practical tools useful in modeling.

Integration. Many proposals for different representations for mathematical programming exist. The authors of each of these proposals (including the author of this book) frequently seem to view their representation proposal as the "best," in some way. Yet, given the variety of representations, it does not seem likely that one representation will come to dominate. Rather, different representations will coexist. Mathematical programming systems need to provide a variety of representations, for input as well as output, in multiple formats. Whether this should be accomplished by using a single underlying representation format, or by combining heterogeneous systems, remains an open question. However, emerging operating system technology is making the exchange of information among different systems far easier, so the heterogeneous approach is becoming more possible. Perhaps all that is necessary are a few components such as spreadsheets, databases, algebraic modeling languages, hypertext, and solution algorithms linked together by facilities provided by computer operating systems.

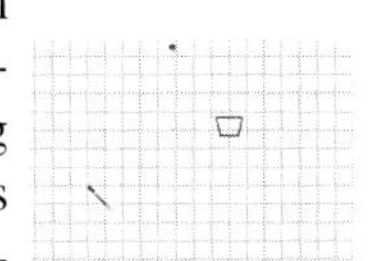

New Visualization Techniques. Newly available visualization techniques, such as animation, sound, computer-supported cooperative work, multimedia [26], [27], and haptic displays, have yet to be applied widely in mathematical programming. Exploratory research that develops novel and interesting uses of these techniques would be of interest. Such research could bring out the "lunatic fringe," but out of that morass could emerge some useful ideas. Of course, those ideas would need to be carefully tested. Researchers interested in pursuing such a direction would be well advised to consult the work of Brooks [28], [29]. Brooks provided an excellent model for research that combines such exploratory investigation with solid experimental justification.

In the information technology business, the future is five minutes away. Predicting trends and events several years into the future is certainly foolhardy. Nevertheless, with gross recklessness, we attempt to make some predictions:

1. All computer software, including software incorporating OR techniques, will use multiple window, mouse driven user interfaces. Those systems that do not provide such capabilities will not survive. As Schultz and Pulleyblank [30] said in a discussion of trends in optimization, "...people now expect much higher quality user interfaces and presentations of results."

2. Spreadsheets will provide increasingly powerful modeling capabilities, especially in the area of multi-dimensional indexing. Spreadsheets will therefore become the delivery vehicle of choice for optimization, at the expense of algebraic modeling languages.

3. A single algebraic modeling language standard will emerge, either through market forces, or through cooperation on the part of members of the mathematical programming community. That standard might perhaps be based on multi-dimensional spreadsheets.

4. A standard paradigm or system to support drawable models or programs will emerge, perhaps based on graphs or networks.

5. Animation will become as widespread as static graphics are today. Applications will be in algorithm animation and animated sensitivity analysis.

6. Hypertext (really hypermedia) will be a standard tool for navigating through spreadsheets, modeling languages, word processors, and any other complex system. The World-Wide Web will become the standard hypertext delivery engine linking together multiple representations across multiple platforms distributed world-wide.

7. Mathematical programming systems will support *multiple* representations including algebraic formulations, block-structured, relational databases, matrix images, graph-based and other graphical representations, spreadsheet, presentation graphics, natural language, hypertext, and animation.

8. More and more optimization systems will be composed of modular building blocks linked together by facilities provided by the computer operating system.

9. Virtual reality will move primarily from the video game stage to see more common application by optimization systems.

10. Multiple user modeling systems will emerge. The basic facilities for allowing multiple users to work together on a project, simultaneously or not, geographically separated or not, will be provided by the operating system. People throughout the world will be able to exchange problem instances and generic models quickly and easily.

11. Research in visualization and OR will continue to be dangerous for untenured faculty. Hence, as Hirshfeld [31] predicted, most of the interesting developments in visualization and optimization will come from industry and not academia.

12. Millions more will make use of optimization, often to great benefit. Even so, many of these new users will not know or care about the heritage of the technology that they are using. If true, this prediction implies that the prominence of optimization as a field will continue to fade, as the techniques it nurtured, ironically, see greater and greater use.

In conclusion, the field of optimization has made great strides in developing techniques to analyze complicated problems. These sophisticated mathematical and computer-based techniques have yielded large benefits. Problem solving in general, however, involves more than building an algorithm. It consists of a process of transforming and understanding various representations or visualizations of the problem at hand. The solution algorithm shares at least equal importance with the problem-solving process and the associated representations and visualizations. If visualization has such importance, then it needs to be studied, researched, and improved.

Unlike mathematics and computer science, however, visualization does not yet constitute a mature, refined science. It remains difficult to prove or predict the relative quality of a particular visualization for specific tasks or users. Yet techniques are emerging based in the fields of cognitive psychology, formal logic, and formal languages that are theoretically grounded and have practical application.

If we improve our ability to visualize our problems as much as we have improved the ability of algorithms to "solve" problems, then we should be able to model and solve our problems much more effectively and efficiently. At the very least, we should see the wider use of the sophisticated techniques that optimization researchers have spent a considerable amount of time developing.

Regardless of the success of academic researchers in providing provable results about the effectiveness of visualization, the market will continue to produce new and better visualization techniques for building, solving, and understanding problems that can be analyzed using optimization.

Bibliography

[1] Bhargava HK, Kimbrough SO. On embedded languages for model management. In Nunamaker JF, editor, *Proceedings of the Twenty-Third Annual Hawaii International Conference on System Sciences*, pages 443–452, Los Alamitos (CA):1990. IEEE Computer Society Press.

[2] Kimbrough SO, Pritchett CW, Bieber MP, Bhargava HK. The coast guard's KSS project. *Interfaces*, 1990;20(6):5–16.

[3] Jones CV. An example based introduction to graph-based modeling. In *Proceedings of the Twenty-Third Annual Hawaii International Conference on the System Sciences*, pages 433–442, Los Alamitos (CA):1990. IEEE Computer Society Press.

[4] Jones CV. An introduction to graph-based modeling systems, part I: Overview. *ORSA Journal on Computing*, 1990;2(2):136–151.

[5] Geoffrion AM. An introduction to structured modeling. *Management Science*, 1987;33:547–588.

[6] Geoffrion AM. A library of structured models. Technical report, UCLA, Los Angeles (CA), 90026:Western Management Science Institute, 1990.

[7] Olson JR, Nilsen E. Analysis of the cognition involved in spreadsheet software interaction. *Human-Computer Interaction*, 1988;3.

[8] Bush V. As we may think. *The Atlantic Monthly*, July 1945;pages 101–108.

[9] Bertin J. *Graphics and Graphic Information-Processing*. Berlin:Walter de Gruyter, 1981. W. J. Berg and P. Scott (Translators).

[10] Bertin J. *Semiology of Graphics: Diagrams, Networks, Maps*. Madison (WI):University of Wisconsin Press, 1983. W. J. Berg (Translator).

[11] Tufte ER. *The Visual Display of Quantitative Information*. Cheshire, Connecticut:Graphics Press, 1983.

[12] Tufte ER. *Envisioning Information*. Cheshire, Connecticut:Graphics Press, 1990.

[13] Kimbrough SO, Pritchett CW, Bieber MP, Bhargava HK. An overview of the coast guard's KSS project: DSS concepts and technology. In Volonino L, editor, *Transactions of DSS-90: Information Technology for Executives and Managers, Tenth International Conference on Decision Support Systems*, pages 63–77, Cambridge (MA):1990.

[14] Krishnan R. A logic modeling language for model construction. *Decision Support Systems*, 1990;6.

[15] Hong SN, Mannino MV, Greenberg B. Measurement theoretic representation of large, diverse model bases: The unified modeling language: L_U. Technical report, Austin (TX):Department of Management Science and Information Systems, The University of Texas at Austin, 1991.

[16] Muhanna WA, Pick RA. Meta-modeling concepts and tools for model management: A systems approach. Technical report, Columbus (OH):Faculty of Accounting and Management Information Systems, Ohio State University, 1990.

[17] Jones CV. An introduction to graph-based modeling systems, part II: Graph-grammars and the implementation. *ORSA Journal on Computing*, 1991;3(3):180–206.

[18] Jones CV. An integrated modeling environment based on attributed graphs and graph-grammars. *Decision Support Systems*, 1993;10:255–275.

[19] Jones CV. Attributed graphs, graph-grammars, and structured modeling. *Annals of OR*, 1992;38:281–324.

[20] Jones CV, D'Souza K. Graph-grammars for minimum cost network flow modeling. Technical report, Burnaby (BC), V5A 1S6 CANADA:Faculty of Business Administration, Simon Fraser University, 1992.

[21] Jones CV, Krishnan R. A visual, syntax-directed environment for automated model development. Technical report, Burnaby (BC), V5A 1S6 CANADA:Faculty of Business Administration, Simon Fraser University, 1992.

[22] Baldwin D. *Principles of Design for a Multiple Viewpoint Problem Formulation Support System.* [dissertation], Lubboch (TX):School of Business Administration, Texas Tech University, 1989.

[23] Greenberg HJ, Murphy FH. Views of mathematical programming models and their instances. *Decision Support Systems*, 1993;13(1):3–34.

[24] Geoffrion AM. The formal aspects of structured modeling. *Operations Research*, 1989;37(1):30–51.

[25] Geoffrion AM. The SML language for structured modeling. *Operations Research*, 1992;40(1):38–75.

[26] Bly SA, Harrison SR, Irwin S. Mediaspaces: Bringing people together in a video, audio, and computing environment. *Communications of the ACM*, 1993;36(1):28–47.

[27] Narasimhalu AD, Christodoulakis S. Multimedia information systems: The unfolding of a reality. *IEEE Computer*, 1991;24(10):6–8.

[28] Brooks FP. Grasping reality through illusion: Interactive graphics serving science. In Soloway E, Frye D, Sheppard SB, editors, *Proceedings of the ACM SIGCHI Human Factors in Computing Systems Conference*, pages 1–1, New York:1988. ACM Press.

[29] Brooks FP, Ouh-Young M, Batter JJ, Kilpatrick PJ. Project GROPE—haptic displays for scientific visualization. *Computer Graphics*, 1990;24(4):177–185.

[30] Schultz H, Pulleyblank W. Trends in optimization. *OR/MS Today*, August 1991;pages 20–25.

[31] Hirshfeld DS. Some thoughts on math programming practice in the '90s. *Interfaces*, 1990;20(4):158–165.

Colophon

Producing this book presented a numerous challenge. The author was responsible for providing camera ready copy to the publisher. This allowed the author complete control over the layout of the book (within page size limits set by the publisher), but also required that the author learn a variety of typesetting and graphical tools to a painful level of detail.

The book was typeset by the author in Times Roman font on an Apple Macintosh IIfx using Textures 1.7 [1], an implementation of TEX [2]. Actually, the book was typeset using LATEX, version 2e, [3] a set of macros that provide higher level access to TEX.

LATEX allows an author to specify "packages" to customize the look of the document. The packages used in this book include [4]:

- `timesmt`, which typesets the document using Times Roman mathematical fonts.
- `fancyheadings`, which creates headings that include a horizontal line.
- `epsfig`, which allows Postscript images to be included in a document.
- `chapterbib`, for putting bibliographies at the end of each chapter.
- `multind`, to allow for multiple indexes.

A variety of custom macros were also written, most notably to include the flip chart in the margins of the odd numbered pages. Although TEX and LATEX are incredibly flexible languages for typesetting text, their limits were reached during the production of this book. In order to allow for larger images in the flip charts, the author wanted to wrap the text around each image at the top of each odd-numbered page. Unfortunately, it does not appear possible to coerce TEX or LATEX to do this.

The pictures that appear in the book were either generated by the author or very generously supplied by the individuals or organizations noted in the text. The images were provided on paper, color slides, and in every digital image format imaginable. Some of the digital image formats included:

- Silicon Graphics RGB format
- Sun Raster File
- Tagged Image File Format (TIFF)
- Macintosh PICT Format
- Postscript. The images from [5] the figure had to be extracted manually from a raw Postscript file using a text editor. The Postscript file contained an entire research paper.
- Encapsulated Postscript
- GIF
- X Bitmap
- PC Paintbrush (PCX)
- Joint Photographic Experts Group (JPEG)
- Motion Picture Expert's Group (MPEG). Snapshots of individual frames of the digital video were extracted using Sparkle [6] a shareware utility for viewing MPEG movies on the Macintosh.
- QuickTime

All images had to be converted either to Encapsulated Postscript (preferred) or to Macintosh PICT format in order to be able to be included in the book. A Macintosh shareware program, GraphicConverter [6] proved invaluable in converting these files.

Moreover, the images frequently had to be converted into black and white, since the book does not include color images. It proved quite challenging to be able to modify the images so that they would print acceptably in black and white. Adobe Photoshop, Version 3.0 [7] helped tremendously in this process.

The cumulative size of the images (over 100Mb) necessitated the purchase of a new hard disk drive. The proof copies were made at WYSIWYG Prepress, 400 West Georgia ST, Vancouver, British Columbia on a 2540 dot-per-inch Linotronic imagesetter.

Bibliography

[1] Metzler M. *Textures Users Guide.* Portland (OR):Blue Sky Research, Inc., 1995.

[2] Knuth DE. *The TEXbook.* Reading (MA):Addison-Wesley, 1986.

[3] Lamport L. *LATEX a Document Preparation System.* Reading (MA):Addison-Wesley, 1986.

[4] Goossens M, Mittelbach F, Samarin A. *The LATEX Companion.* Reading (MA):Addison-Wesley, 1994.

[5] Gay DM. Pictures of Karmarkar's linear programming algorithm. Technical Report 136, Murray Hill (NJ):AT&T Bell Laboratories, Computer Science, 1987.

[6] Lemke T. *GraphicConverter.* Insterburger Str 6, 31228 Peine, Germany.

[7] Adobe Systems Incorporated. *Adobe Photoshop Version 3.0 User Guide.* Mountain View (CA):Adobe Systems Incorporated, 1994.

Author Index

Subject Index